DESIGN OF STEEL STRUCTURES FOR BUILDINGS IN SEISMIC AREAS

ECCS Eurocode Design Manuals

ECCS Editorial Board
Luís Simões da Silva (ECCS)
António Lamas (Portugal)
Jean-Pierre Jaspart (Belgium)
Reidar Bjorhovde (USA)
Ulrike Kuhlmann (Germany)

Design of Steel Structures – 2ND Edition
Luís Simões da Silva, Rui Simões and Helena Gervásio

Fire Design of Steel Strcutures – 2ND Edition
Jean-Marc Franssen and Paulo Vila Real

Design of Plated Structures
Darko Beg, Ulrike Kuhlmann, Laurence Davaine and Benjamin Braun

Fatigue Design of Steel and Composite Structures
Alain Nussbaumer, Luís Borges and Laurence Davaine

Design of Cold-formed Steel Structures
Dan Dubina, Viorel Ungureanu and Raffaele Landolfo

Design of Joints in Steel and Composite Structures
Jean-Pierre Jaspart and Klaus Weynand

Design of Steel Structures for Buildings in Seismic Areas
Raffaele Landolfo, Federico Mazzolani, Dan Dubina, Luís Simões da Silva and Mario D'Aniello

ECCS – SCI EUROCODE DESIGN MANUALS

Design of Steel Structures, UK Edition
Luís Simões da Silva, Rui Simões, Helena Gervásio
Adapted to UK by Graham Couchman

Design of Joints in Steel Structures, UK Edition
Jean-Pierre Jaspart and Klaus Weynand
Adapted to UK by Graham Couchman and Ana M. Girão Coelho

ECCS EUROCODE DESIGN MANUALS – BRAZILIAN EDITIONS

Dimensionamento de Estruturas de Aço
Luís Simões da Silva, Rui Simões, Helena Gervásio, Pedro Vellasco, Luciano Lima

Information and ordering details
For price, availability, and ordering visit our website www.steelconstruct.com.
For more information about books and journals visit www.ernst-und-sohn.de.

Design of Steel Structures for Buildings in Seismic Areas

Eurocode 8: Design of steel structures in seismic areas
Part 1-1 – General rules and rules for buildings

Raffaele Landolfo
Federico Mazzolani
Dan Dubina
Luís Simões da Silva
Mario D'Aniello

ECCS
CECM
EKS

Ernst & Sohn
A Wiley Brand

Design of Steel Structures for Buildings in Seismic Areas

1st Edition, 2017

Published by:
ECCS – European Convention for Constructional Steelwork
publications@steelconstruct.com
www.steelconstruct.com

Sales:
Wilhelm Ernst & Sohn Verlag für Architektur und technische Wissenschaften GmbH & Co. KG, Berlin

All rights reserved. No parts of this publication may be reproduced, stored in a retrieval system, or transmitted in any form or by any means, electronic, mechanical, photocopying, recording or otherwise, without the prior permission of the copyright owner.

ECCS assumes no liability with respect to the use for any application of the material and information contained in this publication.

Copyright © 2017 ECCS – European Convention for Constructional Steelwork

ISBN (ECCS): 978-92-9147-138-6
ISBN (Ernst & Sohn): 978-3-433-03010-3

Legal dep.: 432199/17   Printed in Sersilito, Empresa Gráfica Lda, Maia, Portugal
Photo cover credits: Dan Dubina

TABLE OF CONTENTS

FOREWORD	xiii
PREFACE	xvii

Chapter 1
SEISMIC DESIGN PRINCIPLES IN STRUCTURAL CODES — 1
1.1   Introduction — 1
1.2   Fundamentals of seismic design — 2
    1.2.1   Capacity design — 2
    1.2.2   Seismic design concepts — 6
1.3   Codification of seismic design — 11
    1.3.1   Evolution of seismic design codes — 11
    1.3.2   New perspectives and trends in seismic codification — 19

Chapter 2
EN 1998-1: GENERAL AND MATERIAL INDEPENDENT PARTS — 25
2.1   Introduction — 25
2.2   Performance requirements and compliance criteria — 27
    2.2.1   Fundamental requirements — 27
    2.2.2   Ultimate limit state — 32
    2.2.3   Damage limitation state — 34
    2.2.4   Specific measures — 35
2.3   Seismic action — 36
    2.3.1   The fundamentals of the dynamic model — 36
    2.3.2   Basic representation of the seismic action — 40
    2.3.3   The seismic action according to EN 1998-1 — 46
    2.3.4   Alternative representations of the seismic action — 52
    2.3.5   Design spectrum for elastic analysis — 54
    2.3.6   Combinations of the seismic action with other types of actions — 56
2.4   Characteristics of earthquake resistant buildings — 58
    2.4.1   Basic principles of conceptual design — 58
    2.4.2   Primary and secondary seismic members — 60

	2.4.3 Criteria for structural regularity	61
2.5	Methods of structural seismic analysis	70
	2.5.1 Introduction	70
	2.5.2 Lateral force method	72
	2.5.3 Linear modal response spectrum analysis	75
	2.5.4 Nonlinear static pushover analysis	84
	2.5.5 Nonlinear time-history dynamic analysis	90
2.6	Structural modelling	94
	2.6.1 Introduction	94
	2.6.2 Modelling of masses	96
	2.6.3 Modelling of damping	98
	2.6.4 Modelling of structural mechanical properties	101
2.7	Accidental torsional effects	107
	2.7.1 Accidental eccentricity	107
	2.7.2 Accidental torsional effects in the lateral force method of analysis	109
	2.7.3 Accidental torsional effects in modal response spectrum analysis	110
	2.7.4 Accidental torsional effects in nonlinear static pushover analysis	111
	2.7.5 Accidental torsional effects in linear and nonlinear dynamic time history analysis	114
2.8	Combination of effects induced by different components of the seismic action	114
2.9	Calculation of structural displacements	117
2.10	Second order effects in seismic linear elastic analysis	118
2.11	Design verifications	121
	2.11.1 Safety verifications	121
	2.11.2 Damage limitation	126

Chapter 3
EN 1998-1: DESIGN PROVISIONS FOR STEEL STRUCTURES   129

3.1	Design concepts for steel buildings	129
3.2	Requirements for steel mechanical properties	133
	3.2.1 Strength and ductility	133
	3.2.2 Toughness	135

3.3 Structural typologies and behaviour factors 137
    3.3.1 Structural types 137
    3.3.2 Behaviour factors 141
3.4 Design criteria and detailing rules for dissipative structural behaviour common to all structural types 145
    3.4.1 Introduction 145
    3.4.2 Design rules for cross sections in dissipative members 145
    3.4.3 Design rules for non-dissipative connections 147
    3.4.4 Design rules and requirements for dissipative connections 148
    3.4.5 Design rules and requirements for non-dissipative members 148
3.5 Design criteria and detailing rules for moment resisting frames 149
    3.5.1 Code requirements for beams 149
    3.5.2 Code requirements for columns 152
    3.5.3 Code requirements for beam-to-column joints 153
3.6 Design criteria and detailing rules for concentrically braced frames 158
    3.6.1 Code requirements for braces 158
    3.6.2 Code requirements for beams and columns 162
3.7 Design criteria and detailing rules for eccentrically braced frames 164
    3.7.1 Code requirements for seismic links 164
    3.7.2 Code requirements for members not containing seismic links 171
    3.7.3 Code requirements for connections of the seismic links 172

Chapter 4
DESIGN RECOMMENDATIONS FOR DUCTILE DETAILS 173
4.1 Introduction 173
4.2 Seismic design and detailing of composite steel-concrete slabs 174
4.3 Ductile details for moment resisting frames 182
    4.3.1 Detailing of beams 182
    4.3.2 Detailing of beam-to-column joints 186
    4.3.3 Detailing of column bases 210
4.4 Ductile details for concentrically braced frames 215
    4.4.1 Introduction 215
    4.4.2 Detailing of brace-to-beam/column joints 216
    4.4.3 Detailing of brace-to-beam midspan connections 228
    4.4.4 Detailing of brace-to-brace connections 230
    4.4.5 Detailing of brace-to-column base connections 235

TABLE OF CONTENTS

    4.4.6 Optimal slope, constructional tolerances and local details for braces 236
 4.5 Ductile details for eccentrically braced frames 239
    4.5.1 Detailing of links 239
    4.5.2 Detailing of link lateral torsional restraints 241
    4.5.3 Detailing of diagonal brace-to-link connections 244
    4.5.4 Detailing of link-to-column connections 245

Chapter 5
DESIGN ASSISTED BY TESTING 247
 5.1 Introduction 247
 5.2 Design assisted by testing according to EN 1990 248
    5.2.1 Introduction 248
    5.2.2 General overview of EN 1990 250
    5.2.3 Testing 252
    5.2.4 Derivation of design values 254
 5.3 Testing of seismic components and devices 262
    5.3.1 Introduction 262
    5.3.2 Quasi-static monotonic and cyclic testing 262
    5.3.3 Pseudo-dynamic testing 275
    5.3.4 Dynamic testing 277
 5.4 Application: experimental qualification of buckling restrained braces 278
    5.4.1 Introduction and scope 278
    5.4.2 Test specifications 279
    5.4.3 Test specimens 280
    5.4.4 Test setup and loading protocol for ITT 280
    5.4.5 Results 281
    5.4.6 Fabrication Production Control tests 283

Chapter 6
MULTI-STOREY BUILDING WITH MOMENT RESISTING FRAMES 285
 6.1 Building description and design assumptions 285
    6.1.1 Building description 285
    6.1.2 Normative references 287
    6.1.3 Materials 288

	6.1.4	Actions	289
	6.1.5	Pre-design	292
6.2		Structural analysis and calculation models	293
	6.2.1	General features	293
	6.2.2	Modelling assumptions	296
	6.2.3	Numerical models and method of analysis	297
	6.2.4	Imperfections for global analysis of frames	301
	6.2.5	Frame stability and second order effects	303
6.3		Design and verification of structural members	304
	6.3.1	Design and verification of beams	304
	6.3.2	Design and verification of columns	310
	6.3.3	Panel zone of beam-to-column joints	316
6.4		Damage limitation	319
6.5		Pushover analysis and assessment of seismic performance	320
	6.5.1	Introduction	320
	6.5.2	Modelling assumptions	321
	6.5.3	Pushover analysis	328
	6.5.4	Transformation to an equivalent SDOF system	331
	6.5.5	Evaluation of the seismic demand	333
	6.5.6	Evaluation of the structural performance	334

Chapter 7
MULTI-STOREY BUILDING WITH CONCENTRICALLY BRACED FRAMES — 335

7.1		Building description and design assumptions	335
	7.1.1	Building description	335
	7.1.2	Normative references	337
	7.1.3	Materials	337
	7.1.4	Actions	338
	7.1.5	Pre-design	340
7.2		Structural analysis and calculation models	342
	7.2.1	General features	342
	7.2.2	Modelling assumptions	342
	7.2.3	Numerical models and method of analysis	344
	7.2.4	Imperfections for global analysis of frames	348
	7.2.5	Frame stability and second order effects	349

7.3	Design and verification of structural members	350
	7.3.1 Design and verification of X-CBFs	350
	7.3.2 Design and verification of inverted V-CBFs	357
7.4	Damage limitation	365

Chapter 8
MULTI-STOREY BUILDING WITH ECCENTRICALLY BRACED FRAMES — 369

8.1	Building description and design assumptions	369
	8.1.1 Building description	369
	8.1.2 Normative references	371
	8.1.3 Materials	371
	8.1.4 Actions	372
8.2	Structural analysis and calculation models	374
	8.2.1 General features	374
	8.2.2 Modelling assumptions	375
	8.2.3 Numerical models and method of analysis	376
	8.2.4 Imperfections for global analysis of frames	380
	8.2.5 Frame stability and second order effects	380
8.3	Design and verification of structural members	381
	8.3.1 Design and verification of shear links	381
	8.3.2 Design and verification of beam segments outside the link	384
	8.3.3 Design and verification of braces	384
	8.3.4 Design and verification of columns	385
8.4	Damage limitation	388

Chapter 9
CASE STUDIES — 391

9.1	Introduction	391
9.2	The Bucharest Tower Centre International	393
	9.2.1 General description	393
	9.2.2 Design considerations	397
	9.2.3 Detailing	421
	9.2.4 Construction	422
9.3	Single storey Industrial Warehouse in Bucharest	432
	9.3.1 General description	432

	9.3.2	Design considerations	435
9.4	The Fire Station of Naples		449
	9.4.1	General description	449
	9.4.2	Design considerations and constructional details	456
	9.4.3	The anti-seismic devices	467

REFERENCES 475

FOREWORD

There are many seismic areas in Europe. As times goes by, regional seismicity is better known and the number of places where earthquake is an action to consider in design increases. Of course, there are substantial differences in earthquake intensity between regions and the concern is much greater in many areas of Italy, for instance, than in most places in Northern Europe. However, even in Northern Europe, for structures for which a greater level of safety is required, like Seveso industrial plants, hospitals and public safety facilities, seismic design can be the most requiring design condition.

Designing for earthquake has original features in comparison with design for classical loading like gravity, wind or snow. The reference event for Ultimate Limit State seismic design is rare enough for an allowance to permanent deformations and structural damages, as long as people's life is not endangered. This means that plastic deformations are allowed at ULS, so that the design target becomes a global plastic mechanism. To be safe, the latter requires many precautions, on global proportions of structures and on local detailing. The seismic design concepts are completely original in comparison to static design. Of course, designing for a totally elastic behaviour even under the strongest earthquake remains possible but, outside of low seismicity areas, this option is generally left aside because of its cost.

This book is developed with a constant reference to Eurocode 8 or EN 1998-1:2004; it follows the organization of that code and provides detailed explanations in support of its rather dry expression. Of course, there are many other seismic design codes, but it must be stressed that there is nowadays a strong common thinking on the principles and the application rules in seismic design so that this book is also a support for the understanding of other continents codes.

Chapter 1 explains the principles of seismic design and their evolution throughout time, in particular the meaning, goals and conditions set forward by capacity design of structures and their components, a fundamental aspect of seismic design.

FOREWORD

Chapter 2 explains the general aspects of seismic design: seismic actions, design parameters related to the shape of buildings, models for the analysis, safety verifications. Methods of analysis are explained in an exhaustive way: theoretical background, justifications of limits and factors introduced by the code, interest and drawbacks of each method, together with occasionally some tricks to facilitate model making and combination of load cases.

Chapter 3 focuses on design provisions specific to steel structures: ductility classes, requirements on steel material, structural typologies and design conditions related to each of them; an original insight on design for reparability is also included.

Chapter 4 provides an overview about the best practice to implement the requirements and design rules for ductile details, particularly for connections in moment resisting frames (MRF), concentrically braced frames (CBF) and eccentrically braced frames (EBF), and for other structural components like diaphragms.

Chapter 5 describes the guidance provided for design assisted by testing by EN 1990 and the specific rules for tests, a necessary tool for evaluating the performance characteristics of structural typologies and components in the plastic field and in cyclic/dynamic conditions.

Chapter 6 illustrates and discusses the design steps and verifications required by EN 1998-1 for a multi-storey Moment Resisting Frame.

Chapter 7 and 8 do the same respectively for buildings with CBF's and EBF's.

Chapter 9 presents three very different examples of real buildings erected in high seismicity regions: one tall building, one industrial hall and one design using base isolation. These examples are complete in the sense that they show the total design, where seismic aspects are only one part of the problem. These examples are concrete, because they illustrate practical difficulties of the real world with materials, execution, positioning…

The concepts, design procedures and detailing in seismic design may seem complex. This publication explains the background behind the rules, which clarify their objectives. Details on the design of the different building typologies are given, with reference to international practice and to recent research results. Finally, design examples and real case studies set out the design process in a logical manner, giving practical and helpful advice.

This book will serve the structural engineering community in expanding the understanding and application of seismic design rules, and, in that way, constitute a precious tool for our societies safety.

André Plumier
Honorary Professor, University of Liege

PREFACE

This manual aims to provide its readers with the background and the explanation of the main aspects dealing with the seismic design of steel structures in Europe. Therefore, the book focuses on EN 1998-1 (usually named part 1 of Eurocode 8 or EC8-1) that is the Eurocode providing design rules and requirements for seismic design of building structures. After 10 years from its final issue, both the recent scientific findings and the design experience carried out in Europe highlight some criticisms. In the light of such considerations, this book complements the explanation of the EC8-1 provisions with the recent research findings, the requirements of renowned and updated international seismic codes (e.g. North American codes and design guidelines) as well as the design experience of the Authors. Although the manual is oriented to EC8-1, the book aims to clarify the scientific outcomes, the engineering and technological aspects rather than sticking to an aseptic explanation of each clause of the EC8-1. Indeed, as shown in Chapter 4, the proper detailing of steel structures is crucial to guarantee adequate ductility of seismic resistant structures and the current codes does not give exhaustive guidelines to design ductile details since it only provides the fundamental principles. In addition, the practice of earthquake engineering significantly varies between European regions, reflecting the different layouts of each national seismic code as well as the level of knowledge and confidence with steel structures of each country. With this regard, a large number of European engineers believe that steel structures can withstand severe earthquakes without requiring special details and specifications as conversely compulsory for other structural materials like reinforced concrete and masonry. This belief direct results from the mechanical features of the structural steel, which is a high performance material, being stronger than concrete but lighter (if comparing the weight of structural members) and also very ductile and capable of dissipating large amounts of energy through yielding when subjected to cyclic inelastic deformations. However, although the material behaviour is important, the ductility of steel alone is not enough to guarantee ductile structural response. Indeed, as demonstrated by severe past earthquakes (e.g. Northridge 1994,

Preface

Kobe 1995 and Christchurch 2011) there are several aspects ensuring good seismic behaviour of steel structures, which are related to (i) the conceptual design of the structure, (ii) the overall sizing of the member, (iii) the local detailing and (iv) proper technological requirements as well as ensuring that the structures are actually constructed as designed.

Therefore, this book primarily attempts to clarify all these issues (from Chapter 1 to 4) for European practising engineers, working in consultancy firms and construction companies. In addition, the examples of real buildings (see Chapter 9) are an added value, highlighting practical and real difficulties related to both design and execution.

This design manual is also meant as a recommended textbook for several existing courses given by the Structural Sections of Civil Engineering and Architectural Engineering Departments. In particular, this manual is oriented to advanced students (i.e. those attending MSc programmes) thanks to the presence of various calculation examples (see Chapter 6, 7 and 8) that simplify and speed up the understanding and the learning of seismic design of EC8 compliant steel structures. In addition, research students (i.e. those attending PhD programmes) can find useful insights for their experimental research activities by reading Chapter 5, which provides some guidance and discussion on how performing experimental tests of structural typologies and components in cyclic pseudo-static and dynamic conditions.

The Authors

Raffaele Landolfo
Federico Mazzolani
Dan Dubina
Luís Simões da Silva
Mario D'Aniello

Chapter 1

SEISMIC DESIGN PRINCIPLES IN STRUCTURAL CODES

1.1 INTRODUCTION

Earthquake Engineering is the branch of engineering aiming at mitigating risks induced by earthquakes with two objectives: i) to predict the consequences of strong earthquakes on urban areas and civil infrastructures; ii) to design, build and maintain structures that are able to withstand earthquakes in compliance with building codes.

Researchers and experts working within emergency management organizations (e.g. the civil protection) actively work on the first issue. On the contrary, structural designers focus their attention and efforts on the second objective. With this regard, it should be noted that the seismic design philosophy substantially differs from the design approaches conventionally adopted for other types of actions, raising difficulties to structural engineers less confident with seismic engineering. Indeed, broadly speaking, for quasi-static loads (e.g. dead and live loads, wind, snow, etc.) the structure should behave mostly elastically without any damage until the maximum loads are reached, while in case of seismic design it is generally accepted that structures can experience damage because they should perform in the plastic range for seismic events. The philosophy of structural seismic design establishes the performance levels that properly engineered structures should satisfy for different seismic intensities, which can be summarized as follows:

1. Seismic Design Principles in Structural Codes

- prevent near collapse or serious damage in rare major ground shaking events, which are called in the following Ultimate Limit State seismic action or ULS seismic action;
- prevent structural damage and minimize non-structural damage in occasional moderate ground shaking events;
- prevent damage of non-structural components (such as building partitions, envelopes, facilities) in frequent minor ground shaking events.

Hence, the most meaningful performance indexes for seismic resistant structures are the amount of acceptable damage and the repair costs. Owing to the unforeseeable nature of seismic actions, it is clear that damage control is very difficult to be quantified by code provisions, especially because it is related to acceptable levels of risk. The challenge for efficient design of seismic resistant structures is to achieve a good balance between the seismic demand (namely the effect that earthquakes impose on structures) and the structural capacity (namely the ability to resist seismic induced effects without failure). However, the quantification of different types of damage (structural and non-structural) associated to the reference earthquake intensity (e.g. frequent/minor, occasional/moderate, and rare/major) and the definition of relevant operational design criteria are still open issues that need clarification and further studies.

This chapter describes and discusses the concept of capacity design in the light of existing seismic codes, illustrating the evolution of seismic design principles throughout time, and explains the criteria that form the basis of EN 1998-1:2004 (CEN, 2004a), henceforth denoted as EC8-1.

1.2 FUNDAMENTALS OF SEISMIC DESIGN

1.2.1 Capacity design

It is generally acknowledged that structural safety depends on the ductility that the structural system can provide against the design loads. Indeed, ductility represents the capacity of a mechanical system (e.g. a beam, a structure, etc.) to deform in the plastic domain without substantially reducing its bearing capacity.

In seismic design of structures it is generally not economical or possible to ensure that all the elements of the structure behave in a ductile manner.

1.2 Fundamentals of Seismic Design

Inevitably, a dissipative (ductile) structure comprises both dissipative (ductile) elements and non-dissipative (brittle) ones. In order to achieve a dissipative (ductile) design for the whole structure, the failure of the brittle elements must be prevented. This may be done by prioritizing structural elements strength, which will lead to the prior yielding of ductile structural elements, preventing the failure of brittle structural elements. This principle is known as "capacity design". Capacity design may be explained by considering the chain model, introduced by Paulay and Priestley (1992) and depicted in Figure 1.1a, whereby the chain represents a structural system made of both ductile elements (e.g. the ring "1") and brittle zones (e.g. the ring "i").

According to non-seismic design procedures for quasi-static loads (hereinafter referred to as "direct design"), the design force is the same for all elements belonging to the chain, because the applied force is equal for all rings, being a system in series. Hence, the design resistance $F_{y,i}$ is the same for all elements. Under this assumption, the yield resistance of the ductile chain $F_{y,1}$ is equal or even slightly larger than $F_{y,i}$.

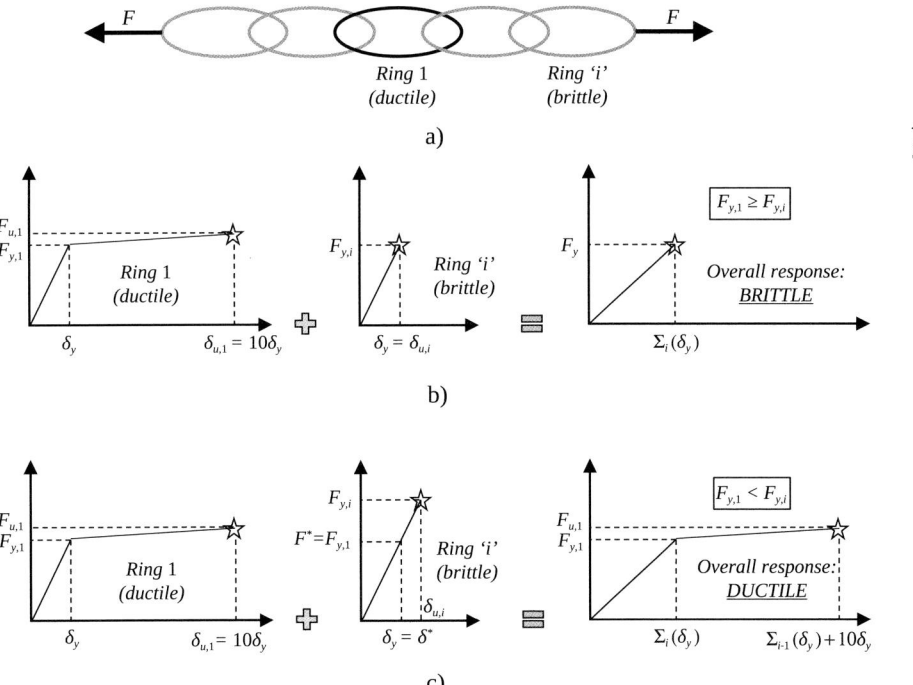

Figure 1.1 – Ductility of a chain with brittle and ductile rings

1. Seismic Design Principles in Structural Codes

As shown in Figure 1.1b, with the direct design approach the system cannot develop strength larger than F_y and the ultimate elongation of the chain is given as

$$\delta_u = \sum_i \delta_y = 5\delta_y \qquad (1.1)$$

According to capacity design principles, in order to improve the ductility of the chain, some rings should be designed with ductile behaviour and lower strength, as is the case of ring "1" in Figure 1.1c. The remaining rings "i" that are brittle should be designed to provide a resistance $F_{y,i}$ larger than the maximum resistance $F_{u,1}$ exhibited by the ring "1" beyond yielding. The ductile ring "1" behaves as a sacrificial element, i.e. a ductile fuse, which filters the external actions and limits the transfer of forces into the brittle elements. Hence, the maximum force that the chain can sustain is equal to the maximum resistance $F_{u,1}$ of the ductile ring "1". It is interesting to observe that the beneficial improvement of the capacity design methodology is the increase of displacement capacity, given as follows:

$$\delta_u = \sum_{i-1} \delta_y + 10\delta_y = 4\delta_y + 10\delta_y = 14\delta_y \qquad (1.2)$$

Comparing equations (1.1) and (1.2), it can be easily recognized that the collapse displacement of the chain is significantly larger than that obtained by adopting the direct design approach.

This trivial example allows to understand that the brittle elements represent protected zones that must be designed to resist larger forces than those supported by the ductile elements. Those larger forces do not directly depend on the external applied loads but they are obtained from the maximum capacity of the connected ductile elements. However, it should be emphasized that the external forces are used to design the dissipative elements, which establish the threshold of structural strength.

Concerning the practical application to building structures, this methodology leads the structural designers to work on two different schemes for the same structure, as follows:

1) elastic behaviour with the calculation of the relevant internal forces F_{Ed} to design the dissipative elements. Hence, following an elastic analysis, the ductile structural elements should satisfy the following check:

$$F_{ductile,Rd} \geq F_{Ed} \tag{1.3}$$

In addition to strength, the ductile elements must possess a ductility corresponding to the chosen ductility class. The ductility is provided by using appropriate structural details and different materials and specific design principles for specific types of structures;

2) inelastic response with design of non-dissipative (i.e. brittle) elements on the basis of the plastic strength of the connected dissipative parts. Hence, in order to prevent their failure, brittle elements must be sized so that they present an over strength with respect to the capacity of the ductile elements, as follows:

$$F_{brittle,Rd} \geq \Omega F_{ductile,Rd} \tag{1.4}$$

where Ω is a coefficient (> 1.0) that takes into account different aspects that may lead to ductile elements strengths larger than the design ones (strain hardening phenomena, material strength larger than the nominal values, etc.).

This twofold approach is the basic characteristic of capacity design and represents the main distinctive difference with respect to direct design for quasi-static actions. The example shown in Figure 1.1 also allows understanding that the common belief of non-seismic designers, which consider that the excess of strength is always beneficial and safe, may dramatically impair the non-linear response of a structure either by overdesigning the fuse elements or, with more serious consequences, by inaccurate quality control of the material properties that results in larger strength for the dissipative elements (e.g. a steel element conceived as fuse with grade S355 is supplied with higher grade as S460). The consequence of such events is clear, namely the failure of the system because the hierarchy of resistance is not complied with.

In case of steel structures the best way to dissipate energy is to exploit the tensile capacity of the material, which can be obtained by enforcing plasticity into specific zones called plastic hinges that can involve either flexural, tensile or shear mechanisms depending on the type of adopted structural scheme (e.g. moment resisting frame, concentrically or eccentrically braced frame), while preserving the rest of the structure from damage.

1.2.2 Seismic design concepts

Two substantially different concepts can be used to design structures located in seismic areas, which correspond to two different structural behaviours:

– Concept (a): low-dissipative (and/or non-dissipative) behaviour;
– Concept (b): dissipative behaviour.

The difference between dissipative and non-dissipative behaviours is dictated by both the ductility and energy dissipation capacity that the structure can provide. The ductility represents the capacity to deform in the plastic domain without substantially reducing its bearing capacity. However, there are other properties that significantly influence the seismic performance, namely the displacement and dissipative capacity. These properties are not synonyms, but all of them contribute towards a satisfactory seismic behaviour. Some examples may be helpful to clarify the differences between ductility, displacement and dissipative capacity.

Figure 1.2 shows the load-deflection response curves of two different frames subjected to monotonically increasing horizontal loads. The maximum strength F_y of the frame corresponds to the yield strength and/or stability limit load, and the deformation capacity δ_u corresponds to the sudden decrease of the strength that can be caused by the rupture of steel material, global and/or local buckling of steel members and/or crushing of concrete. Even though the strength of both frames is identical, the one with the response curve shown in Figure 1.2a represents a ductile behaviour, which is substantially different from that of Figure 1.2b that corresponds to a brittle performance. Indeed, the first structure is characterized by a larger ductility $\mu = \delta_u/\delta_y$ and also a larger displacement capacity δ_u, which is the capacity of the structural system to experience large ultimate displacements. Also, the amount of energy absorbed by the frame shown in Figure 1.2a before it reaches the limit deformation δ_u is larger than that of the frame shown in Figure 1.2b. In light of the remarks in section 1.2.1, the response of the frame shown in Figure 1.2a is more efficient for an earthquake resisting structure.

1.2 FUNDAMENTALS OF SEISMIC DESIGN

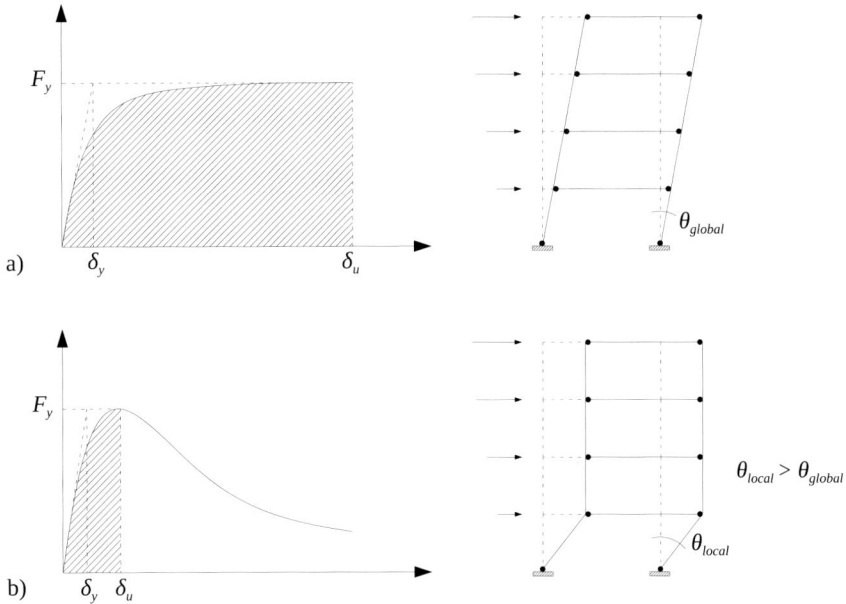

Figure 1.2 – Ductility of frames: a) high and b) poor displacement capacity

However, adequate seismic behaviour also depends on the shape of the cyclic response of both the structure and the dissipative zones. With this regard, Figure 1.3 shows two examples of hysteresis loops of frames under repeated horizontal load, having the same monotonic response and displacement capacity δ_u. In these cases, in addition to the effects indicated above, the shape of the hysteresis loops also depends on the number of loading cycles, since deformation phenomena associated with fatigue caused by the repeated loading may have some effect on it. The frame shown in Figure 1.3a dissipates larger energy before failure than the one in Figure 1.3b, thus providing a better seismic performance, the energy being the area within a loop. Hence, dissipative capacity can be defined as the ability to dissipate energy by means of stable and compact hysteretic loops.

Ductile and dissipative structures are very convenient because they avoid brittle phenomena and lead to less expensive constructions. In order to exploit the ductility, ductile structures are generally designed to resist seismic forces substantially smaller than those needed to obtain an elastic response under seismic action corresponding to the Ultimate Limit State (ULS). However, plastic deformations imposed by the seismic action must not exceed the deformation capacity of the structure in the plastic domain, in order to prevent excessive damage that may compromise the stability against gravity loads and/or make unfeasible a subsequent refurbishment. Thus, the minimum strength F_y of the structure against lateral forces that is needed to avoid excessive damage is directly related to the structure's deformation capacity in the plastic domain. For the ULS seismic action, different strength/ductility combinations can be determined that satisfy the design demands.

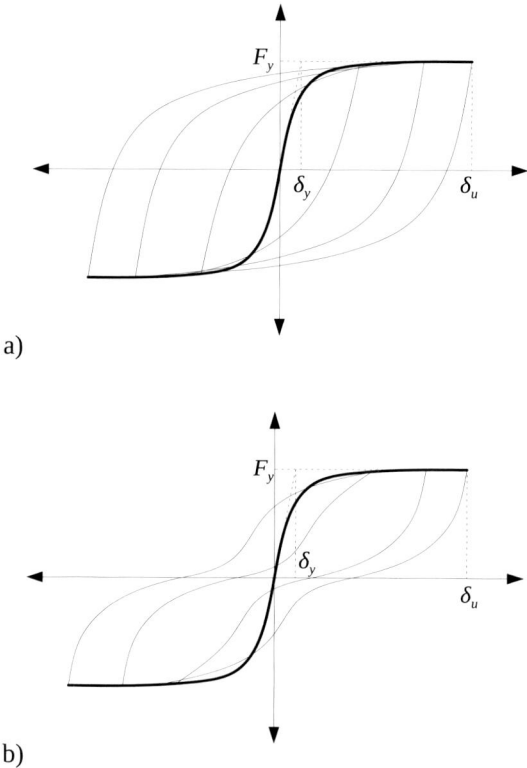

Figure 1.3 – Dissipative capacity of frames: a) high and b) poor energy absorption

The fundamental relationship between the strength of the structure to lateral forces (F_y) and the displacement demand (δ_{Ed}) imposed to the structure by a given level of the seismic action is presented in Figure 1.4a.

1.2 FUNDAMENTALS OF SEISMIC DESIGN

For the same the same displacement capacity, the lower is the strength of the structure to lateral forces (F_y), the higher is the ductility demand ($\mu_{Ed,i} = \delta_{Ed}/\delta_{y,i}$) imposed to the structure. Thus, the more ductile and dissipative structures may be designed to lower lateral forces that can be determined by scaling the elastic forces by the so-called behaviour factor q, which strictly depends on the structural system (see Figure 1.4b).

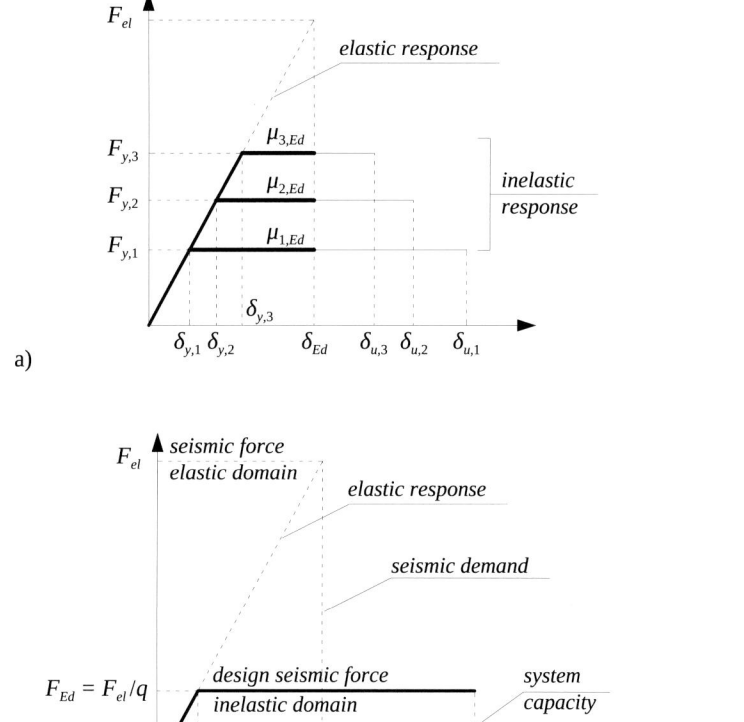

Figure 1.4 – Strength vs. displacement demand relationship

Modern codes like EC8-1 give the possibility to choose different ductility levels for a structure, providing different ductility classes. It is understandable that choosing a ductility class instead of another has direct consequences on the design process. In case of EC8-1 there are at least two major features. The first is the value of the design seismic load, which is obtained by scaling the elastic design forces by a behaviour factor q (see Figure 1.4b). The structures

1. Seismic Design Principles in Structural Codes

that are designed to behave in a more ductile way (i.e. on a higher ductility class) have higher values of the behaviour factor q, and, consequently, lower design seismic forces. The second consequence of choosing a ductility class is the necessity of providing a certain ductility level to the structure. To achieve this purpose, the codes provide specific detailing and design requirements for all structural materials (e.g. steel, reinforced concrete, timber, etc.) and relevant types of structures (e.g. moment resisting and braced frames, structural walls, etc.) compliant with each ductility class.

This approach cannot be adopted for structural typologies that do not provide any ductility and/or dissipative capacity, such as the so-called low-dissipative (brittle) structures. Indeed, because the force exhibits a sudden decrease beyond their elastic limit, these structures must be designed to remain almost elastic under the ULS seismic action. This corresponds to using a behaviour factor q close to unity. Because these structures do not exploit any plastic behaviour, their design may be carried out according to the direct design procedures used for non-seismic conditions. Therefore, seismic design provisions (for example EC8-1) are used only to determine the seismic loading, and the ULS structural checks are carried out according to general structural design standards (for example, the EN 1993 series in case of steel structures).

Designing a structure as dissipative or low-dissipative is a decision of the structural engineer. Fundamentally, any structure can be designed according to one of the two concepts. Generally, choosing the design concept accounts for economical aspects, depending on the type of the structure and the seismic area. With this regard, it should be noted that structural details and design demands necessary to provide ductility and dissipative behaviour may lead to higher constructional and design effort. Therefore, if the elastic (non-reduced) seismic forces acting on the structure are relatively small and the design is mainly governed by non-seismic load conditions, the low-dissipative design principle of the structure can be economically used. By omitting the design demands meant to provide a ductile global behaviour, the design process will be simplified and will lead to reduced material consumption.

However, for many types of structures, the seismic action represents a very severe design action, more critical than the other loading conditions, and providing an elastic response of the structure under the effect of the design seismic action at ULS will lead to excessive size of the structural elements and, consequently, to an excessive material consumption. Hence, in those cases, the dissipative design concept of the structure should be adopted, exploiting the structural ductility by both designing the structure for reduced seismic forces smaller than those corresponding to an elastic response and detailing the dissipative zones in order to provide the required larger ductility.

Consequently, it can be stated that the low-dissipative design concept is reasonably suitable and economic for small seismic forces, while a dissipative design one is more economic and effective for large seismic forces.

An interesting consideration can be drawn considering the nature of seismic forces. Indeed, actions induced by earthquakes are inertial forces, generated by the acceleration upon the structure masses as a result of the seismic motion imposed to the base of the structure. Therefore, the seismic forces will have smaller values for light structures. On the contrary, the seismic forces will have important values for structures with larger mass. An example of a light structure, for which the low-dissipative design principle is suitable, are single storey steel warehouses. These constructions are characterized on one hand by a relative small self weight and on the other hand by small live loads. In contrast, typical examples of structures for which the dissipative design principle is suitable are multi-storey buildings, because of the large masses resulting from permanent (reinforced concrete slabs) and live loads.

1.3   CODIFICATION OF SEISMIC DESIGN

1.3.1   Evolution of seismic design codes

Seismic engineering is a relatively new branch of structural engineering, since the first criteria were developed only at the beginning of the 20th century, while the most important modern concepts were established in the last 50 years (Gioncu and Mazzolani, 2002). In Europe, the first seismic design concepts were introduced by Gustave Eiffel at the beginning of last century, who modelled the earthquake action through an equivalent wind load. San Francisco, in California, was rebuilt after the 1906 great seismic event with this model by assuming a 1.4 kPa equivalent wind pressure to estimate seismic actions.

In Europe, the first quantitative seismic code was developed by an Italian Government Commission following the 1908 Messina–Reggio earthquake,

1. SEISMIC DESIGN PRINCIPLES IN STRUCTURAL CODES

which killed 160000 people. A report was issued giving a procedure that, for the first time, proposed to estimate the forces induced by the earthquake on a structure as a percentage of its weight. Accordingly, the first floor earthquake equivalent force was estimated to be 1/12 of the weight above, changing to 1/8 of the weight for the upper floors. This method promoted an equivalent static approach, which is still in use nowadays in most design codes.

In Japan, after the 1923 earthquake in Kanto, which killed 140000 people, the Home Office of Japan adopted a design regulation that stipulated the use of horizontal equivalent static forces equal to 10 % of the building weight, limiting also the height of the buildings.

In the USA, the concept of lateral seismic forces proportional to mass was introduced after the Santa Barbara earthquake in 1925 and the 1933 Long Beach earthquake and the buildings then had to be designed to carry lateral forces equal to 7.5 % and 10 % of their dead load for rigid and soft soils, respectively. The influence of structural flexibility and the number of floors on the design forces was recognized by the Los Angeles city code in 1943. The San Francisco recommendations gave the first provisions to take into account the influence of the dynamic properties of a structure by relating the seismic forces to the fundamental period of vibration. These simple concepts were based on grossly simplified physical models, engineering judgment and a number of empirical coefficients. For many years, the standard design methodology was based on models where the structures were considered as elastic systems and the earthquake actions as static loads.

The modern concepts of response spectrum and plastic deformation were introduced by Benioff (1934) and Biot (1941). The concept of ductility and energy dissipation capacity was proposed for the first time by Tanabashi in 1935, according to whom the earthquake resistance capacity of a structure should be measured by the amount of energy that the structure can dissipate before collapse. The first attempts to combine these two aspects, namely the response spectrum and the dissipation of seismic energy through plastic deformations, were made by Housner (1956, 1959), who used the velocity response spectra of the elastic system to have a quantitative evaluation of the total amount of energy input that contributes to the building response, by assuming the hypothesis that the energy input, responsible for the damage in the elastic-plastic system, is identical to that in the elastic system (Akiyama, 1985). The first studies on inelastic spectra were carried out by Velestos and Newmark (1960) who obtained the maximum response deformation for an elastic-perfectly plastic structure. At that stage, the response spectrum became a standard measure of the demand of the ground motion. Despite being based on a simple Single-Degree-Of-Freedom (SDOF) linear system, the concept of the response spectrum was extended to Multi-Degree-Of-Freedom systems (MDOF), non-linear elastic systems and inelastic hysteretic systems. Indeed, the response spectrum represents a powerful design tool because it gives a simple and direct indication of the overall displacement and acceleration demands of the earthquake ground motion, for structures having different period and damping characteristics, without the need to perform detailed numerical analysis. Newmark and Hall proposed a new concept in 1969 (Newmark and Hall, 1969, 1982), by constructing spectra based on accelerations, velocities, and displacements, in short, medium and long period ranges, respectively. More recently, different design methodologies have been elaborated for near-field and far-field regions. Indeed, the ground motions in near-field regions are qualitatively different from those of the commonly used far-field earthquake ground motions. For near-field earthquakes the importance of the higher vibration modes cannot be neglected; therefore, the use of the equivalence of multi-degree-of-freedom systems with only one degree-of-freedom gives inaccurate results. In the early 1970s, an important change in seismic design took place thanks to advent of personal computers and the availability of a large number of softwares able to perform static and dynamic analyses in the elastic and plastic ranges. This new technology allows to obtain more refined results, giving researchers the possibility to improve the design spectra methodology by means of a more correct calibration of design values. At the same time, the seismic behaviour of structures may be evaluated in a more precise way thanks to time-history methodologies, by using real recorded accelerograms.

Consequently, by the end of 1970s, the second generation of seismic design codes was developed, which started to take into account both the dynamic amplification and the energy dissipation properties in the estimation of the statically equivalent seismic design forces. However, the design and calculation procedures remained quite rough and did not allow for particularities between the behaviour of structures made of different materials and of different lateral force resisting systems (Bisch, 2009).

The significant economic losses and human casualties that resulted from the Northridge and Kobe earthquakes (that occurred in 1994 and 1995, respectively), even if the no-collapse objective had been met for many

1. SEISMIC DESIGN PRINCIPLES IN STRUCTURAL CODES

structures (Bommer and Pinho, 2006), led to the development of a new generation of prescriptive seismic design codes (i.e. the third one) with the aim to improve the criteria for overstrength and ductility and to qualify structural details in dissipative zones. Moreover, significant progress was done in using advanced methodologies of structural analysis, such as non-linear static and, particularly, dynamic analyses.

This category of codes established minimum requirements for safety through the specification of prescriptive criteria that regulate acceptable materials of construction. Moreover, they identified structural and non-structural systems approved for seismic applications, specifying the required minimum levels of strength and stiffness and controlling the relevant details. Although these prescriptive criteria were intended to result in buildings capable of providing certain levels of performance, the actual performance of individual building designs is not assessed as part of the traditional code design process. As a result, the performance capability of some buildings designed to these prescriptive criteria could be better than anticipated by the code, while the performance of others could be worse.

However, the added value of these codes was the introduction of a novel design philosophy that is known as "Performance-Based Seismic Design (PBSD)", which synthesises the significant concept that multi-level design criteria have to be applied to achieve a set of design objectives. The main peculiarity consists in correlating the structural performance at various limit states to the probability of occurrence of the earthquake action that reaches the intensity required to induce the corresponding failure modes (Mazzolani and Gioncu, 2000). The combination of the structure performance level with the specific level of ground motion represents the performance design objective. The aim of this new approach is to provide criteria for selecting the appropriate structural system and for detailing both structural and non-structural components, so that for specified levels of earthquake intensity the structural damage will be constrained within pre-defined limits in order to achieve a good balance between adequate safety levels and economy (Mazzolani and Piluso, 1996).

PBSD has been developed within the activities of SEAOC Vision 2000 Committee (SEAOC, 1995). The acceptability of varying levels of damage was determined on the basis of the consequences of this damage to the user community and the frequency of occurrence of such damage (see Figure 1.5). The four following performance levels were proposed:

- Fully operational, in which no damage occurs and the consequences to the building and its user community are negligible;
- Operational, in which moderate damage to non-structural elements and contents, and light damage to structural elements may occur; however, the damage does not compromise the safety of the building for occupancy;
- Life safe (damage state), in which moderate damage to both structural and non-structural elements occurs; nevertheless, both lateral stiffness and strength of the structure to resist additional lateral loads is reduced, but some margin against collapse remains;
- Near collapse (extreme state), in which the lateral and vertical load resistance of the building are substantially compromised; aftershocks could result in partial or total collapse of the structure.

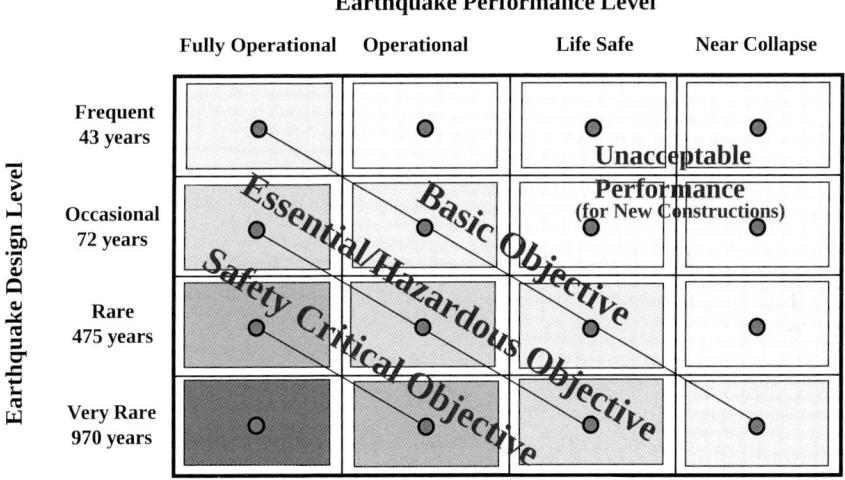

Figure 1.5 – Seismic performance design objective matrix
(SEAOC, 1995)

In addition, SEAOC Vision 2000 specified four earthquake design levels: frequent, occasional, rare and very rare, which are characterised by return periods equal to 43, 72, 475 and 970 years, respectively. It is clear that it is accepted that structures may fail at more severe seismic intensities (Bertero 1996).

Three design levels are defined in Figure 1.5. The "Basic Objective" applies to the majority of buildings. For more critical structures, higher performance objectives would be the reference, like the "Essential/Hazardous Objective" or even higher, the "Safety Critical Objective".

In contrast to prescriptive design approaches, PBSD ideally provides a systematic methodology for assessing the performance capability of a building, system or component. Indeed, PBSD explicitly evaluates how a structure will likely perform considering the potential hazard it is likely to experience, accounting for both uncertainties related to the quantification of the potential hazard and random and epistemic uncertainties that are related to the assessment of the actual building response. The use of PBSD allows to design both new buildings or to upgrade existing buildings with an accurate and realistic understanding of the risks and economic loss that may occur in case of future earthquakes. In addition, PBSD provides a framework to determine the levels of safety and property protection with the corresponding costs, thus allowing to evaluate the thresholds acceptable by building owners, tenants, lenders, insurers, regulators and other decision makers based upon the specific needs of a project.

However, this framework was far from being entirely implemented in seismic codes, owing to its complexity and the lack of guidelines. An initial suggestion by Bertero (1996) was to carry out a preliminary design of structures taking into account only two performance levels, such as the operational and the life safety, then check the structure for all the intermediate levels in order the assess the design acceptability. In this way a compromise was reached between the traditional design philosophy and this new philosophy, which was convenient considering the widespread use of the traditional design practice.

About ten years later, the fourth generation of seismic codes was developed. Indeed, FEMA 445 (2006) opened the way to the full implementation of Performance Based Seismic Design (PBSD) methods in current design practice, which represents a significant improvement with respect to the previous seismic codes.

FEMA 445 started a work plan devoted to cover the shortcomings and criticisms of the previous third generation codes, with a threefold action, as follows:

- To revise the discrete performance levels (as those defined in SEAOC Vision 2000), to develop new performance measures (e.g. repair costs, casualties, and time of occupancy interruption) that more efficiently relate to the decision-making needs of stakeholders,

and that communicate these losses in a way that is more meaningful to stakeholders;
- To develop accurate guidelines to carry out both analytical and numerical procedures for the prediction of actual building response; and to create procedures for estimating probable repair costs, casualties, and time of occupancy interruption, for both new and existing buildings;
- To develop a framework for performance assessment that properly accounts for, and adequately communicates to stakeholders, the limitations in the ability to accurately predict response and uncertainty in the level of earthquake hazard.

This activity culminated in two new codes, i.e. FEMA 695 (Quantification of Building Seismic Performance Factors) and FEMA 750 (Recommended Seismic Provisions for New Buildings and Other Structures), that are at the present time the most advanced seismic regulations.

In particular, FEMA 695 provides a standard procedural methodology that allows quantifying the inelastic response characteristics and performance of typical structures and verifying the adequacy of the structural system provisions to meet the design performance objectives. Such a methodology directly accounts for the potential variations in structural configurations of structures designed, and for the variation in ground motion to which these structures may be subjected. In addition, the behavioural characteristics of structural elements are validated on the basis of available experimental data.

Within this continuously progressing codification process, Eurocode 8 bridges in-between the third and fourth generation of codes. At the present time, also in Europe there is a six years action plan devoted to update the current version of Eurocode 8. With this regard, CEN/TC250/SC8, which is the official body devoted to the maintenance and development of all parts of Eurocode 8, established specific working groups (WG), one per chapter of the code, aiming to overcome the criticisms to the code and to implement the recent outcomes from scientific research. One crucial specific aspect relates to the quantification of the seismic hazard and the improvement and harmonization of the European Seismic Zonation (Solomos *et al*, 2008). In light of Figure 1.5, it should be emphasized that while many parts of central and northern Europe present low seismicity with low peak ground accelerations for a rare seismic event (475 years return period or, equivalent, 10 % probability of exceedance over 50 years), see Figure 1.6, crucial infrastructures such as nuclear power plants should be designed for 10000 years return periods. In this case, a major part of Europe may be affected. Seismic design provisions for critical structures (e.g. dams, nuclear power plants, etc.) are out of the scope of both EC8-1 and this handbook. However, the seismic zonation provided by EC8 may be considered as the basis for the definition of the seismic input.

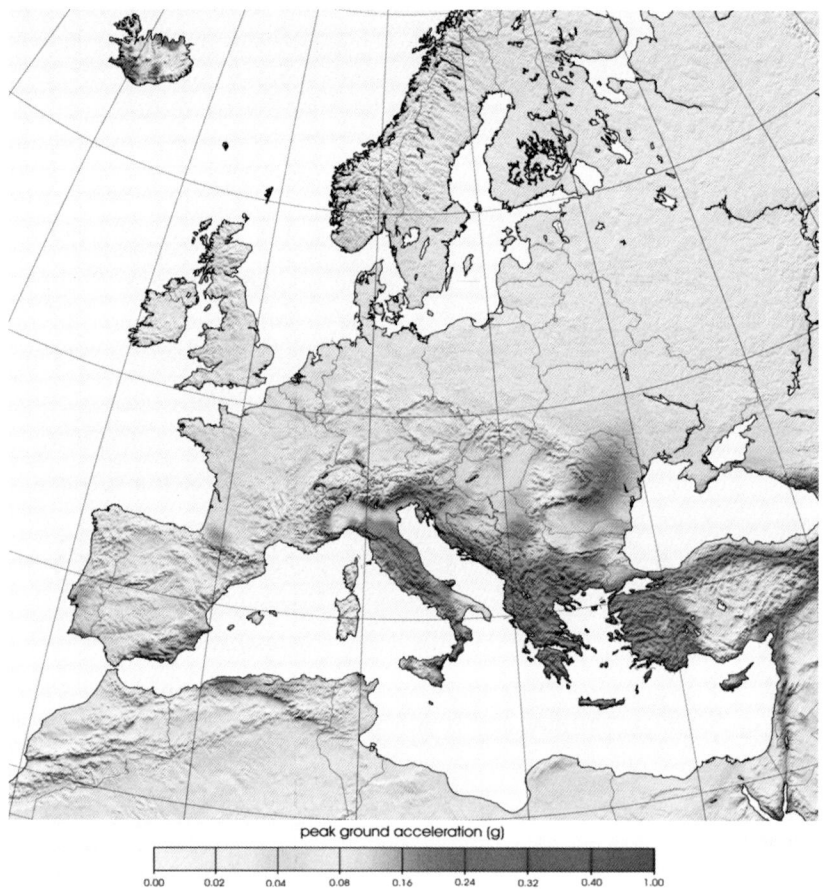

Figure 1.6 – ESC-SESAME European-Mediterranean seismic hazard map for the peak ground acceleration with 10 % probability of exceedance in 50 years for rock site (Solomos *et al*, 2008)

It should be noted that the seismic zonation prescribed by EC8 defines zones for which the reference peak ground acceleration hazard on a "rock" site (a_{gR}) is assumed as uniform. This approach differs from many existing seismic codes, which define that hazard directly for the specific site under consideration (e.g., NEHRP 2003 and 2009, NBCC 2005, NTC-Italy 2008, MOC 2008), or allowing for interpolation between contoured levels of uniform hazard (e.g. NZS 1170.5, 2004). Recently, the European research project SHARE (i.e. "Seismic Hazard Harmonization in Europe", developed within the 7th Framework Program of the European Commission, redefined the European seismic zonation for the application of Eurocode 8. Figure 1.7 depicts the updated version of the European Seismic Hazard Map.

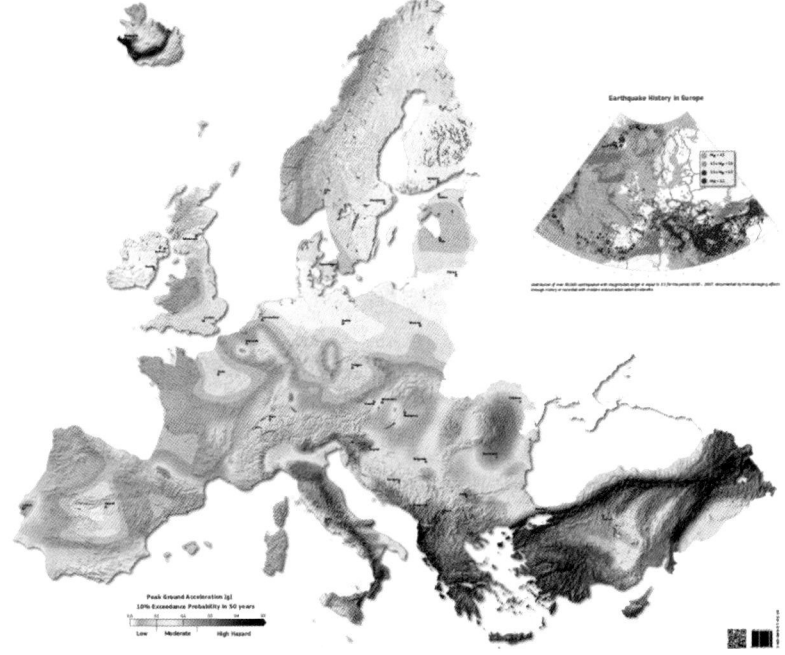

Figure 1.7 – ESHM13 European Seismic Hazard Map for the peak ground acceleration with 10 % probability of exceedance in 50 years for rock site (SHARE, 2013)

1.3.2 New perspectives and trends in seismic codification

The latest seismic events showed that the degree of seismic protection is unsatisfactory. This was evident from what happened during recent earthquakes in Iran, Turkey, China, Italy, Chile and New Zeeland. Under

1. SEISMIC DESIGN PRINCIPLES IN STRUCTURAL CODES

severe or even moderate earthquake activity, buildings have suffered extensive damage and even total collapse. As a consequence, the building design codes increase constantly the seismic demands, to cover the uncertainty of hazard quantification by improving the structural response capacity through accuracy of design and enhanced technical solutions. On the other hand, the design methodology includes enhanced models of analysis and calculation tools, associated with more relevant performance criteria, in order to obtain a better prediction and control of structural response. There are three practical and efficient strategies to reduce the seismic vulnerability, i.e.:

(1) reducing seismic design forces;
(2) enhancing structural damping;
(3) adjusting the structural response to seismic demand (see Figure 1.8).

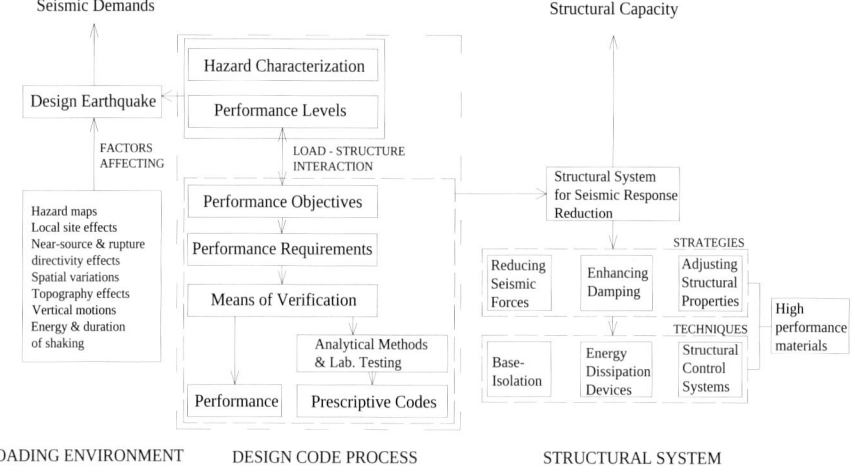

Figure 1.8 – Seismic protection strategies (Rai, 2000)

The first approach, which implies the method of structural isolation, is very efficient, but expensive for existing buildings. The principle behind isolation is to change the natural period of the structure, substantially decoupling a structure from the ground motion input and therefore reducing the resulting inertia forces that the structure must resist. This is done by the insertion of devices allowing a relative displacement between the structure base and the ground or between the upper part of the structure (the roof for instance) and the bottom part of the structure. These devices can be for instance energy absorbing material such as rubber bearings. They will reduce the amount of seismic forces transmitted to the structure.

The second strategy applies passive energy dissipation devices and has shown great potential for seismic hazard mitigation for civil engineering structures. They can significantly enhance the structural performance by reducing inelastic deformation demands on the primary lateral load resisting system and the drift, acceleration and velocity demands on non-structural components. Passive devices can be categorized as rate independent devices (e.g., hysteretic or friction systems) and rate dependent devices (e.g., viscous or viscoelastic systems), where the force output of the latter type of devices is dependent on the rate of applied deformation. In the USA, the NEHRP recommendations allow the structural engineer to utilize passive damping devices to attain performance similar to that of conventional lateral load resisting systems. The design methods used for structures with passive energy dissipation systems are usually based on an approximate or iterative approaches. Experiments and load tests are often required to evaluate and validate these design methods.

The third strategy is in contrast to the approaches mentioned earlier. Indeed, both base isolation and passive devices do not allow changing and adjusting the structural response evolutively with the seismic signal. On the contrary, there is an innovative expanding class of systems referred to as 'smart' or active control systems that allow modifying the vibrational response with the variation of seismic excitation. Different smart techniques have been proposed in recent years that involve adjusting lateral strength, stiffness and damping of the structure during the earthquake to reduce the structural response. Many studies and some field applications have emphasized their potential in reducing the structural response. However, many serious problems are still far from being solved such as the time delay in the control actions, modelling errors, inadequacy of sensors and controllers, structural nonlinearities and reliability, and finally the high operational costs. Therefore, this design strategy is not robust enough and still requires research to be validated and to be implemented in seismic codes.

EC8-1 allows the design of buildings with base isolation (in its chapter 10), but this part of Eurocode 8 is still more informative than applicative, and requires additional guidelines. In Europe, there is also a code specifically devoted to the design, characterization and the acceptance criteria of anti-

1. Seismic Design Principles in Structural Codes

seismic devices, namely EN 15129 (CEN, 2009). This standard specifies the functional requirements and general design rules for the seismic situation, material characteristics, manufacturing and testing requirements, as well as evaluation of conformity, installation and maintenance requirements, but does not give guidelines for structural design and/or criteria for retrofitting intervention with seismic devices on existing buildings. Therefore, in Europe there are surely needs for further codification and technical support documents for design, in order to support their use in practice.

In light of recent scientific findings and the current status of North American codes previously described, it is clear that EC8-1 needs to be updated and improved. With this regard, the Joint Research Centre (JRC) in 2007 elaborated a EN document (Pinto et al, 2007), which summarises the current needs to achieve improved Design Guidelines for seismic protection in Europe. The following objectives of further engineering research aiming to improve European seismic regulations were identified:

- Development of a common methodology to evaluate the earthquake hazard in Europe;
- Development of an assessment and strengthening methodology for more economical and safe solutions for the seismic retrofit of the existing building stock in European earthquake prone areas;
- Development of strengthening techniques of low intrusive effect for application in monuments, historical buildings and other structures;
- Seismic design and upgrading of mechanical, electric and other types of equipment used in the lifelines and industry.

More recently, with respect to steel structures, the Technical Committee 13 - Seismic Design, of the European Convention for Constructional Steelwork (ECCS), issued the document P131/2013 entitled "Assessment of EC8 provisions for seismic design of steel structures" that identifies and proposes the needs and subjects that should be addressed by the drafting team of the ongoing revision of the eurocodes and chapter 6 – Steel structures of EC8-1 in particular. As a non-exhaustive list, the following items were identified for development and inclusion in the future version of EC8-1, chapter 6:

- New criteria for choice of material, in terms of overstrength and toughness;
- Better definition of local ductility: relevant criteria, consideration of class 4 sections;
- Connections in dissipative zones: prequalification criteria;
- New structural systems: new typologies, definition of q factors for them; dissipative components working as fuse devices (such as Buckling Restrained Braces, removable links in Eccentric Braced Frames); systems with re-centering capacity;
- Seismic design of structures in low seismicity zones.

Chapter 2

EN 1998-1:
GENERAL AND MATERIAL INDEPENDENT PARTS

2.1 INTRODUCTION

EN 1998, also denoted as Eurocode 8 or EC8, represents the ensemble of European codes for "Design of structures for earthquake resistance". EC8 applies to the design and construction of buildings and civil engineering works in seismic regions (clause 1.1.1 of EN 1998-1 or EC8-1). It covers common structures, although its provisions are of general validity. It should be pointed out that special structures such as nuclear power plants, large dams or offshore structures are beyond its scope. The objectives of seismic design in accordance with Eurocode 8 are explicitly stated. Its purpose is to ensure that in the event of earthquakes the following design objectives are fulfilled:

- human lives are protected;
- damage is limited;
- structures important for civil protection remain operational.

These objectives are present throughout the code and condition the principles and application rules therein included.

Eurocode 8 is composed of 6 parts dealing with different types of buildings and civil engineering works or subjects, namely:

- EN 1998-1: General rules, seismic actions and rules for buildings;

2. EN 1998-1: General and Material Independent Parts

- EN 1998-2: Bridges;
- EN 1998-3: Assessment and retrofitting of buildings;
- EN 1998-4: Silos, tanks and pipelines;
- EN 1998-5: Foundations, retaining structures and geotechnical aspects;
- EN 1998-6: Towers, masts and chimneys.

Part 1 is the leading document, since it includes also the general provisions for the other parts of EC8, covering the following aspects: seismic performance levels, types of seismic action, types of structural analysis, general concepts and rules which should be applied to all types of structures and different structural materials beyond those generally used for buildings.

It is subdivided into 10 sections as follows:

- Section 1 reports on general aspects about scope, normative references and symbols;
- Section 2 provides the seismic performance requirements and compliance criteria;
- Section 3 gives the rules to represent the seismic actions and their combination with other design actions;
- Section 4 describes the general design rules specifically conceived for buildings;
- From sections 5 to 9 the code provides specific rules for buildings made of each type of building materials (concrete, steel, composite steel-concrete, timber and masonry);
- Section 10 gives the fundamental requirements and design rules for base isolation of buildings.

The main aspects of the material independent sections of EC8-1 (CEN, 2004a) are described in this chapter. The requirements for the seismic design of steel buildings will be explained and discussed in the following chapters.

2.2 PERFORMANCE REQUIREMENTS AND COMPLIANCE CRITERIA

2.2.1 Fundamental requirements

In EC8-1 two performance levels should be accounted for the seismic design of building structures (clause 2.1(1)P), which correspond to the following objectives:

- *No-collapse requirement*: the protection of human lives under rare seismic actions, by preventing the local or global collapse of the structure and preserving the structural integrity with a residual load capacity;
- *Damage limitation requirement*: the limitation of both structural and non-structural damage in case of frequent seismic events (namely seismic action having a larger probability of occurrence than the design seismic action), also without associated limitations of use, the costs of which would be disproportionately high in comparison with the costs of the structure itself.

The first performance level is achieved by applying capacity design rules based on the hierarchy of strength concept (see chapter 1). In the framework of the Structural Eurocodes (namely in accordance with EN 1990 – Basis of Design), that uses the concept of Limit States, this performance requirement is associated with the Ultimate Limit State (ULS) since it deals with the safety of people or the whole structure.

The second performance level is accomplished by limiting the lateral interstorey drifts of the structure within levels acceptable for the integrity of both non-structural (i.e. infill walls, claddings, plants, etc.) and structural elements. It is clear that this requirement is related to the reduction of economic losses in frequent earthquakes. Under such events, the structure should not have evident permanent deformations and its elements should retain their original strength and stiffness and hence should not need structural repair. However, some damage to non-structural elements is acceptable but it should not impose significant limitations of use and should be repairable economically.

2. EN 1998-1: General and Material Independent Parts

Consistently with the previous performance level, this requirement is associated with the Serviceability Limit State (SLS), since it deals with the use of the building, comfort of the occupants and economic losses.

In line with the concept of Performance Based Seismic Design (PBSD) discussed in the previous chapter, these two performance levels have to be verified against two different levels of the seismic action, correlated with the seismicity of the region and the soil conditions. According to the European Union treaties allocating national competence on issues of safety and economy, Eurocode 8 refers to National Annexes for the determination of the hazard levels corresponding to the two performance levels described above. However, as also reported in Figure 2.1a for ordinary structures, EC8 recommends the following:

- *Design (ULS) earthquake*: it corresponds to a seismic action having 10 % exceedance probability in 50 years (namely a mean return period equal to 475 years) for the ultimate limit state (i.e. collapse prevention). At this limit state the design seismic action for structures of ordinary importance over rock ground conditions is termed "reference" seismic action;
- *Damage limitation (SLS) earthquake*: it corresponds to a seismic action having 10 % exceedance probability in 10 years (namely a mean return period equal to 95 years) for the control of both structural and non-structural damage.

Damage Limitation Requirement	No Collapse Requirement
Damage Limitation State (DLS)	**Ultimate Limit State (ULS)**

• For ordinary structures this requirement should be met for a seismic action with 10% probability of exceedance in 10 years
 	
• Return Period = 95 years
• Performance level required:
  - Resist the design seismic action without damage
  - Avoid limitations of use and high repair costs
	• For ordinary structures this requirement should be met for a reference seismic action with 10% probability of exceedance in 50 years
	
	• Return Period = 475 years
	• Performance level required:
	  - Resist the design seismic action without local or global collapse
	  - Preserve structural integrity and residual load bearing capacity after the seismic event

a)

Figure 2.1 – Performance levels according to EC8-1

2.2 PERFORMANCE REQUIREMENTS AND COMPLIANCE CRITERIA

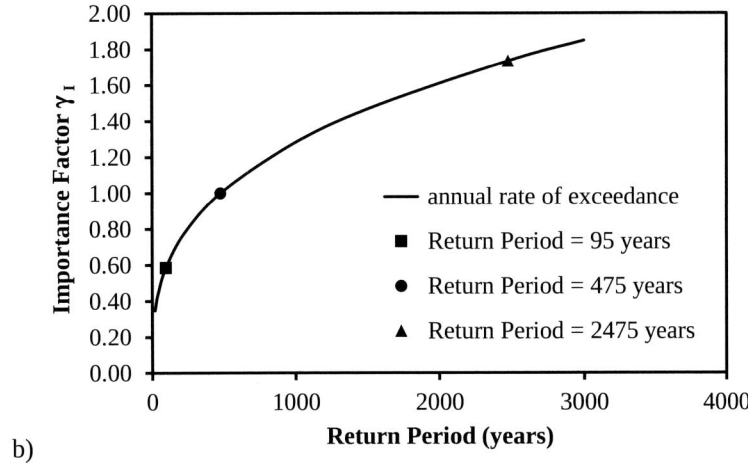

b)

Figure 2.1 – Performance levels according to EC8-1 (continuation)

With this regard, it is useful to clarify shortly the concept of mean return period which is the inverse of the mean (annual) rate of occurrence of a seismic event exceeding a certain threshold. Therefore, assuming a Poisson statistical distribution for the occurrence of earthquakes, the value of the probability of exceedance P_R in T_L years (i.e. the reference time period) of a specific level of the seismic action is related to the mean return period T_R of this level of the seismic action in accordance with the expression $T_R = -T_L / \ln(1 - P_R)$. Thus, for a given T_L (e.g. $T_L = 50$ years) the seismic action may equivalently be specified either via its mean return period T_R (e.g. 475 years) or its probability of exceedance P_R (e.g. 10 %) in T_L years (e.g. $T_L = 50$ years).

In case of essential or large occupancy facilities the code recommends to guarantee an enhanced performance when compared to ordinary structures. This objective is achieved by modifying the hazard level (namely the mean return period) for both collapse prevention and damage limitation, in clause 2.1(3)P, which prescribes that:

"Reliability differentiation is implemented by classifying structures into different importance classes. An importance factor γ_I is assigned to each importance class. Wherever feasible this factor should be derived so as to correspond to a higher or lower value of the return period of the seismic event (with regard to the reference return period) as appropriate for the design of the specific category of structures"

2. EN 1998-1: General and Material Independent Parts

The different levels of reliability are obtained by multiplying the reference seismic action by the importance factor, γ_I, which, in case of using linear analysis, may be applied directly to the action effects obtained with the reference seismic action.

Although EC8-1 (and also the other Parts of EC8) presents recommended values for the importance factors, this is a nationally determined parameter, since it depends not only on the global policy for seismic safety of each country, but also on the specific characteristics of its seismic hazard.

EC8-1 classifies building structures in 4 importance classes depending on: i) the consequences of collapse for human life; ii) their importance for public safety and civil protection in the immediate post-earthquake period and; iii) the social and economic consequences of collapse that correspond roughly to the consequence classes defined in Annex B of EN 1990 (Simões da Silva *et al*, 2016). In particular, at the collapse prevention level, the recommended value of γ_I ranges from 1.4 (for essential buildings) to 1.2 (for large occupancy buildings). In addition, the code allows using $\gamma_I = 0.8$ for buildings of reduced importance for public safety. The recommended values of γ_I for the main building categories are reported in Table 2.1 (Table 4.3 of EC8-1).

Table 2.1 – Importance factors for each building category

Importance class	Buildings	γ_I
I	Buildings of minor importance for public safety, e.g. agricultural buildings, etc.	0.8
II	Ordinary buildings, not belonging in the other categories.	1.0
III	Buildings whose seismic resistance is of importance in view of the consequences associated with a collapse, e.g. schools, assembly halls, cultural institutions, etc.	1.2
IV	Buildings whose integrity during earthquakes is of vital importance for civil protection, e.g. hospitals, fire stations, power plants, etc.	1.4

EC8-1 provides some additional guidance on the determination of the importance factor γ_I in a note of its section 2.1, which clarifies the relationship between the annual rate of exceedance $H(a_{gR})$ and the reference peak ground acceleration a_{gR} for most European seismic sites. This relationship may be taken as: $H(a_{gR}) \cong k_0 \, a_{gR}^{-k}$, with the value of the exponent k depending on seismicity. The value of the exponent k is the slope of the hazard curve in logarithmic scale and it is directly related to the level of seismicity of the site. In particular, the k values are close to 2 for zones where the Peak Ground Acceleration or PGA of very rare earthquake is significantly larger than that corresponding to frequent seismic excitation. In sites where the PGA of very rare and frequent earthquakes have slight difference the k values are larger than 4. In case of regions of high seismicity in Europe like Italy the exponent k is generally assumed equal to 3. Then, if the seismic action is defined in terms of the reference peak ground acceleration a_{gR}, the value of the importance factor γ_I can be considered as the multiplier that amplifies or reduces a_{gR} to achieve the same probability of exceedance in T_L years as in the T_{LR} years for which the reference seismic action is defined. Hence, γ_I may be computed as $\gamma_I \cong (T_{LR}/T_L)^{-1/k}$. Alternatively, the value of the importance factor γ_I that needs to multiply the reference seismic action to achieve a value of the probability of exceeding the seismic action P_L in T_L years other than the reference probability of exceedance P_{LR}, over the same T_L years, may be estimated as $\gamma_I \cong (P_L/P_{LR})^{-1/k}$. This relation is depicted in Figure 2.1b for the seismicity exponent k equal to 3.

Additionally, section 2.2 of EC8-1 requires compliance with a number of pertinent specific measures in order to limit the uncertainties and to promote a good behaviour of structures under seismic actions more severe than the design seismic action.

Hereinafter, these measures are described and commented in the relevant sections. It should be noted that the use of additional prescriptions is implicitly equivalent to account for a third performance level that intends to prevent global collapse under very strong and rare earthquakes (i.e with return period in the order of 2475 years, more severe than the design earthquake). After such an earthquake, the structure may be heavily damaged, with large permanent drifts and having lost significantly its lateral stiffness and resistance but it should still keep a minimal load bearing capacity to prevent global collapse. More details about this third performance level can be found in EN 1998-3 (CEN, 2005c), where specific measures and control are recommended for the seismic assessment of existing structures. However, the application of part 3 of EC8 is outside the scope of this design manual.

2. EN 1998-1: General and Material Independent Parts

2.2.2 Ultimate limit state

This limit state requires the verification of both lateral resistance and energy-dissipation capacity. This implies that the fulfilment of the no-collapse requirement does not require that the structure remains elastic under the design seismic action. On the contrary, it allows/accepts the development "of significant inelastic deformations in the structural members or their connections, provided that the integrity of the structure is kept. It also relies on the (stable) energy dissipation capacity of the structure to control the amount of plastic engagement into specific zones that, otherwise, would result in excessively large demand into the structure.

Besides the verification of the individual structural elements (for resistance and ductility), in accordance with specific rules for the different structural materials, the ULS verification entails the checking of: i) the overall stability of the structure (overturning and sliding); ii) the foundations and the bearing capacity of the soil; iii) the influence of second order effects; and iv) and iv) the non-detrimental effects of non-structural elements.

The core concept of the Eurocode 8 approach is the possible trade-off between resistance and ductility, that is at the base of the introduction of Ductility Classes, and the use of behaviour factors. This is explained as follows in clause 2.2.2(2):

"The resistance and energy-dissipation capacity to be assigned to the structure are related to the extent to which its non-linear response is to be exploited. In operational terms such balance between resistance and energy-dissipation capacity is characterised by the values of the behaviour factor q and the associated ductility classification, which are given in the relevant Parts of EN 1998. As a limiting case, for the design of structures classified as low-dissipative, no account is taken of any hysteretic energy dissipation and the behaviour factor may not be taken, in general, as being greater than the value of 1.5 considered to account for overstrengths. For steel or composite steel concrete buildings, this limiting value of the q factor may be taken as being between 1.5 and 2 (see Note 1 of Table 6.1 or Note 1 of Table 7.1, respectively). For dissipative structures the behaviour factor is taken as being greater than these limiting values accounting for the hysteretic energy dissipation that mainly occurs in specifically designed zones, called dissipative zones or critical regions."

Notwithstanding these basic concepts, the provisions and calculations required by EC8-1 to verify the fulfilment of this limit state are force-based, essentially in line with all the other Eurocodes. It should be noted that, exactly on the contrary, the physical feature of the seismic action corresponds to the application of (rapidly changing) displacements at the base of the structures and not to the application of forces. It is trivial that in perfectly linear systems there is a complete equivalence in representing the action as imposed forces or imposed displacements. However, in non-linear systems, the application of force controlled or displacement controlled actions may lead to substantially different structural responses. Accordingly, the ability of structures to survive earthquakes depends basically on their aptitude to develop both local and global deformations adequate to the seismic demand, keeping its load bearing capacity, which is fundamentally different to the simple capacity to resist lateral forces. Notwithstanding this inconsistency, the use of force-based design is well established and, as mentioned above, is adopted in EC8-1 as the reference method, because most of the other non-seismic actions that structural designers have to handle are forces imposed to the structures. Hence, within the overall design process the use of a force-based approach is very practical and attractive even for seismic actions. Furthermore, analytical methods for a displacement based approach in seismic design are not fully developed and not familiar to many designers.

In the last twenty years, structural engineers and researchers have demonstrated a growing interest in displacement-based design (Priestley, 2000) because, as discussed above, displacements describe more rationally the structural response than forces. The displacement design procedure is based on the definition of a substitute structure that models an inelastic system as an equivalent elastic system. In such a way the inelastic structural system can be designed and analysed using elastic response spectra. The main difficulties are hidden in the determination of both the design displacement, the dynamic characteristics for the substitute structure, and the displacement spectra. Nevertheless, it should be noticed that EC8-1 opens the possibility to use displacement-based approaches as alternative design methods, for which it presents an Informative Annex with operational rules to compute the target displacements for Nonlinear Static (Pushover) Analysis. Very recently, in the framework of the European project DISTEEL (2015), a set of practical displacement-based design guidelines for steel moment-resisting frame structures was developed and performance criteria accounting for different beam-column joint typologies have been provided. Further detailed information on displacement-based design may be found in the freely available

2. EN 1998-1: General and Material Independent Parts

final report of the European project DISTEEL – Displacement based seismic design of steel moment resistant frame structures (Calvi *et al*, 2015).

2.2.3 Damage limitation state

The performance requirements associated with this Limit State require the structure to withstand frequent earthquakes without significant damage or loss of operationality. Damage is mainly expected in non-structural elements and its occurrence depends on the interaction with the deformation of the structure under the earthquake motion. The loss of operationality of systems and networks depends also on the same interaction even though in some equipments accelerations may also be responsible for significant damage.

Indeed, non-structural components in buildings can be categorized as drift-sensitive and acceleration-sensitive, and they include a large variety of different architectural, mechanical and electrical components such as parapets, cabinets, ornaments, general mechanical, manufacturing and process machinery, lighting fixtures, contents of the buildings etc. (HAZUS, 1997). The non-structural components, which constitute the major portion of the economic value of buildings (Ayers and Sun, 1973; Whitman *et al*, 1973; Rihal, 1992), are generally acceleration-sensitive components. Therefore, they are mainly damaged due to the large floor acceleration demands under the seismic effects.

Unfortunately, EC8-1 does not provide any requirements to verify non-structural elements against acceleration demand, while it stipulates verifications for structural lateral displacements and interstorey drifts that should be smaller than a set of deformation limits depending on the characteristics of the non-structural elements.

For instance, EC8-1 establishes for buildings the following limits of the interstorey drift ratios (relative displacement divided by the interstorey height) due to the frequent earthquake (serviceability seismic action):

- 0.5 % for buildings having non-structural elements of brittle materials attached to the structure;
- 0.75 % for buildings having ductile non-structural elements;
- 1.0 % for buildings without non-structural elements or with non-structural elements fixed in a way such that they do not interfere with structural deformations.

Additional requirements may be imposed in structures important for civil protection so that the function of the vital services in the facilities is maintained. These requirements should lead to resilient structures, i.e., structures that are able to absorb or avoid damage and to ensure their capability to function during a seismic event and to recover quickly after an earthquake (i.e. to re-establish their operativeness), thus reducing risk to the lives of persons. Very recently, the US president issued an Executive Order ("Establishing a Federal Earthquake Risk Management Standard") that addresses these objectives (Obama, 2016).

2.2.4 Specific measures

As previously stated, EC8-1 implicitly aims at providing the satisfaction of a third performance level that intends to prevent global collapse (e.g. Collapse Prevention Limit State – CPLS) during a very strong and rare earthquake.

This is not fulfilled by imposing specific requirements for a higher threshold of the seismic action but rather by imposing some specific measures to be accounted for throughout the entire design process. These specific measures (section 2.2.4 of EC8-1) aim at increasing the structural reliability by imposing the following prescriptions:

- To the extent possible, structures should have simple, compact and regular shapes both in plan and elevation;
- Brittle failure or the premature formation of unstable plastic mechanisms should be avoided. Therefore, in order to guarantee overall dissipative and ductile behaviour, hierarchy of resistances should be adopted among the primary and secondary structural components, trying to enforce ductile failure modes into plastic fuses, but also verifying the existence of secondary ductile mechanisms (e.g. in case of overstrength shear connections it is advisable to design for bearing capacity or net tearing strength lower than bolt shear strength, because the latter mechanism is brittle). In any case, every measure for ensuring a suitable plastic mechanism and for avoiding brittle failure modes should be envisaged;
- Special care should be addressed in the design of the dissipative zones, which should be detailed for a large and stable hysteretic

response and to maintain their capacity to transmit the required forces under cyclic conditions;
- The analysis should be based on adequate structural models and, if necessary, it should also take into account the influence of soil deformability and of non-structural elements;
- The strength and stiffness of the foundations should be adequate for transmitting the actions received from the superstructure to the ground as uniformly as possible;
- The design documents should be well detailed and include all relevant information regarding materials properties, dimensions of all members, constructional details and erection instructions, if appropriate;
- Provisions for the necessary quality control should also be given in the design documents, specifying also the verification methods to be used for the elements of primary structural importance;
- In regions of high seismicity and in highly important structures, formal quality system plans, covering design, construction, and use, additional to the control procedures prescribed in the other relevant Eurocodes, should be used.

2.3 SEISMIC ACTION

2.3.1 The fundamentals of the dynamic model

The seismic action is characterized by ground vibration under the base of structures. The acceleration, velocity and displacement imposed to the base of the structure vary with the duration of the motion. It is thus clear that the seismic action has dynamic properties that interact with those of the striken building. Indeed, due to ground vibrations, deformations occur in the structure, which in turn generate restoring forces (due to the structure's stiffness), damping forces (due to internal friction, external dissipation and essentially by the damage of the dissipative zones) and inertial forces (due to the mass of the structure). These forces act in dynamic equilibrium. In the case where all dynamic degrees of freedom of the structure have the same direction as the seismic action, the dynamic equilibrium of the structure can be expressed by equation (2.1), as follows:

2.3 SEISMIC ACTION

$$[m]\{a(t)\}+[c]\{v(t)\}+[k]\{u(t)\}=-[m]\{1\}a_g(t), \tag{2.1}$$

where $[m]$ is the mass matrix, $[c]$ is the damping matrix, $[k]$ is the stiffness matrix, $\{a\} = \{ü\}$ is the relative acceleration vector, $\{v\} = \{\dot{u}\}$ is the relative velocity vector, $\{u\}$ is the relative displacement vector, $\{1\}$ is the unit vector, and $a_g(t)$ is the ground acceleration.

The structural analysis of the seismic action requires: (a) a structural model, (b) a seismic action model and (c) a structural analysis method. As it can be seen from the equations of motion (2.1) and Figure 2.2, the structural model requires the establishment of a mass model, a damping model and the mechanical properties of the structural model (stiffness and possibly strength).

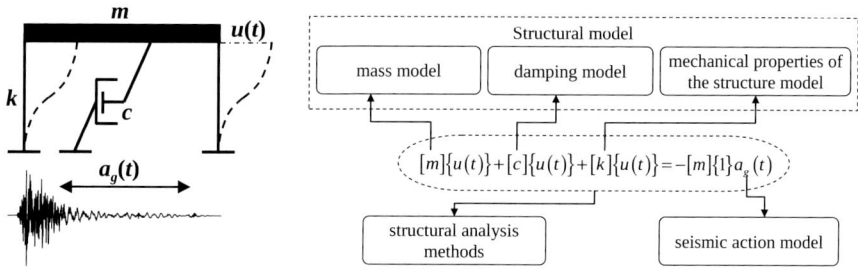

Figure 2.2 – Fundamental chart of the necessary elements for the structural analysis

Equation (2.1) shows that the structural response under dynamic loading, as is the case of the seismic action, substantially depends on the fundamental dynamic properties of the system such as the fundamental period of vibration and the effective mass participation factor. In order to understand the meaning of these quantities, consider the case of a single-degree-of-freedom system (SDOF) under free vibration (namely with $a_g(t) = 0$) and undamped conditions (namely damping $c = 0$), which can be written as:

$$m \cdot \ddot{u}(t) + k \cdot u(t) = 0 \tag{2.2}$$

where k is the stiffness of the system and m is the relevant mass. The solution of equation (2.2) is an harmonic function with the following general solution:

$$u(t) = A\cos\omega t + B\sin\omega t \tag{2.3}$$

2. EN 1998-1: General and Material Independent Parts

where A and B are constants of integration and ω is the circular frequency, defined in equation (2.4) that represents the natural frequency of the mass (therefore it is expressed in terms of rad/sec). It physically represents the frequency at which the mass will move regardless of the amplitude of the motion as long as the spring in the system (i.e the structural system) behaves in the elastic range.

$$\omega = \sqrt{(k/m)} \tag{2.4}$$

The cyclic frequency (i.e. the number of times the mass experiences a complete oscillation per second) is expressed in Hertz, as follows:

$$f_n = \frac{1}{2\pi}\sqrt{\left(\frac{k}{m}\right)} \tag{2.5}$$

For structural design, the inverse of equation (2.5) is mostly used and it is called the fundamental period T_n, given by:

$$T_n = \frac{1}{f_n} = \frac{2\pi}{\omega} = 2\pi\sqrt{\left(\frac{m}{k}\right)} \tag{2.6}$$

The physical meaning of equation (2.3) is that the SDOF system under free and undamped vibration oscillates indefinitely with the same shape of displacement, and regardless of the amplitude of the displacement the time to have a complete reversal of motion is constant. This property characterizing the shape of the displacements is called the fundamental mode of vibration.

Considering now a SDOF system under free but damped vibration, equation (2.1) becomes

$$m \cdot a(t) + c \cdot v(t) + k \cdot u(t) = 0 \tag{2.7}$$

whose general solution for values of damping lower than the critical damping c_{crit} is given by:

$$u(t) = e^{-\frac{c}{2m}t} \cdot \{A \cdot \sin\omega_D t + B \cdot \cos\omega_D t\} \tag{2.8}$$

Equation (2.8) shows that, similarly to the undamped case, the SDOF system oscillates indefinitly but with a decaying amplitude. Damping is often expressed as the damping ratio ξ, defined as:

$$\xi = \frac{c}{c_{crit}} = \frac{c}{2\sqrt{k \cdot m}} \quad (2.9)$$

It can be demonstrated that the results briefly summarized for a SDOF system can be extended to the case of multi-degree of freedom systems (MDOF). In particular, the solution to the homogeneous form of equation (2.1) can be found in terms of eigenvalues and eigenvectors. The eigenvalues are the fundamental frequencies, while the eigenvectors represent the vibration modes. This implies that a generic MDOF with "N" degrees of freedom is characterized by "N" fundamental periods with corresponding "N" fundamental modes. Each mode of vibration can be treated as an equivalent SDOF having the natural frequency ω_n equal to that of the "N^{th}" mode (Chopra, 2011).

To clarify this issue, it is convenient to transform the undamped and unforced form of equation (2.1) into modal coordinates. Considering $u = \phi\eta$ and $\ddot{u} = \phi\ddot{\eta}$, where $\{\phi\}$ is the mode shape vector and η is the modal coordinate, replacing into equation (2.1) and premultiplying by ϕ^T, leads to:

$$\{\phi\}^T[m]\{\phi\}\ddot{\eta}(t) + \{\phi\}^T[k]\{\phi\}\eta(t) = 0 \quad (2.10)$$

In general, the modal coordinate may vary with time whereas the mode shape vector $\{\phi\}$ does not vary. In free vibration, the modal coordinate *i* varies like a simple harmonic with a frequency equal to the natural frequency of the *i*-th mode, as follows:

$$\eta_i(t) = A_i \cos\omega_i t + B_i \sin\omega_i t \quad (2.11)$$

As it can be easily recognized, equation (2.11) is similar to equation (2.3), and it is possible to rearrange the non-trivial solution of equation (2.10) as follows:

$$\{u(t)\} = \{\phi\}[\eta(t)] = \{\phi\}_i \left[A_i \cos\omega_i t + B_i \sin\omega_i t\right] \quad (2.12)$$

2. EN 1998-1: General and Material Independent Parts

Hence, equation (2.12) allows treating each mode of vibration as an equivalent SDOF system.

Starting from equation (2.10) it is also possible to define the system's generalized mass matrix $\hat{m} = \{\phi\}^T [m] \{\phi\}$. Let $\bar{r} = \{1\}$ be the influence vector, which represents the displacements of the masses resulting from static application of a unit ground displacement (it should be noted that the influence vector induces a rigid body motion in all modes), it is possible to introduce the coefficient vector $\bar{L} = \{\phi\}^T [m] \bar{r} = \{\phi\}^T [m] \{1\}$.

With some manipulations it is possible to define two fundamental dynamic properties, namely the modal participation factor Γ_i and the effective modal mass $m_{eff,i}$ for mode "i", which are, respectively, given by:

$$\Gamma_i = \frac{\bar{L}_i}{\hat{m}_{ii}} \quad (2.13)$$

$$m_{eff,i} = \frac{\bar{L}_i^2}{\hat{m}_{ii}} \quad (2.14)$$

The modal participation factor Γ_i mass represents the ratio between the mass participating in the unitary forcing function and mass participating in inertia effects for mode "i".

The effective modal mass $m_{eff,i}$ for mode "i" provides another measure of the amount of system mass participating in a particular mode, namely it represents the work done by the mass of the system oscillating with mode "i" and excited by a uniform unitary acceleration. Therefore, $m_{eff,i}$ is a very important parameter because it characterizes the actual contribution that each mode of vibration gives to the overall dynamic response of the system. It is also clear that the sum of the effective modal masses from "N" modes of vibration is equal to the total mass of the structure, namely $\Sigma m_{eff,i} = M_{tot}$.

Further information about dynamics of structures may be found in classical textbooks (Clough and Penzien, 1993).

2.3.2 Basic representation of the seismic action

The seismic response can be expressed in terms of displacement, velocity or acceleration by means of the so-called response spectrum. A response spectrum is a plot of the peak response (displacement, velocity or acceleration) of a series of oscillators (i.e. single degree of freedom systems) with varying natural frequency, which are forced into motion by the same seismic input (namely an accelerogram). For illustration, Figure 2.3 depicts the shapes of the response spectra using a damping ratio of $\xi = 2\ \%$ for the Castelnuovo-Assisi station seismic record of the Umbria-Marche earthquake (26 September 1997).

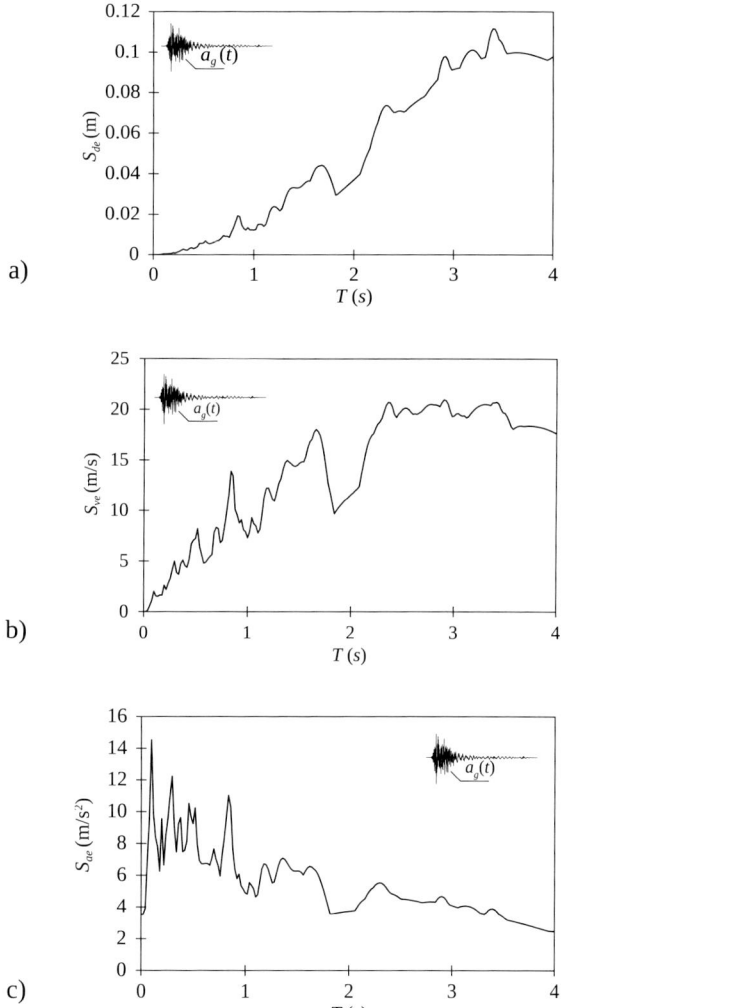

Figure 2.3 – Displacement (S_{de}), pseudo-velocity (S_{ve}) and pseudo-acceleration (S_{ae}) elastic response spectra for the Castelnuovo-Assisi station seismic record of the Umbria-Marche earthquake (26 September 1997)

2. EN 1998-1: GENERAL AND MATERIAL INDEPENDENT PARTS

The principle to determine the displacement elastic spectrum for the same record is reported in Figure 2.4. Indeed, as previously mentioned, in order to build a displacement spectrum it is necessary to determine the seismic response $u(t)$ of SDOF systems having different values of the period of vibration (T) and a fixed damping ratio ξ (that is assumed equal to 2 % in this example), under the action of the accelerogram $a_g(t)$. Only the peak values (maximum absolute positive or negative values) of the displacement $u(t)$ are kept, which are indicated by $S_{de}(T)$ and represent the points of the displacement spectrum as a function of the period of vibration T. Both velocity and acceleration spectra are obtained in the same manner.

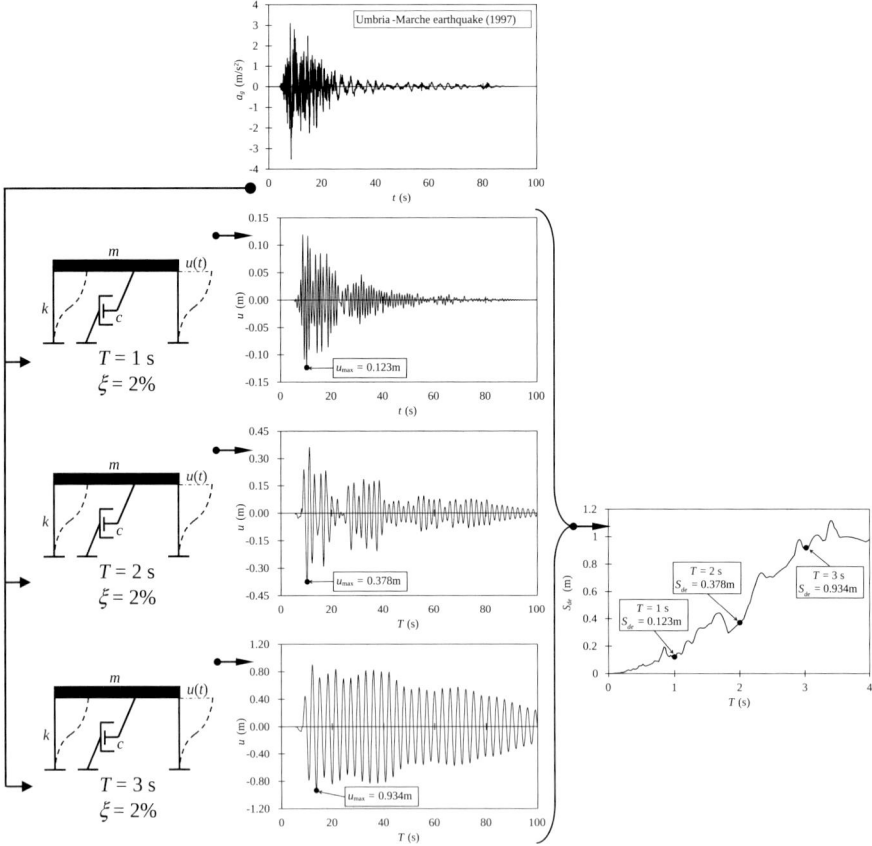

Figure 2.4 –The schematic representation of the procedure to derive the displacement elastic spectrum for the Castelnuovo-Assisi station seismic record of the Umbria-Marche earthquake (26 September 1997)

2.3 Seismic Action

The resulting plot can be used to obtain the response of any linear system, given its natural frequency of oscillation. With reference to a building structure under seismic excitation, once the natural period of the structure is known, then the peak response of the building can be estimated by reading the value from the ground response spectrum for the appropriate frequency or period. In Eurocode 8, this value is the basis for calculating the seismic design forces.

Once the elastic spectra are known the peak seismic response (displacements, velocities, accelerations) of a single dynamic degree of freedom system having a period of vibration T and damping ratio ξ, can be determined. However, velocity and acceleration of the structural masses, which can be determined from the velocity and acceleration spectra, do not have a practical direct use. Instead, other notions are of interest for the seismic action structural analysis, namely the pseudo-velocity and the pseudo-acceleration.

The pseudo-velocity $S_{ve}(T)$ corresponding to the period of vibration T is obtained by multiplying the spectral displacement $S_{de}(T)$ by the vibration circular frequency $\omega = 2\pi/T$:

$$S_{ve} = \frac{2\pi}{T} S_{de} \qquad (2.15)$$

The pseudo-velocity S_{ve} has velocity units, but the prefix "pseudo" is used because it differs from the peak velocity of the single dynamic degree of freedom system. The pseudo-velocity S_{ve} is directly related to the peak value of the deformation energy.

Similarly, the pseudo-acceleration $S_{ae}(T)$ corresponding to the period of vibration T is obtained by multiplying the spectral pseudo-velocity $S_{ve}(T)$ by the vibration circular frequency $\omega = 2\pi/T$:

$$S_{ae} = \frac{2\pi}{T} S_{ve} = \left(\frac{2\pi}{T}\right)^2 S_{de} \qquad (2.16)$$

The pseudo-acceleration $S_{ae}(T)$ has acceleration units, but differs from the peak acceleration of a single dynamic degree of freedom system. It corresponds to the acceleration at the peak displacement of the seismically excited mass.

A more convenient solution to determine seismic actions on a structure consists in using the concept of equivalent static force. For a single dynamic

2. EN 1998-1: GENERAL AND MATERIAL INDEPENDENT PARTS

degree of freedom system, an equivalent static force F that generates the displacement $|u(t)|_{max} \equiv S_{de}(T)$ can be determined, see Figure 2.5, as follows:

$$F = k \cdot S_{de} \qquad (2.17)$$

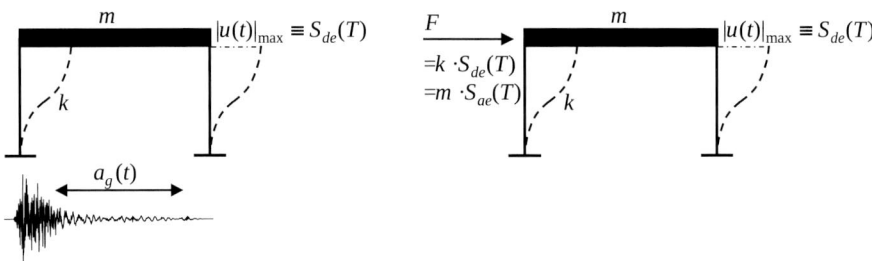

Figure 2.5 – Concept of equivalent static force

Taking into account the definition of the circular frequency (equation 2.4) and its relationship with the period of vibration, $\omega = 2\pi/T$, the equivalent static force can be expressed as follows:

$$F = k \cdot S_{de} = m \cdot \left(\frac{2\pi}{T}\right)^2 \cdot S_{de} = m \cdot S_{ae} \qquad (2.18)$$

Equation (2.18) shows that for a dynamic single degree of freedom system the equivalent static force F can be determined by multiplying the mass m of the system with the spectral pseudo-acceleration S_{ae}. From practical considerations, the use of equivalent static forces and response spectra to determine the structural seismic response is particularly convenient. Firstly, the response spectra for each accelerogram should be determined by laborious dynamic analysis. Once the spectra are available, the structural response (internal forces, deformations, etc.) can be obtained from a static structural analysis. Secondly, equivalent static forces and response spectra analysis directly provide the peak structural response, necessary to design the structure.

It should be noted that the seismic excitation is a phenomenon characterized by a high degree of uncertainty. For this reason, it is unfeasible to design structures to resist solely seismic action based on the response spectrum of a single accelerogram, because it does not provide an adequate level of safety. Therefore,

2.3 Seismic Action

the codified elastic response spectra from seismic design provisions are obtained by processing comprehensive sets of existing seismic records from seismic hazard analyses, representing an individual smooth spectra "envelope". Moreover, the elastic response spectra used for structural analysis and design are idealized in order to have a convenient mathematical description, as shown in Figure 2.6 that illustrates the shape of displacement (S_{de}), pseudo-velocity (S_{ve}) and pseudo-acceleration (S_{ae}) elastic response spectra compliant with EC8 recommendations.

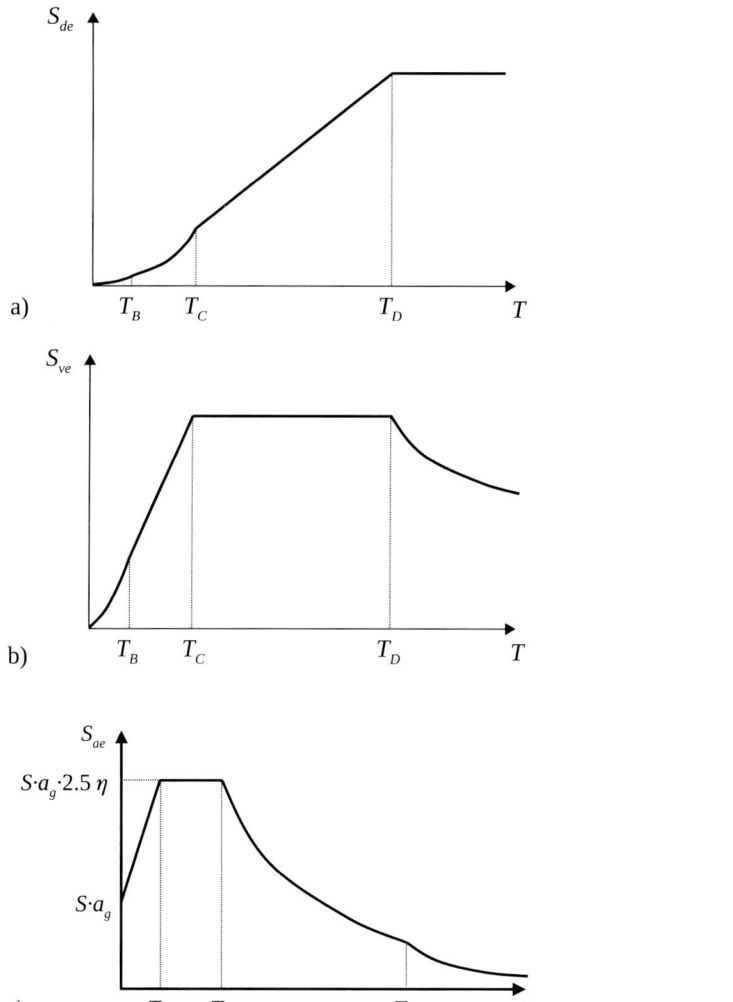

Figure 2.6 – The shape of displacement (S_{de}), pseudo-velocity (S_{ve}) and pseudo-acceleration (S_{ae}) elastic response spectra in EC8

2. EN 1998-1: General and Material Independent Parts

2.3.3 The seismic action according to EN 1998-1

Depending on the distance from the fault, two types of ground motions can occur, namely near- and far-fault earthquakes. The former are characterized by short duration and consist of one or more pulses of motions, which expose the structures to high input energy (i.e. large displacement and strength demands basically in one direction) at the beginning of the earthquake and large permanent ground displacements.

Records from near fault earthquakes indicate that this pulse signal is a narrow band pulse whose period increases with magnitude, namely the period at the peak of the corresponding response spectra increases with magnitude, such that the acceleration spectra of near-fault ground motions from moderate magnitude earthquakes may exceed those of larger earthquakes at intermediate periods (Somerville *et al*, 1997; Somerville 2003). This feature implies that the influence of higher modes of vibrations can substantially increase with respect to far-fault motions as well as the ductility demand. In addition, near-fault ground motions are often characterized by high vertical peak ground acceleration a_{vg}, that may reach a similar value as the horizontal peak ground acceleration a_g. Under this combined action, the axial forces acting in the columns may be substantially modified, thus producing very severe ductility demands and premature failure.

An example of collapse induced by a near fault earthquake is the well-known reinforced concrete Olive View Hospital, which sustained significant damage during the San Fernando (California) earthquake in 1971 (Chopra and Chintanapakdee 2001). Severe structural damage due to near-fault motions also occurred during Northridge (1994), Kobe (1995), Chi-Chi (1999) and L'Aquila (2009) earthquakes, where the combined effect of pulse-like and vertical motion induced severe damage and collapse of both industrial and residential buildings.

EC8-1 considers solely far-fault earthquakes and it does not characterize the seismic action for near-source effects. Hence, no provision for near-fault effects is given for the design of buildings. However, EN 1998-2 recommends a site-specific spectra requirement for the design of bridges within the fault zone. In this case, clause 3.2.2.3 (1)P of EN 1998-2 clearly states that "*…site-specific spectra considering near source effects shall be used, when the site is located within 10 km horizontally of a known active seismotectonic fault that may produce an event of Moment Magnitude higher than 6.5*".

2.3 SEISMIC ACTION

In EC8-1 (section 3.2.2) the seismic action is represented by an elastic acceleration response spectra at a given point of the ground surface. Neglecting the rotational components, the seismic excitation of a point on the ground surface is characterized by three translational components: two orthogonal components in the horizontal plane and one component in the vertical plane ($S_{ae,x}$, $S_{ae,y}$ and $S_{ae,z}$ respectively, depicted in Figure 2.7). The two horizontal components of the seismic action ($S_{ae,x}$ and $S_{ae,y}$) are considered independent and are represented by the same spectral shape. The vertical component of the seismic action can be neglected for most of the current structures. According to clause 4.3.3.5.2, the vertical component of the seismic motion a_{vg} must be taken into consideration solely when the vertical peak ground acceleration exceeds 0.25g (i.e. about 2.5 m/s^2) and the structure has at least one of the following features:

- it includes horizontal elements with spans larger than 20 m;
- it includes cantilever elements with lengths larger than 5 m;
- it includes prestressed horizontal elements;
- it includes columns supported by beams;
- it is base isolated.

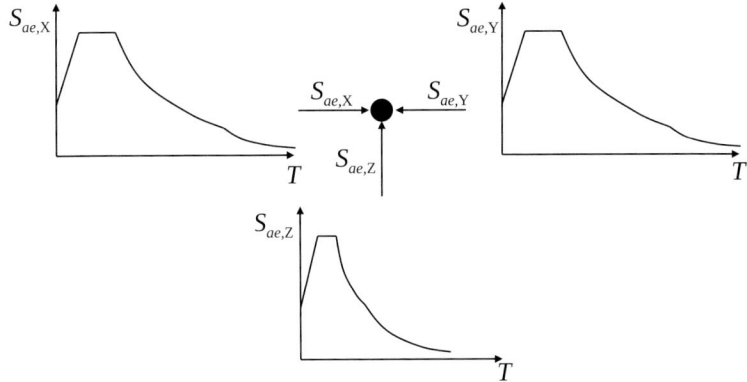

Figure 2.7 – EC8 representation of seismic action acting upon a point on the ground surface: two horizontal response spectra ($S_{ae,x}$ and $S_{ae,y}$) and the vertical response spectrum ($S_{ae,z}$)

EC8 assumes the same spectral shape for both damage limitation and collapse prevention limit states (clause 3.2.2.1(2)). In order to consider different hazard levels, a factor ν is provided to obtain the seismic demand at the

2. EN 1998-1: GENERAL AND MATERIAL INDEPENDENT PARTS

serviceability limit state (clause 4.4.3.2(2)). The latter is derived multiplying by ν the ordinates of the elastic acceleration response spectrum at ultimate limit state. The value of ν depends on two aspects: i) the local seismic hazard conditions; ii) the protection of property objective. The recommended values for ν are equal to 0.4 and 0.5 for ordinary and large-occupancy buildings, respectively.

The elastic response spectrum of the design seismic action (namely that corresponding to collapse prevention) is characterized by the reference ground acceleration on rock, a_{gR}, which should be provided by national seismic zoning maps. The shape of the EC8 pseudo-acceleration spectrum is shown in Figure 2.6c and it is constituted by three main regions having constant properties, namely the spectral acceleration (for period T from T_B to T_C), the spectral pseudo-velocity (for period T from T_C to T_D) and spectral displacement (for period $T > T_D$). The amplitude and the period of these zones depend on the soil type. In EC8-1, five standard soil types are considered (Table 3.1 in EC8-1), as follows:

- Type A: rock, with an average shear wave velocity v_s in the top 30 m larger than 800 m/s;
- Type B: very dense sand or gravel, or very stiff clay, with v_s ranging from 360 to 800 m/s;
- Type C: medium-dense sand or gravel, or stiff clay, with v_s ranging from 180 m/s to 360 m/s;
- Type D: loose-to-medium sand or gravel, or soft-to-firm clay, with v_s less than 180 m/s;
- Type E: 5m to 20m thick soil with v_s less than 360 m/s, underlain by rock.

The entire elastic spectrum is anchored to the mapped "reference" acceleration on rock (clause 3.2.1(2)) multiplied by: i) the importance factor γ_I (clause 3.2.1(3), already defined in section 2.2); ii) the soil factor $S \geq 1$ accounting for the dynamic amplification effects on spectral values due to soil conditions (clause 3.2.2.2(1)P); iii) a damping correction factor η (clauses 3.2.2.2(1)P and 3.2.2.2(3)) equal to:

$$\sqrt{10/(5+\xi)} > 0.55, \qquad (2.19)$$

where ξ is the viscous damping ratio expressed as a percentage.

2.3 SEISMIC ACTION

Assuming $\gamma_I = 1$, the elastic acceleration response spectra are given as follows:

$$0 < T < T_B \quad S_e(T) = a_g \cdot S \cdot \left(1 + \frac{T}{T_B} \cdot (\eta \cdot 2.5 - 1)\right)$$

$$T_B < T < T_C \quad S_e(T) = a_g \cdot S \cdot \eta \cdot 2.5$$

$$T_C < T < T_D \quad S_e(T) = a_g \cdot S \cdot \eta \cdot 2.5 \cdot \left(\frac{T_C}{T}\right)$$

$$T > T_D \quad S_e(T) = a_g \cdot S \cdot \eta \cdot 2.5 \cdot \left(\frac{T_C \cdot T_D}{T^2}\right)$$

(2.20)

The corner periods (T_A, T_B, T_C and T_D) and the corresponding amplification factors depend on the spectral shape of the full hazard spectrum. According to EC8-1, two earthquake scenarios are considered to which correspond two different types of spectra with the same spectral shape described by equations (2.20):

- Type 1 (see Figure 2.8a) for moderate to large magnitude earthquakes, namely with with surface wave magnitude M_s larger than 5.5;
- Type 2 (see Figure 2.8b) for low magnitude earthquakes with M_s less than 5.5.

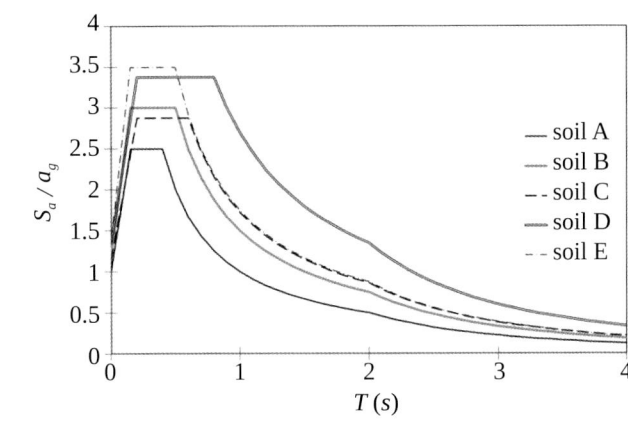

a)

Figure 2.8 – EC8 Elastic acceleration response spectra: Type 1 (a) and 2 (b)

2. EN 1998-1: General and Material Independent Parts

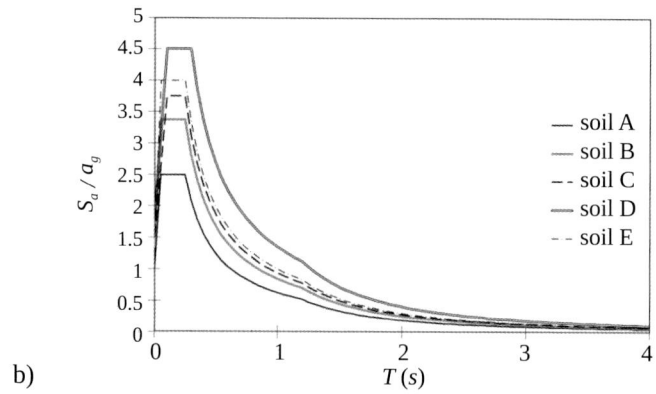

b)

Figure 2.8 – EC8 Elastic acceleration response spectra: Type 1 (a) and 2 (b) (continuation)

The values to be ascribed to T_B, T_C, T_D and S for each ground type and type (shape) of spectrum to be used in a country may be found in its National Annex (clause 3.2.2.2(2)P). However, EC8-1 provides recommended values for the spectral parameters (see Table 2.2, or Tables 3.2 and 3.3 of EC8-1).

Table 2.2 – EC8 recommended values of the parameters describing both Type 1 and Type 2 elastic response spectra

Elastic Response Spectra	Ground Type	S	T_B (s)	T_C (s)	T_D (s)
Type 1	A	1.0	0.15	0.4	2.0
	B	1.2	0.15	0.5	2.0
	C	1.15	0.20	0.6	2.0
	D	1.35	0.20	0.8	2.0
	E	1.4	0.15	0.5	2.0
Type 2	A	1.0	0.05	0.25	1.2
	B	1.35	0.05	0.25	1.2
	C	1.5	0.10	0.25	1.2
	D	1.8	0.10	0.30	1.2
	E	1.6	0.05	0.25	1.2

2.3 SEISMIC ACTION

The elastic displacement response spectrum, $S_{de}(T)$, shall be obtained by direct manipulation of equation (2.16) expressed as a function of the elastic acceleration response spectrum, as follows:

$$S_{de} = \left(\frac{T}{2\pi}\right)^2 S_e \qquad (2.21)$$

EC8 recommends to use equation (2.21) for vibration periods not exceeding 4.0 s.

In addition, EC8 provides analytical expressions to determine the vertical component of the seismic action (clause 3.2.2.3(1)P), which shall be represented by an elastic response spectrum, $S_{ve}(T)$, as follows:

$$
\begin{aligned}
0 < T < T_B &\quad S_{ve}(T) = a_{vg} \cdot \left[1 + \frac{T}{T_B} \cdot (\eta \cdot 3.0 - 1)\right] \\
T_B < T < T_C &\quad S_e(T) = a_{vg} \cdot \eta \cdot 3.0 \\
T_C < T < T_D &\quad S_e(T) = a_{vg} \cdot \eta \cdot 3.0 \cdot \left(\frac{T_C}{T}\right) \\
T > T_D &\quad S_e(T) = a_{vg} \cdot \eta \cdot 3.0 \cdot \left(\frac{T_C \cdot T_D}{T^2}\right)
\end{aligned}
\qquad (2.22)
$$

As for the horizontal acceleration response spectra, also for vertical acceleration EC8 considers two spectral types, namely Type 1 and Type 2. The values to be assumed in equation (2.22) for T_B, T_C, T_D and a_{vg} for each type of vertical spectrum may be found in the National Annexes. EC8 recommends the following values (Table 2.3, or Table 3.4 of EC8-1).

Table 2.3 – EC8 recommended values of parameters describing the vertical elastic response spectra

Elastic Response Spectra	a_{vg}/a_g	T_B (s)	T_C (s)	T_D (s)
Type 1	0.90	0.05	0.15	1.0
Type 2	0.45	0.05	0.15	1.0

2. EN 1998-1: GENERAL AND MATERIAL INDEPENDENT PARTS

2.3.4 Alternative representations of the seismic action

EC8-1 (section 3.2.3) allows adopting an alternative representation of the seismic motion in terms of ground acceleration time-histories (accelerograms) and related quantities (velocity and displacement).

Generally, for a spatial (i.e. three-dimensional) structural analysis, three components of ground motion are necessary: two orthogonal components in horizontal plane and one component in vertical direction. With this regard, EC8 mandates to use two different accelerograms along both horizontal directions (clause 3.2.3.1.1(2)P). For a planar (i.e. two-dimensional) structural analysis, two accelerograms should be used: one in the horizontal and another in the vertical direction, respectively.

It is important to highlight that accelerograms recorded during a real seismic events have different shape and frequency contents in the three orthogonal directions. In addition, the values of the peak acceleration are generally recorded at different time periods. Therefore, attention is drawn upon the fact that the accelerograms used to represent the seismic action on the three orthogonal directions must be different.

The selection of accelerograms to be used in structural analysis should consider the following aspects: (1) the source of the accelerograms (i.e. by artificial generation, from past seismic events or by simulation); (2) the compatibility between the accelerogram characteristics and the source characteristics, propagation environment and local soil conditions that control the site seismicity; (3) the match between the target response spectrum and the accelerograms response spectrum, accounting for the structural properties; (4) the number of used accelerograms and the relevant implications on interpretation of results.

EC8-1 (clause 3.2.3.1) considers three different types of accelerograms for seismic time history analysis, namely artificial, recorded and simulated records whose main features are briefly summarized hereinafter.

Artificial accelerograms are obtained by generating mathematically a signal whose response spectrum matches the EC8 elastic acceleration response spectrum. One of the most common procedures to generate artificial spectrum-compatible accelerograms is implemented in the software SIMQKE (Gasparini and Vanmarcke, 1979). The approach employed in SIMQKE is to generate sinusoidal signals having random phase angles and amplitudes. The sinusoidal motions are subsequently summed and by means of an iterative procedure are combined to improve the match with the target response spectrum, by calculating the ratio between the target and actual response ordinates at selected frequencies. Artificial signals are attractive because it is possible to obtain acceleration time-series that are almost completely compatible with the elastic design spectrum, which is mostly the only information available to designers and structural analysts regarding the type of earthquake to be investigated. However, it is now widely accepted that the use of such artificial records, particularly for non-linear analyses, raises several problems (Bommer and Acevedo, 2004). In particular, the main criticism with artificial records is that these signals are unrealistic and generally have a number of cycles excessively larger than natural accelerograms and, consequently, they possess unreasonably high energy content, thus being more damage demanding than natural records for the same values of peak ground acceleration. However, artificial records are generally able to produce results that present relatively lower dispersion than natural records. Hence, the use of simulated accelerograms allows to limit the influence of the aleatory uncertainty on the predicted seismic response parameters.

The second category of seismic signals is synthetic accelerograms that are obtained by physical simulated source and seismic waves propagation mechanisms and can include local site effects modelling. Several methods to generate synthetic records have been developed (e.g. Zeng *et al*, 1994; Beresnev and Atkinson, 1998; Boore, 2003) and are freely available. However, their application is very difficult for structural engineers because the definition of the physical parameters requires to characterise the earthquake source thus needing the services of a specialist consultant in engineering seismology. EC8 allows using synthetic records provided that the samples used adequately represent the seismogenetic features of the sources and the soil conditions appropriate to the site, and their values are scaled to the value of "$S \cdot a_g$" for the seismic area under consideration.

The third category of records is real accelerograms recorded during real earthquakes. This type of signals is, by definition, free from the problems affecting artificial spectrum-compatible records. Most engineers prefer recorded accelerograms over the artificial or simulated ones, because of two main aspects: i) these signals represent real seismic events: ii) structural analysts do not often have sufficient background knowledge to generate artificial/simulated accelerograms, requiring specialist engineering seismologists;

iii) nowadays, real natural records are easily accessible thanks to several available databases. With this regard, the access to recorded accelerograms has been permanently improving during the last decades due to fast development of digital seismic networks and the availability of databases containing severe seismic motions. Among the seismic records sources of interest for European countries the following databases are worth to be cited: the European Strong-Motion Database (http://www.isesd.cv.ic.ac.uk); ITACA database (http://itaca.mi.ingv.it); INCERC database (http://www.incerc2004.ro/accelerograme.htm and http://www.incerc.ro/download.htm).

According to EC8, whichever type of suite of accelerograms is used, the records have to comply with the following conditions (clauses 3.2.3.1.2(3) and 3.2.3.1.2(4)):

1) the minimum duration T_s of the stationary part of the accelerograms should be equal to 10 s, if site-specific data are not available;
2) the minimum number of accelerograms should be equal to 3;
3) the mean value of the zero period spectral response acceleration (calculated from each time histories) should not be smaller than the value of "$S \cdot a_g$" for the given site;
4) in the range of periods between $0.2T_1$ and $2T_1$, where T_1 is the fundamental period of the structure in the direction where the accelerogram will be applied, the ordinates of the mean elastic acceleration response spectrum calculated from all time histories should not be smaller than 10 % of the corresponding values belonging to the target EC8 elastic spectrum.

It is finally emphasized that the selection and use of accelograms, using real records and/or artificial or semi-artificial is in itself a topic, specific guidance being found in (Baker and Cornell, 2006; Hancock *et al*, 2008; Araújo *et al*, 2016). The issue of the number of accelerograms to be used according to EC8-1 is further discussed in section 2.5.5.

2.3.5 Design spectrum for elastic analysis

EC8 takes into account the capacity of structures to dissipate energy through inelastic deformations, so that the design seismic forces can be smaller than those related to a linear elastic response, but sophisticated non-linear structural analyses can also be avoided at the design stage. In fact, an elastic analysis based on a response spectrum reduced with respect to the elastic one, called the "design spectrum", can be performed. The reduction is accomplished by introducing the behaviour factor q (3.2.2.5(2)), which can be approximately interpreted as the ratio between the seismic forces that a single degree of freedom system equivalent to the real structure would experience if its response would be completely elastic (with 5 % equivalent viscous damping) and the seismic forces that may be used in the design (clause 3.2.2.5(3)), directly accounting for the energy dissipation capacity of the system. The values of the behaviour factor q are given for various materials and structural systems according to the relevant ductility classes within EC8.

In any case, once the behaviour factor q suitable for the structure to be calculated is fixed, the design response spectrum is obtained from the elastic spectrum (see equation (2.20)) using the following equations:

$$0 < T < T_B \qquad S_e(T) = a_g \cdot S \cdot \left(\frac{2}{3} + \frac{T}{T_B} \cdot \left(\frac{2.5}{q} - \frac{2}{3} \right) \right)$$

$$T_B < T < T_C \qquad S_e(T) = a_g \cdot S \cdot \frac{2.5}{q}$$

$$T_C < T < T_D \qquad S_e(T) \begin{cases} = a_g \cdot S \cdot \dfrac{2.5}{q} \cdot \dfrac{T_C}{T} \\ \geq \beta a_g \end{cases} \qquad (2.23)$$

$$T > T_D \qquad S_e(T) \begin{cases} = a_g \cdot S \cdot \dfrac{2.5}{q} \cdot \left(\dfrac{T_C \cdot T_D}{T^2} \right) \\ \geq \beta a_g \end{cases}$$

The parameter β is the lower bound factor for the horizontal design spectrum, whose appropriate value should be provided by the National Annexes. EC8 recommends to assume $\beta = 0.2$.

Figure 2.9 compares elastic and design spectra, the latter illustrated for two different q factors.

2. EN 1998-1: General and Material Independent Parts

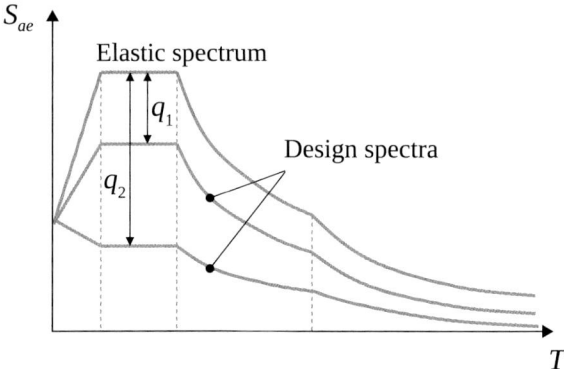

Figure 2.9 – Elastic vs. design acceleration spectrum according to EC8

2.3.6 Combinations of the seismic action with other types of actions

EC8-1 (clause 3.2.4.1(P)) states that the design value E_d of the effects of actions in the seismic design situation should be determined by adopting the load combinations for structural checks in accordance with EN 1990 (CEN, 2002a), which recommends in its clause 6.4.3.4 the following load combination at the ultimate limit state:

$$\sum_{j\geq 1} G_{k,j} + P + A_{Ed} + \sum_{i\geq 1} \psi_{2,i} Q_{k,i} \qquad (2.24)$$

where:
- $G_{k,j}$ is the characteristic value of the permanent load "j";
- $Q_{k,i}$ is the characteristic value of the variable load "i";
- A_{ed} is the design value of the seismic action, $A_{ed} = \gamma_I \cdot A_{ek}$;
- A_{ek} is the characteristic value of the seismic action for the reference return period (e.g. 475 years at ULS);
- P is the pretensioning action representative value (if present);
- $\psi_{2,i}$ is the coefficient for the quasi-permanent value of the variable load Q_i, whose values are also reported in Table 2.4;
- γ_I is the importance coefficient.

It is important to highlight that in the ULS load combination, the permanent loads have their characteristic value. On the contrary, the variable loads are considered only with the quasi-permanent fraction of the characteristic loading. This approach reflects the smaller probability of occurrence at ULS that the mean return period earthquake might occur simultaneously with the full value of the variable load. The variable actions that are accounted for in the seismic combination are both imposed and snow loads. The wind and temperature variation variable loads are not considered in the same load combination with the seismic action (i.e. $\psi_{2,i} = 0$ in this case).

Table 2.4 – Recommended values for $\psi_{2,i}$ coefficient used to determine the quasi-permanent fraction of the variable action according to EN 1990 (CEN, 2002a)

Type of variable action	ψ_2
Category A : domestic, residential areas	0.3
Category B : office areas	0.3
Category C : congregation areas	0.6
Category D : shopping areas	0.6
Category E : storage areas	0.8
Category F : traffic area, vehicle weight ≤ 30 kN	0.6
Category G : traffic area, 30 kN < vehicle weight ≤ 160 kN	0.3
Category H : roofs	0
Snow loads on buildings (see EN 1991:1-3) Finland, Iceland, Norway, Sweden	0.2
Remainder of CEN Member States, for sites located at altitude $H > 1000$ m a.s.l.	0.2
Remainder of CEN Member States, for sites located at altitude $H \le 1000$ m a.s.l.	0
Wind loads on buildings (see EN 1991:1-4)	0
Temperature (non-fire) in buildings (see EN 1991:1-5)	0

In order to evaluate the effects of the seismic action upon a structure it is also necessary to evaluate the inertial properties, which are associated to the structural masses that correspond to the gravitational (permanent and variable) loads given in the seismic load combination (see equation (2.24)). In detail, in accordance with clause 3.2.4(2)P, the inertial effects in the seismic design situation have to be evaluated as follows:

2. EN 1998-1: GENERAL AND MATERIAL INDEPENDENT PARTS

$$\sum_{j\geq 1} G_{k,j} + \sum_{i\geq 1} \psi_{E,i} Q_{k,i} \qquad (2.25)$$

where $\psi_{E,i}$ is the combination coefficient for variable action "i", which takes into account the likelihood of the loads $Q_{k,i}$ not to be present over the entire structure during the earthquake, as well as a reduced participation in the motion of the structure due to a non-rigid connection with the structure. According to clause 4.2.4(2)P, the combination coefficient $\psi_{E,i}$ should be computed from the following expression:

$$\psi_{E,i} = \varphi \cdot \psi_{2,i} \qquad (2.26)$$

Values to be ascribed to φ may be found in the National Annexes. The Eurocode recommended values for φ are listed in Table 2.5 (Table 4.2 of EC8-1).

Table 2.5 – EC8 Recommended φ coefficient values used to determine $\psi_{E,i}$

Type of variable action	Storey	φ
	Roof	1.0
Categories (*) A-C	Storeys with correlated occupancies	0.8
	Independently occupied storeys	0.5
Categories(*) D-F and Archives		1.0
(*) Categories according to EN 1991:1-1 (CEN, 2002b)		

2.4 CHARACTERISTICS OF EARTHQUAKE RESISTANT BUILDINGS

2.4.1 Basic principles of conceptual design

EC8-1 requirements aim to mitigate seismic vulnerability within acceptable costs. The governing design principles are hereafter summarized (clause 4.2.1(2)):

- structural simplicity (clause 4.2.1.1(1)): it consists in realizing clear and direct paths for the transmission of the seismic forces, thus allowing the modelling, analysis, detailing and construction

of simple structures. It directly implies a simplified morphology of both the structural plan and elevation. In such a way, uncertainties in the structural behaviour are limited;
- uniformity, symmetry and redundancy (clause 4.2.1.2): uniformity is characterised by an even distribution of the structural elements both in-plane and along the height of the building, allowing short and direct transmission of the inertia forces and eliminating the occurrence of sensitive zones where concentration of stress or large ductility demands might prematurely cause collapse. Moreover, if the building configuration is symmetrical, a symmetrical layout of structural elements is envisaged. In addition, the use of distributed structural elements may increase redundancy and allow a redistribution of action effects and widespread energy dissipation across the entire structure;
- bi-directional resistance and stiffness (clause 4.2.1.3): the building structure must be able to resist horizontal actions in any direction. Consequently, the structural elements should be arranged in an orthogonal in-plane structural pattern, ensuring similar resistance and stiffness characteristics in both main directions;
- torsional resistance and stiffness (clause 4.2.1.4(1)): building structures should possess adequate torsional resistance and stiffness in order to limit torsional motions which tend to stress the structural systems in a non-uniform way;
- diaphragmatic behaviour at storey level (clause 4.2.1.5): the floors (including the roof) should act as horizontal diaphragms that collect and transmit the inertia forces to the vertical structural systems and ensure that those systems behave together in resisting the horizontal seismic action. In order to guarantee this behaviour, floors should be provided with in-plane stiffness and resistance and with effective connection to the vertical structural systems;
- adequate foundation (clause 4.2.1.6): the foundations have a key role, because they have to ensure that the whole building is subjected to a uniform seismic excitation.

By satisfying these basic principles of conceptual design allows to obtain a regular building both in plan and in elevation, which is a fundamental requirement to achieve a high seismic performance and a reliable structural model.

2. EN 1998-1: GENERAL AND MATERIAL INDEPENDENT PARTS

2.4.2 Primary and secondary seismic members

EC8-1 (section 4.2.2) explicitly makes a distinction between primary and secondary structural members in relation to their role to withstand seismic action. The elements contributing to the seismic resistance of the structure are considered as primary, while those soley devoted to resist gravity loads are intended as secondary. Therefore, their contribution to the seismic capacity must be neglected provided that they are able to preserve their ability to support gravity loads under the design earthquake. This requirement should be fulfilled by controlling that the forces developing into those elements under the most unfavourable seismic condition, considering also the influence of 2nd order effects (i.e. P-Δ effects), do not exceed their design strength. In addition, since secondary elements should behave elastically, they do not need to conform to any requirements of capacity design and non-seismic structural details can be used.

An example of secondary elements in a building frame is shown in Figure 2.10, where the perimeter moment-resisting frames (i.e. bold lines) are devoted to resist seismic action and its members (i.e. beams and columns) must be considered as primary, namely designed for high ductility, while the remaining frames resist vertical loads and their members should be considered as secondary members.

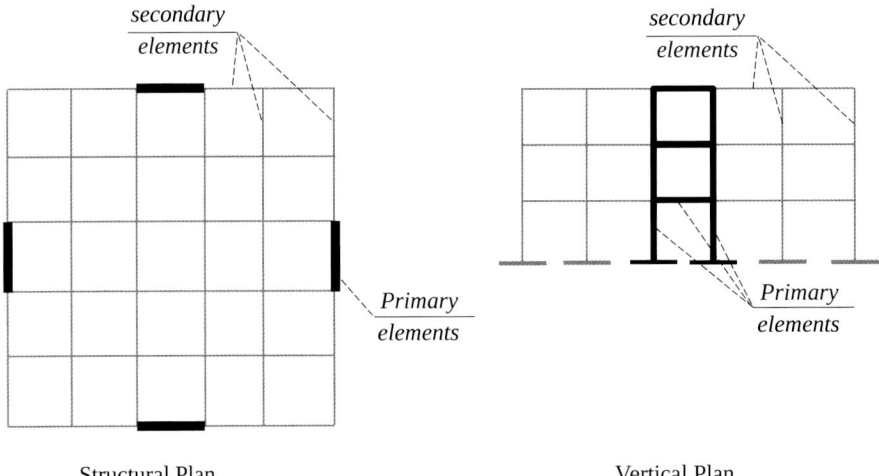

Figure 2.10 – Multi-storey building with perimeter primary moment-resisting frames (i.e. bold lines) and secondary members

It should be also considered that in complex structures with primary members located in restricted parts (e.g. moment-resisting frames positioned in a limited number of bays) the resulting contribution of the non-seismic zones (e.g. stairwell, gravity spans, etc.) can influence the lateral stiffness and the relevant dynamic response of the building. Therefore, clause 4.2.2(4) mandates that the total contribution to lateral stiffness of all secondary seismic members should be less than 15 % of the stiffness provided by all primary seismic members.

2.4.3 Criteria for structural regularity

2.4.3.1 Introduction

Past earthquakes clearly showed that "seismically regular" structures experienced a more satisfactory seismic response than "irregular" structures. Although it is easy to understand the meaning of regular structure, it is very complex to determine accurately the "regularity degree". In order to avoid misinterpretation of this concept, EC8-1 (section 4.2.3) specifies a number of fundamental criteria to be respected in order to obtain regular structures, distinguishing between plan and elevation regularity.

It is important to highlight that a structural classification based on regularity has implications for the following aspects of the seismic design (section 4.2.3.1 of EC8-1):

- the structural model, that can be two or three-dimensional;
- the method of structural analysis, which can be either a simplified response spectrum analysis (lateral force procedure) or a modal one;
- the value of the behaviour factor q that must be reduced in case of vertically irregular structures.

Following this approach, EC8-1 does not exclude the design of irregular structures, but it discourages the use of irregular configurations, imposing more sophisticated design methods and making them more expensive due to higher design forces (i.e. by imposing reduced behaviour factors).

An irregular structure in plan exhibits torsional deformations when subjected to horizontal forces due to the eccentricity between the centre of mass and the centre of stiffness at each floor level. Figure 2.11 illustrates

2. EN 1998-1: GENERAL AND MATERIAL INDEPENDENT PARTS

typical examples of irregular structures that are sensitive to torsional effects, therefore requiring 3D analysis.

3D analysis is also essential to identify the significance of higher vibration modes, as these may become critical in case their period of vibration coincides with the period T_C of the design spectrum.

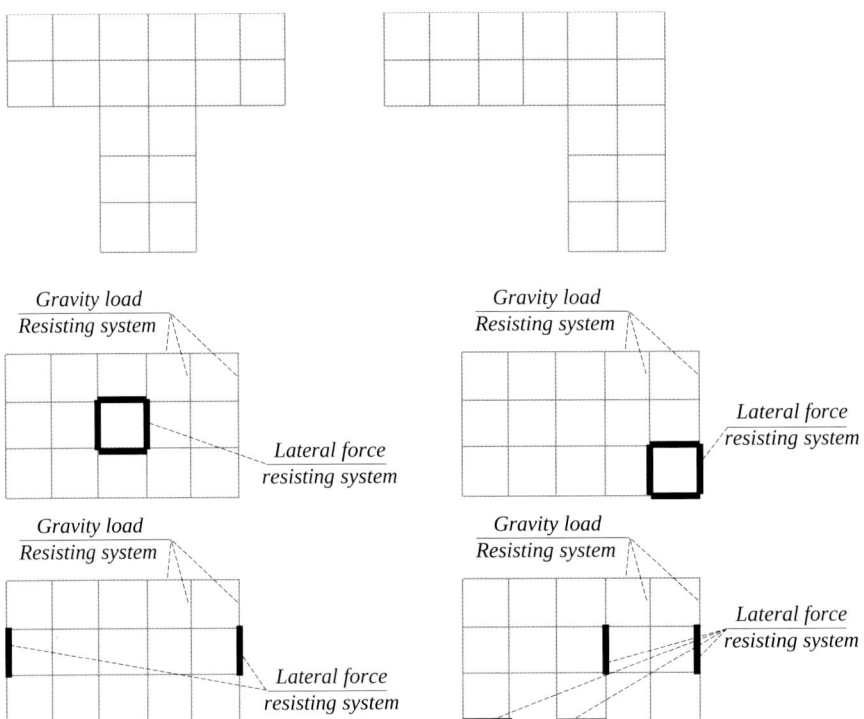

Figure 2.11 – Typical examples of irregular structures sensitive to torsional deformations

2.4.3.2 Criteria for regularity in plan

According to EC8-1 (section 4.2.3.2), in order to be considered regular in plan a building structure must have a symmetric planar distribution of stiffness and mass with respect to two orthogonal axes. In addition, the plan configuration must be compact, namely each floor should be delimited by a polygonal convex line.

In those cases where set-backs (re-entrant corners or edge recesses) exist, EC8 considers that regularity in plan may still be obtained, provided that these setbacks do not affect the floor in-plane stiffness and that, for each set-back, the area between the outline of the floor and a convex polygonal line enveloping the floor does not exceed 5 % of the floor area (see Figure 2.12a).

In order to distribute seismic forces among lateral forces resisting systems, the in-plane stiffness of each floor must be sufficiently large in comparison with the lateral stiffness of the vertical structural elements; this allow their modelling as rigid diaphragms. The slenderness of the building in plan, defined as $\lambda = L_{max}/L_{min}$ must not be greater than 4, where L_{max} and L_{min} are the largest and the smallest in-plan dimensions of the building, measured in orthogonal directions (see Figure 2.12b).

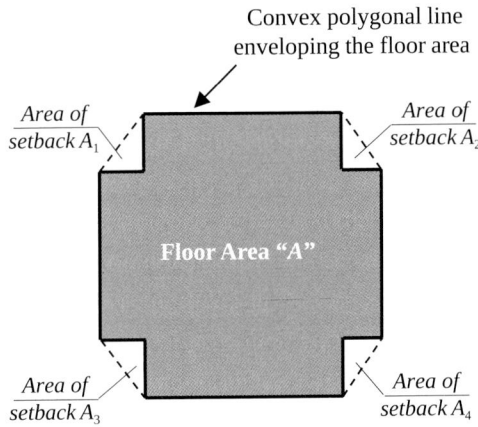

a)

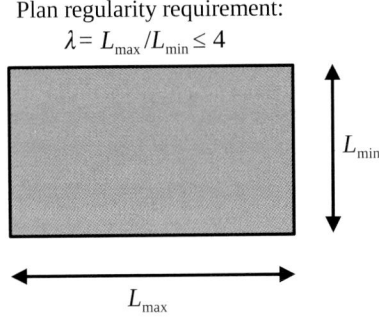

b)

Figure 2.12 – Plan regularity requirement for setback area (a); Plan regularity requirement for plan slenderness of building (b)

2. EN 1998-1: General and Material Independent Parts

EC8-1 (clause 4.2.3.2(6)) imposes an additional requirement in order to classify a building as regular in plan, namely that at each storey and for each main building plan direction x and y, the eccentricity e_0 and the torsional radius r should satisfy the following conditions:

$$e_{0x} \leq 0.30 \cdot r_x \text{ and } e_{0y} \leq 0.30 \cdot r_y \qquad (2.27)$$

$$r_x \geq l_s \text{ and } r_y \geq l_s \qquad (2.28)$$

where:

e_{0x}, e_{0y} are the distances between the centre of stiffness and the centre of mass, measured perpendicularly (i.e. x and y) to the direction of analysis considered (i.e. y and x, respectively);

r_x, r_y are torsional radius defined as the square root of the ratio between the torsional stiffness and the lateral stiffness of the structure in the direction of the analysis;

l_s is the radius of gyration of the floor masses in plan defined as the square root of the ratio between the polar moment of inertia of masses in plan with respect to the centre of mass of the floor and floor mass.

It should be noted that for most building structures regular in plan, the first two modes of vibration represent translational shapes in the two main horizontal directions of the structure (x and y). This optimal behaviour is mainly guaranteed by condition (2.28), which avoids torsional coupling with translational response that is considered uncontrollable and potentially very dangerous (Fardis et al, 2005). It is noted that whenever the periods of the structure are determined with a 3D modal analysis, condition (2.28) is satisfied for the building as a whole if the period of the first torsional mode of vibration is lower than the periods of the translational modes in the two horizontal directions. It is important to observe that the appropriate position of seismic resisting members implicitly satisfies condition (2.28), as shown in Figure 2.13a where the lateral loads resisting systems (e.g. braced frames or structural walls) are placed on the plan perimeter. On the contrary, if the lateral loads resisting systems are concentrated near the centre of the floor (see Figure 2.13b), the condition (2.28) might not be fulfilled and must be explicitly verified.

2.4 CHARACTERISTICS OF EARTHQUAKE RESISTANT BUILDINGS

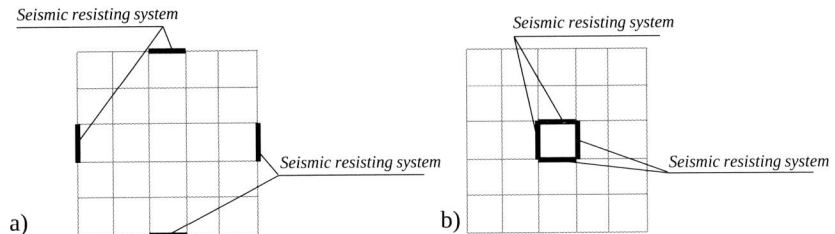

Figure 2.13 – Favourable (a) and unfavourable (b) plan position of lateral forces resisting systems

For buildings with a single storey, the centre of stiffness is defined as the centre of the lateral stiffness of the seismic forces resisting system (clause 4.2.3.2(7)). It is important to note that the centre of lateral stiffness is the point in plan that satisfies the following condition: any set of lateral forces applied to that point on the floor of the building generates only floor translations, without any torsion around the vertical axis. The torsional radius r is defined as the square root of the ratio of the global torsional stiffness with respect to the centre of the lateral stiffness and the global lateral stiffness, in one direction, taking into consideration all of the primary seismic members in that direction.

The stiffness centre may be uniquely determined solely for one storey buildings. For multi-storey buildings, the centre of lateral stiffness cannot be uniquely determined, because its position depends on the shape and distribution of the lateral forces (clause 4.2.3.2(8)). A simplified definition for classification of the structural regularity in plan and for the approximate analysis of torsional effects is possible if the following two conditions are satisfied:

- all lateral load resisting systems, such as reinforced concrete (RC) cores, structural walls, or frames are continuous from the foundation to the top of the building;
- the deformed shapes of the resisting systems are similar to the shape of internal lateral load distribution. This condition may be considered satisfied in case of frame or structural walls systems, while in general it is not satisfied for dual systems (e.g. frames and RC walls working in parallel).

For frames and slender walls with mainly bending deformations, the position of the centre of stiffness and the torsional radius of each floor may be

2. EN 1998-1: General and Material Independent Parts

computed as the centre of the moments of inertia of the cross sections of the vertical elements as follows:

$$x_{cs} = \frac{\sum(x \cdot EI_y)}{\sum(EI_y)} \qquad y_{cs} = \frac{\sum(y \cdot EI_x)}{\sum(EI_x)} \qquad (2.29)$$

$$r_x = \sqrt{\frac{\sum(x^2 \cdot EI_y + y^2 \cdot EI_x)}{\sum(EI_y)}} \qquad r_y = \sqrt{\frac{\sum(x^2 \cdot EI_y + y^2 \cdot EI_x)}{\sum(EI_x)}} \qquad (2.30)$$

where EI_x and EI_y are the bending stiffnesses of the vertical elements in a vertical plan parallel to x and y axis, respectively. For other structural systems, for example braced frames, equations (2.29) and (2.30) can be adapted by replacing the terms EI_x and EI_y with the lateral stiffness K_x and K_y of the bracing systems in the x and y directions, respectively.

2.4.3.3 Criteria for regularity in elevation

EC8-1 (section 4.2.3.3) provides the following criteria to classify building structures as being regular in elevation:

- lateral forces resisting systems must continue without interruptions from foundations to the top storey of the structure;
- lateral stiffness and mass of the structure must be constant or gradually decrease with height;
- in framed buildings, the ratio between the effective resistance of each storey and the necessary design resistance should not vary disproportionately between adjacent storeys.

When setbacks are present, EC8-1 provides the following additional conditions to be applied (clause 4.2.3.3(5)):
    a) for setbacks gradually distributed along the building height preserving vertical symmetry of the structure, the setback at any floor should be less than 20 % of the dimension of the floor beneath in the direction of the setback (see Figure 2.14a and 2.14b);

b) for a single setback located within the lower 15 % of the total height of the main structural system, the setback should be smaller than 50 % of the dimension of the floor beneath (see Figure 2.14c). In this case, the structure of the base zone within the vertically projected perimeter of the upper storeys should be designed to resist at least 75 % of the horizontal shear forces that would develop in that zone in a similar building without the base enlargement;

c) if the setbacks do not preserve vertical symmetry of the building, in each face the sum of the setbacks at all storeys should be smaller than 30 % of the plan dimension at the ground floor above the foundation or above the top of a rigid basement, and the individual setbacks should be smaller than 10 % of the dimension of the floor beneath (see Figure 2.14d).

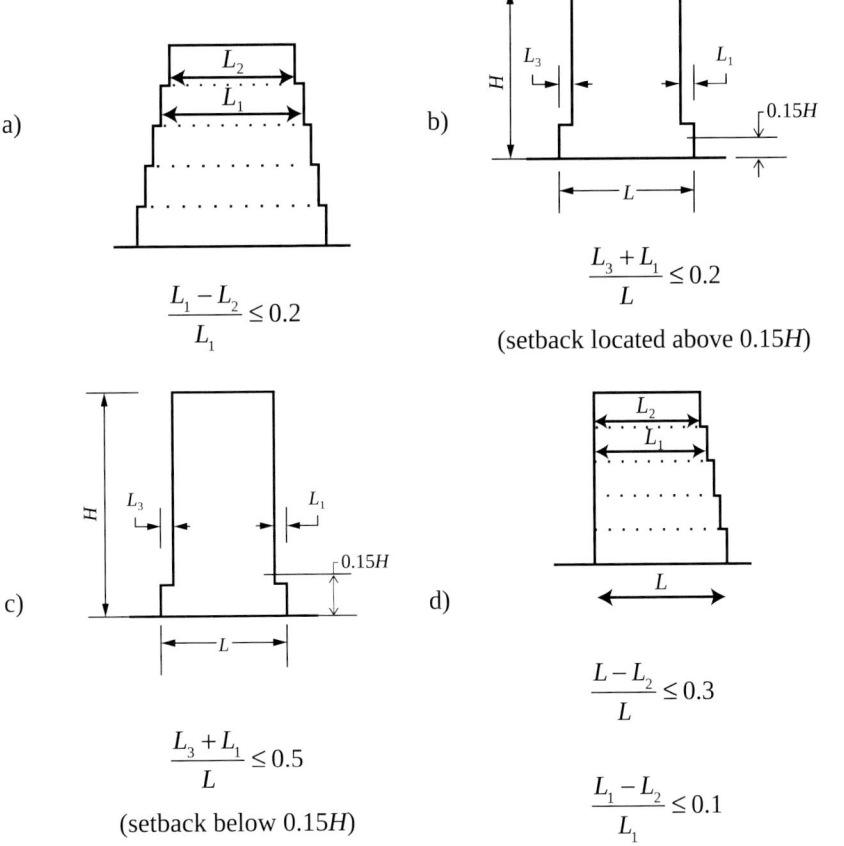

Figure 2.14 – Criteria for regularity of buildings with setbacks

2. EN 1998-1: GENERAL AND MATERIAL INDEPENDENT PARTS

2.4.3.4 Consequences of structural regularity on seismic analysis

Regularity both in plan and elevation significantly affect the structural behaviour. Plan regularity of the structure implies small eccentricities between the centre of mass (CM) and the centre of stiffness (CS), resulting in minor torsional effects. In this case, the seismic forces acting in a certain direction are sustained by the seismic resisting systems located in the same direction that are equally loaded (see Figure 2.15a). This allows using planar models to analyse each seismic resisting system. Structures that are irregular in plan involve important torsional effects and, consequently, seismic resisting systems located in the same direction of the acting seismic forces are loaded non-uniformly. Additionally, the seismic actions acting in a certain direction also impose forces into the resisting systems located in the perpendicular direction (see Figure 2.15b). Under these conditions, it is not possible to determine the distribution of seismic forces in the structure and to identify the contribution of each resisting system by simulating the structural behaviour with planar models. Hence, the most appropriate approach is to determine the structural response using a three-dimensional model of the structure.

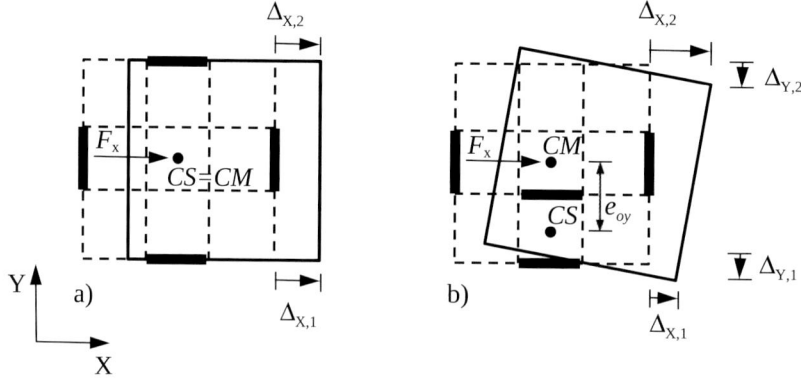

Figure 2.15 – Consequences of plan regularity: regular (a) and irregular (b) structure

Structures regular in elevation that are not very flexible (the period of vibration T_1 must be smaller than the minimum between $4T_C$ and 2.0 s), mainly respond in the fundamental mode of vibration. Consequently, these structures may be analysed using the lateral force method. On the contrary, building structures irregular in elevation are characterized by a lateral response affected by higher modes of vibration. Consequently, these structures must be analysed using a modal response spectrum analysis.

It should be noted that regularity in plan and in elevation do not absolutely compel the structural model or the analysis method. Indeed, any structure may be analysed using a three-dimensional model and a modal response spectrum analysis. However, due to plan and elevation regularity the structural engineer may adopt simplifications, such as simple (planar) models and/or more simple analysis methods (lateral force method).

Table 2.6 (Table 4.1 of EC8-1) synthetically presents the relationship between structural regularity (plan and elevation) and the simplifications accepted in structural analysis, and also the necessity of reducing the behaviour factor q, according to clause 4.2.3.1(3)P.

Table 2.6 – Consequences of structural regularity on seismic analysis

Regularity		Accepted analysis simplification		Behaviour factor (q) for linear elastic analysis
Plan	Elevation	Model	Linear elastic analysis	
Yes	Yes	Plan	Lateral force *	Reference value
Yes	No	Plan	Modal	Reduced value ($0.8q$)
No	Yes	Spatial	Lateral force *	Reference value
No	No	Spatial	Modal	Reduced value ($0.8q$)

*Only if the structure has the period of vibration $T_1 < \min(4T_C; 2.0 \text{ s})$

The behaviour factor q reflects the ductility, dissipative capacity and also the redundancy and over-strength of the structure. Structures that are not regular in elevation are susceptible to concentrations of plastic deformations in some parts of the structure (i.e. non-uniform ductility demand distribution), which is equivalent to a reduced ductility in the whole structure. This implies the necessity of using a reduced behaviour factor q with respect to the reference value. In this case, according to clause 4.2.3.1(7), the decreased value of the behaviour factor is given by the reference value multiplied by 0.8.

Seismic action analysis methods currently used in design are: (a) lateral force method and (b) modal response spectrum analysis method. Therefore

2. EN 1998-1: General and Material Independent Parts

EC8-1 refers only to these two analysis methods when it discusses the implications of plan and elevation regularity upon the structural model and the analysis method. However, the standard provisions may be easily extended also to other methods of analysis. Thus, similarly to the lateral force method, nonlinear static analysis is based on the hypothesis that the structural response is governed by the fundamental mode of vibration. Therefore, nonlinear static analysis can only be used for vertically regular structures that have a period of vibration T_1 smaller than the minimum of $4T_C$ and 2.0 s. Irregular structures in elevation must be analysed using more sophisticated methods, like linear or nonlinear dynamic analysis.

2.5 METHODS OF STRUCTURAL SEISMIC ANALYSIS

2.5.1 Introduction

The system of second order non-homogeneous differential equations (2.1) describes the dynamic response of multi-degree of freedom systems under seismic excitation. The direct solution of the equations of motion by analytical processes is not possible due to the fact that: (i) the dynamic action (seismic motion) cannot be described by an analytical function, and (ii) the structural response is generally in plastic domain. Therefore, various simplifications are often adopted in order to simplify the structural analysis. Practically, instead of solving the equations of motion, different methods for structural seismic analysis can be adopted. According to EC8-1 (section 4.3.3) the following types of structural analysis may be performed to calculate and verify building structures:

- lateral force method of analysis (namely a linear static analysis with a lateral distribution of forces);
- linear modal response spectrum analysis;
- nonlinear static pushover analysis;
- nonlinear time-history dynamic analysis.

The methods of structural seismic analysis commonly adopted for calculating the seismic-induced effects on structures are the lateral force method and modal response spectrum analysis, because of their computational simplicity and high efficiency. It should be noted that both methods of analysis assume elastic structural response and the seismic action is combined with gravitational loads using the principle of superposition of effects.

The lateral force method is the easiest structural analysis method and it is generally preferred by structural engineers because of the familiarity with the basic calculation approaches used in structural mechanics, allowing also for hand computation for simple and low-redundant structural schemes. However, this method can be applied only to structures whose dynamic response is basically dominated by the fundamental mode of vibration. This type of analysis cannot be applied in case of complex structures that are not vertically regular because of the non-uniform stiffness, strength or mass distribution (see Table 2.6).

Modal response spectrum analysis is generally considered as the reference method to calculate design forces for building structures, because it accounts for the dynamic properties of the structure. Indeed, if a sufficient number of modes of vibration is considered, the calculated elastic seismic structural response closely approximates the real elastic behaviour of the bare building.

Although both the lateral force and the modal response spectrum method are conventionally adopted in design, these types of analysis have two major limitations. The first limitation relates to the nature of the seismic action, being a dynamic action that varies in time. The results obtained using these two methods of analysis represent the envelope of response features (e.g. internal forces, displacements, etc.), without providing any information about their time variation. The second major limitation is that both the lateral force and the modal response spectrum methods are elastic analyses. In reality, structures designed using the dissipative behaviour principle should guarantee plastic ductile response under the seismic action. However, the structural ductility is not directly checked, being fictitiously considered in a simplified manner through behaviour factors q.

Due to these disadvantages, there are situations requiring more advanced methods, such as static or dynamic non-linear analyses. These situations are manifold and include the design of special structures (large spans or heights, or with complex appearance), or the evaluation of the seismic performance of existing buildings that do not comply with the seismic demands imposed by modern design provisions.

2. EN 1998-1: GENERAL AND MATERIAL INDEPENDENT PARTS

In the case of the non-linear static analysis method, the seismic capacity of the structure is derived by applying to the structural model a monotonically increasing distribution of lateral forces. This analysis aims at obtaining the response curve of the structure, generally expressed as the base shear (i.e. the resultant of all applied horizontal forces at the base of the building) with respect to the roof displacement. This capacity curve is fictitiously intended to reproduce the backbone response curve of the structure produced by the cyclic actions.

The nonlinear dynamic analysis method is defined through accelerograms applied to a structural model where the nonlinear (plastic) behaviour of the structure is explicitly modelled. It represents the analysis method that allows the most accurate modelling of the seismic structural response, but it is also the most complex and time-consuming.

Hereinafter, the fundamentals of all methods of analysis addressed by EC8 are described and discussed.

2.5.2 Lateral force method

The lateral force method is characterized by a linear static analysis to be performed by applying a pre-defined lateral distribution of forces on the two main horizontal directions of the structure (e.g. x and y as shown in the example reported in Figure 2.15), which are proportional to the shape of the 1st translational vibration mode in each horizontal direction. The results of the lateral force analysis represent peak values of internal forces and structural displacements. As it can be easily recognized this approach corresponds to a simplified modal response spectrum analysis that considers only the contribution of the fundamental mode of vibration to the structural response.

The main advantage of this method of analysis is its simplicity, and it can be applied to both two-dimensional and three-dimensional structural models. However, its easiness does not allow to satisfactorily predict the structural behaviour for most practical applications (e.g. complex structural configurations, in the presence of irregularities, etc.). With this regard, EC8-1 (section 4.3.3.2) imposes that this method may be used only for buildings where the effects of higher modes are negligible. Clause 4.3.3.2.1(2) stipulates that this condition is met whenever the building is regular both in plan and in elevation and its fundamental period of vibration T_1 is smaller than min($4T_C$; 2.0 s) for the two main horizontal directions of the building. In addition, since the lateral force method models only the horizontal components of the seismic action, it cannot be used if vertical component of the seismic action should be accounted for.

The seismic forces are determined in two steps: 1) calculation of the base shear force (i.e. the resultant of all seismic forces at the basis of the building); 2) distribution of the lateral forces along the height of the structure according to the shape of the fundamental mode (i.e. the first mode in that direction).

The base shear force F_b corresponding to the fundamental mode, per main horizontal direction considered in the structural analysis, is given by (clause 4.3.3.2.2(1)P):

$$F_b = S_d(T_1) m \lambda \tag{2.31}$$

where:

$S_d(T_1)$ is the *ordinate* of the design spectrum (see equation (2.23)) at period T_1 as defined in section 3.2.2.5 of EC8-1;

T_1 is the fundamental *period* of vibration of the building for lateral motion in the direction considered;

m is the total mass of the building, above the foundation or above the top of a rigid basement, computed in accordance with equation (2.25) (clause 3.2.4(2));

λ is the correction factor, which takes into account the fundamental mode through the associated modal mass, whose values are: $\lambda = 0.85$ if $T_1 \leq 2T_C$ and the building has more than two storeys, or $\lambda = 1.0$ for all other cases. The factor λ accounts for the fact that in buildings with at least three storeys and translational degrees of freedom in each horizontal direction, the effective modal mass of the 1st (fundamental) mode is smaller, on average by 15 %, than the total building mass.

Regarding the fundamental period T_1, the structural designer may perform an eigenvalue analysis to determine it (clause 4.3.3.2.2(2)). However, in line with the simplified approach, clause 4.3.3.2.2(3) provides an empirical equation for T_1

$$T_1 = C_t H^{(3/4)} \tag{2.32}$$

C_t depends on the structural typology (e.g. $C_t = 0.085$ for steel moment-resisting frames, $C_t = 0.05$ for eccentrically steel braced buildings), and H is the total building height expressed in m. Generally, the period of vibration determined according to equation (2.32) is smaller than the period of vibration determined from an eigenvalue analysis, thus being more conservative (i.e. smaller periods of vibration correspond to larger spectral ordinates, namely larger lateral forces). Once T_1 is known it is possible to evaluate $S_d(T_1)$ and to calculate F_b.

Clause 4.3.3.2.3(1) recommends to distribute the horizontal seismic forces consistently with the fundamental mode shapes in the horizontal directions of analysis of the building. Therefore, the fundamental mode of vibration should be known and an eigenvalue analysis should be carried out to evaluate the modal shapes (ϕ). However, it allows to approximate the fundamental modal shape by means of horizontal displacements increasing linearly along the height of the building.

In the first case, the equivalent static horizontal seismic force F_i acting at the storey "i" in each main horizontal direction of the building may be computed using the following expression (clause 4.3.3.2.3(2)):

$$F_i = F_b \frac{m_i \phi_i}{\sum m_j \phi_j} \qquad (2.33)$$

where:
- F_b is the base shear force corresponding the fundamental mode;
- ϕ_i, ϕ_j are the displacements in the direction of the translational dynamic degree of freedom at storey i, namely "j" of masses m_i and m_j;
- m_i, m_j is the mass at storey "i", namely "j".

In the second case, the horizontal forces are given by:

$$F_i = F_b \frac{m_i z_i}{\sum m_j z_j} \qquad (2.34)$$

where:
- z_i, z_j are the heights of the masses m_i and m_j above either the foundation or a rigid basement.

The horizontal seismic forces are applied as lateral forces to each slab considered non deformable in its plane (i.e. if diaphragmatic behaviour is guaranteed). Figure 2.16 schematically illustrates the typical distribution of horizontal forces obtained from the lateral force method. It should be mentioned that the "inverted triangular" distribution of lateral forces (height proportional) represents a simplified shape of the fundamental mode of vibration. Lateral forces, being proportional to the storey "i" mass, will have this distribution only if the floor masses are equal.

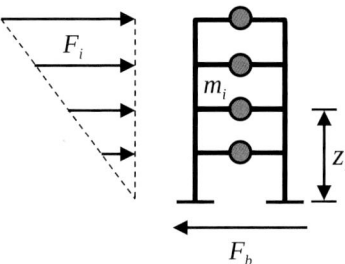

Figure 2.16 – The schematic representation of horizontal forces determined by lateral force method

2.5.3 Linear modal response spectrum analysis

Linear modal response spectrum analysis is the reference method to carry out structural analysis according to EC8-1 (section 4.3.3.3). This type of analysis is based on decoupling the equations of motion (see equation 2.10) of a dynamic multi degree of freedom system. In other words, solving a system with N differential equations is replaced with solving N independent equations, which significantly simplifies the problem.

Decoupling the system of equations is possible due to the application of the principle of superposition of modal responses. Thus, to perform a modal response spectrum analysis it is necessary to determine the modes of vibration and the periods of vibration, solving the eigenvalue problem:

$$\left([k] - \omega_n^2 [m]\right)\{\phi\}_n = \{0\} \tag{2.35}$$

where ω_n^2 is the vibration circular frequency of mode "n", and $\{\phi\}_n$ is the deformed shape in the "n" mode of vibration. Hence, for a structure having

2. EN 1998-1: GENERAL AND MATERIAL INDEPENDENT PARTS

"n" dynamic degrees of freedom, "n" modes of vibration are determined. The fundamental mode of vibration corresponds to that having the largest period of vibration T.

It is clear that for practical application it is necessary to use computational tools to calculate the dynamic properties of each structure. However, most structural analysis software are able to determine the modes of vibration, the periods of vibration and the effective modal masses of the structure.

Anyway, irrespective of the computational platform, the modal response spectrum analysis implies a sequence of steps, described in Figure 2.17 and summarized as follows:

1. Definition of the structural mass [m], stiffness [k] and damping ratio ξ_i;
2. Calculation of the circular frequencies ω_i (and the corresponding period $T_i = 2\pi/\omega_i$) and the relevant modes of vibration $\{\phi\}_i$;
3. Calculation of the modal participation factor Γ_i and the effective modal masses m_{eff} for all modes of vibration in each main direction of the structure by means of equations (2.13) and (2.14);
4. Calculation of the structural response per mode as follows:
   - for period T_i and damping ratio ξ_i, from the pseudo-acceleration spectrum, determination of the spectral ordinate $S_d(T_i)$;
   - computation of the equivalent static forces $\{f\}_i$;
   - calculation of the response $E_{E,i}$ from forces $\{f\}_i$, for all requested response features (internal forces, displacements, etc.).
5. Combination of the modal contributions $E_{E,i}$ in order to obtain the total response E_E.

The fifth step of this procedure is very important. Indeed, the peak response of the modal contributions $V_{bi}(t)$ given by the i-th mode occurs at different periods, namely the maxima per mode are not contemporary. In addition, the modal response spectrum analysis does not provide the response time variations from the modes of vibration $E_{E,i}(t)$, but only the peak values of the response $E_{E,i} = \max|E_{E,i}(t)|$. Consequently, various approximations are used in order to estimate the total peak response r based on the modal peak responses $E_{E,i}$.

Clause 4.3.3.3.2(2) recommends combining modal responses by means of the *square root of sum of squares* (SRSS) method, which gives the maximum value E_E of a seismic action effect as:

2.5 METHODS OF STRUCTURAL SEISMIC ANALYSIS

$$E_E = \sqrt{\sum_{i=1}^{n} E_{E,i}^2} \qquad (2.36)$$

This modal responses combination method provides an adequate approximation of the total response only for structures that have distinct periods

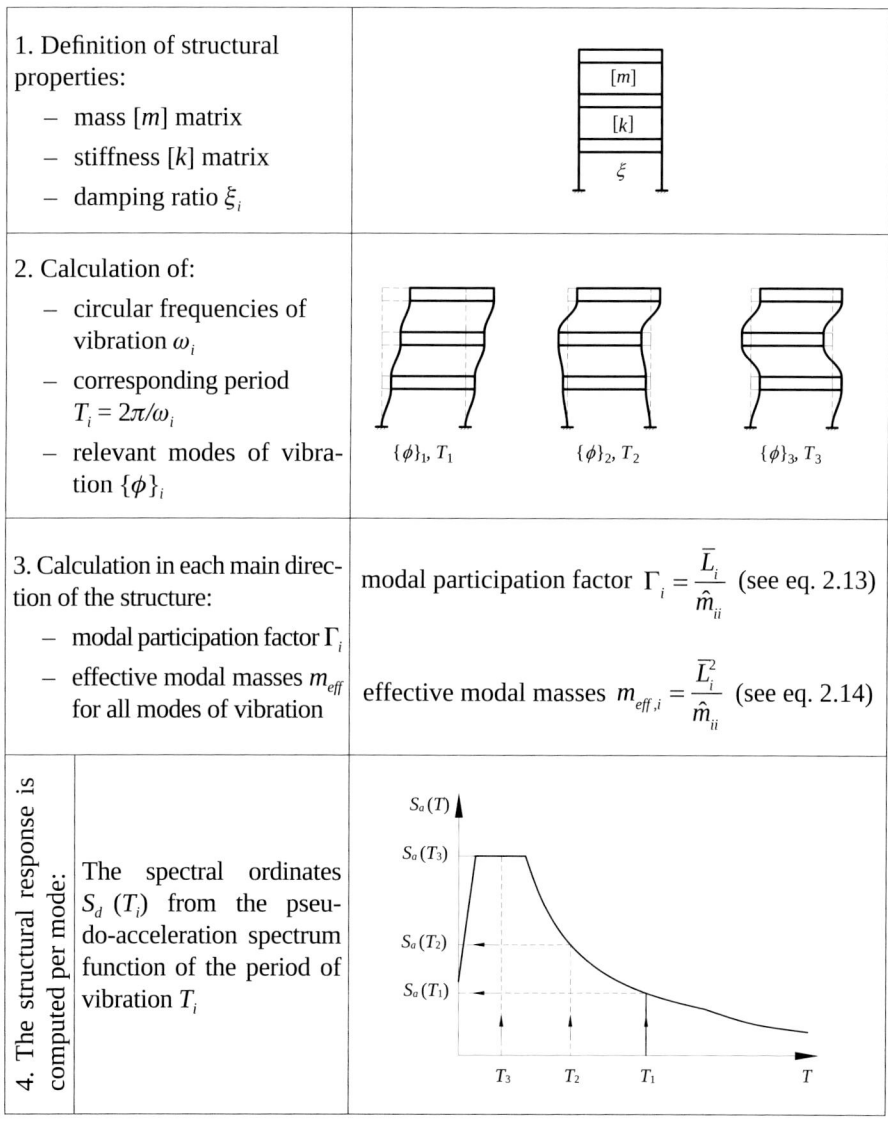

Figure 2.17 – The fundamental scheme of the modal response spectrum analysis

2. EN 1998-1: General and Material Independent Parts

4. The structural response is computed per mode:	The equivalent static forces: $\{f\}_i = m_{eff,i} \cdot S_a(T_i) \cdot \{\phi\}_i$	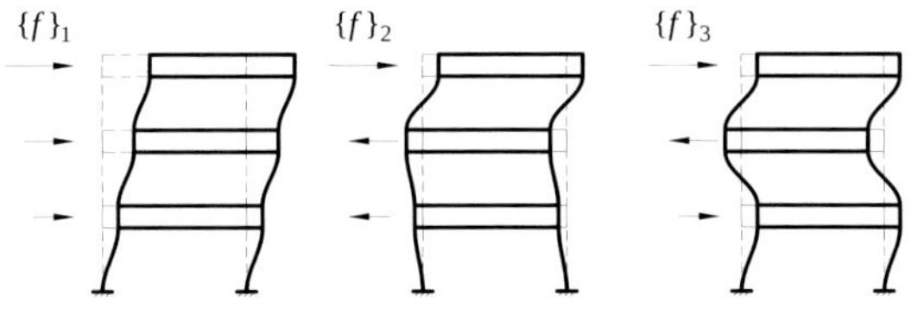 $\{f\}_1$ $\{f\}_2$ $\{f\}_3$
	The response $E_{E,i}$ from forces $\{f\}_i$, all requested response feature (internal forces, displacements, etc.)	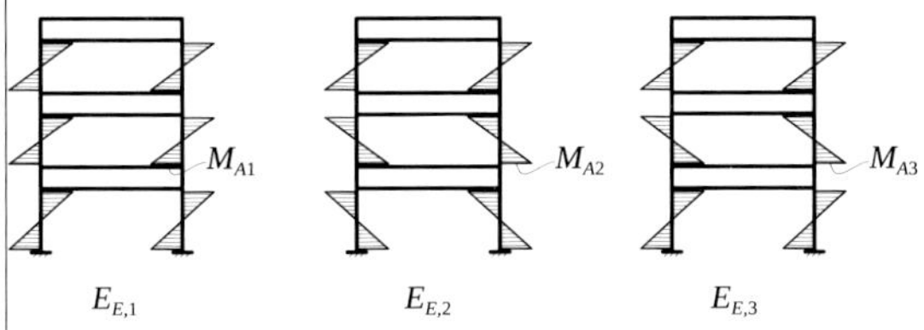 M_{A1}, $E_{E,1}$ M_{A2}, $E_{E,2}$ M_{A3}, $E_{E,3}$
5. The total response E_E is computed by combining the modal contributions $E_{E,i}$ (for example using the SRSS method)		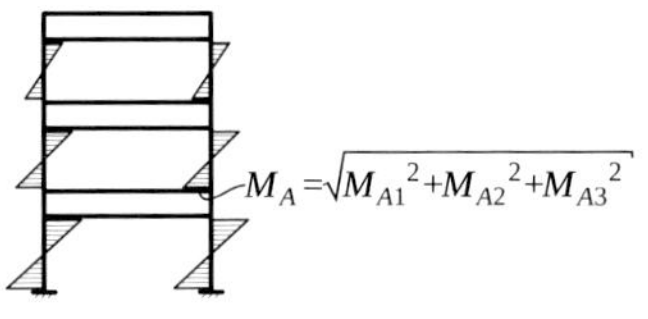 $M_A = \sqrt{M_{A1}^2 + M_{A2}^2 + M_{A3}^2}$, E_E

Figure 2.17 – The fundamental scheme of the modal response spectrum analysis (continuation)

of vibration (independent modes of vibration). According to clause 4.3.3.3.2(1), modal responses for two modes of vibration, "j" and "i" are considered independent if the periods of vibration T_j and T_i (with $T_j \leq T_i$) satisfy the following condition:

$$T_j \leq 0.9 T_i \qquad (2.37)$$

It should be noted that SRSS can lead to a large underestimation of the seismic demand in some cases, e.g. when a three-dimensional structure is highly asymmetric (Wilson *et al*, 1981) or if the condition expressed by equation (2.37) is not satisfied. In the latter case, clause 4.3.3.3.2.(3)P recommends to use the *complete quadratic combination (CQC),* which is a more flexible method of combining modal responses that can be applied both to structures with close modes of vibration and to structures with distinct modes of vibration. The peak response in this method is given by:

2.5 METHODS OF STRUCTURAL SEISMIC ANALYSIS

$$E_E = \sqrt{\sum_{i=1}^{n} \sum_{j=1}^{n} \rho_{ij} E_{E,i} E_{E,j}} \qquad (2.38)$$

Each term under the square root of equation (2.38) represents the product between the correlation coefficient ρ_{ij} and the modal responses peak values $E_{E,i}$ and $E_{E,j}$. The correlation coefficient varies between 0 and 1, having the unit value for $i = j$: $\rho_{ij} = 1$. Thus, equation (2.38) can be rewritten as:

$$E_E = \sqrt{\sum_{k=1}^{n} E_{E,k}^2 + \underbrace{\sum_{i=1}^{n} \sum_{j=1}^{n} \rho_{ij} E_{E,i} E_{E,j}}_{i \neq j}} \qquad (2.39)$$

The first term under the square root is identical with SRSS method, and the second term includes all the factors ($i \neq j$), representing the "correction" for the modes that are not distinct.

The correlation factor ρ_{ij} depends on the damping ratio and the modal circular frequencies ratio $\beta_{ij} = \omega_i / \omega_j$. In case of damping ratios differing for all modes, ρ_{ij} can be determined as follows:

$$\rho_{ij} = \frac{8(\xi_i \cdot \xi_j)^{1/2} \cdot (\xi_i + \beta_{ij} \cdot \xi_j) \cdot \beta_{ij}^{3/2}}{(1-\beta_{ij}^2)^2 + 4\xi_i \cdot \xi_j \cdot \beta_{ij} \cdot (1+\beta_{ij}^2) + 4(\xi_i^2 + \xi_j^2) \cdot \beta_{ij}^2} \qquad (2.40)$$

For structural systems having the same damping ratios for all modes (i.e. $\xi_i = \xi_j = \xi$), equation (2.40) becomes:

$$\rho_{ij} = \frac{8\xi^2 \cdot (1+\beta_{ij}) \cdot \beta_{ij}^{3/2}}{(1-\beta_{ij}^2)^2 + 4\xi \cdot \beta_{ij} \cdot (1+\beta_{ij}^2)} \qquad (2.41)$$

Figure 2.18 shows that for distinct modes (with values $\omega_i \neq \omega_n$) the correlation factor rapidly decreases with the increase of the ratio between circular frequencies, so that the CQC method reduces to the SRSS method. For structures with close modes ($\omega_i \cong \omega_n$), the correlation factor tends to 1

2. EN 1998-1: GENERAL AND MATERIAL INDEPENDENT PARTS

and the corresponding total response is larger than the one obtained using the SRSS method.

The SRSS and CQC methods are based on stochastic (random) vibration theory. Therefore, these combination methods of structural responses allow to estimate the structural response that matches the actual seismic motions for a wide band of frequencies and long period. However, the modal response spectrum analysis method is not appropriate in case of pulse-type earthquakes or those close to a harmonic motion.

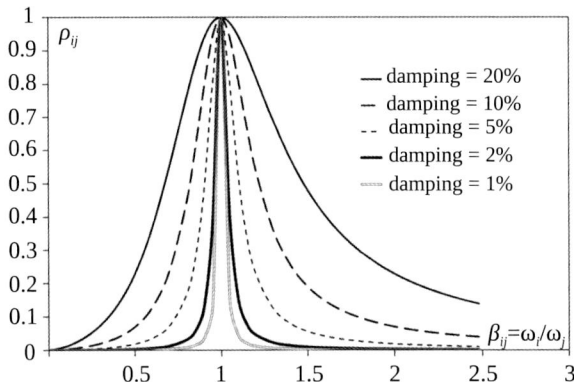

Figure 2.18 – The correlation factor ρ_{ij} vs. the modal circular frequencies ratio $\beta_{ij} = \omega_i/\omega_j$

Rigorously, the modal response spectrum analysis implies taking into account all the modes of vibration of a structure. In the case of a more complex structure with a large number of dynamic degrees of freedom this approach can become excessively difficult and time consuming, even when structural design software is used. However, solutions close enough to the "exact" results may be obtained if only few modes of vibration are taken into account. According to clause 4.3.3.3.1(2)P, the modes with a significant contribution to the total response have to be considered in the analysis. This condition is accomplished if either of the following conditions can be demonstrated (clause 4.3.3.3.1(3):

- the sum of effective modal masses for the considered modes is at least 90 % of the total structural mass, or
- all the modes having an effective modal mass larger than 5 % of the total mass are considered in the analysis.

Generally, the first condition (the sum of effective modal masses to be larger than 90 % of the total structural mass) is easily verified, being a reference method in establishing the minimum number of modes of vibration that need to be considered in the analysis.

In case of three-dimensional models, the above requirements need to be verified for each analysis direction (clause 4.3.3.3.1(4)). As an exception, if the previous conditions cannot be satisfied for a sufficiently large number of modes of vibration (for example, in case of buildings with a significant contribution of torsional modes), the minimum number of modes that should be considered in a three-dimensional analysis must satisfy the following conditions (clause 4.3.3.3.1(5)):

$$k \geq 3\sqrt{n} \quad \text{and} \quad T_k \leq 0.20 \text{ s} \qquad (2.42)$$

where:
- k is the minimum number of modes that needs to be considered in the analysis;
- n is the number of stories above ground level;
- T_k is the period of vibration of the last mode of vibration considered.

As previously described, the modal response spectrum analysis is the EC8 reference method for structural analysis at design stage. This is due to the fact that it combines the advantage of a structural seismic response close enough to the "actual" behaviour with satisfactory numerical efficiency. However, it should be noted that the modal response spectrum analysis method is an approximate method and it has some restrictions.

The approximations of this method are represented by the combination of the modal responses and the combination of the effects induced by the seismic action. Indeed, due to combination of modal responses using either SRSS or CQC, the results loose their sign, thus leading to diagrams of internal forces or structural deformed shapes apparently inconsistent with physical principles. In order to clarify this issue, Figure 2.19 compares the bending moment diagram of a moment resisting frame obtained using the lateral force method (Figure 2.19a) and a modal response spectrum analysis (Figure 2.19b). It can be observed that the bending moment diagram obtained using the modal response spectrum analysis seems incorrect. Actually, the values of the internal forces are correct in magnitude, but they all have positive sign. It is the responsibility of the structural engineer to read correctly the information supplied by the design software.

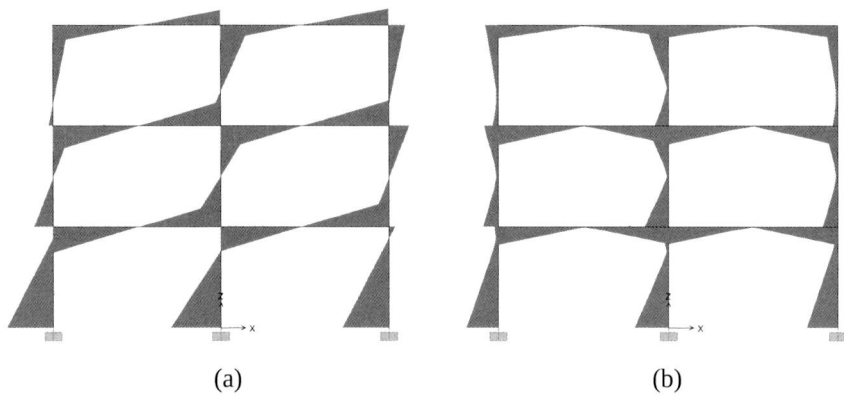

(a)   (b)

Figure 2.19 – Moment resisting frame – bending moment diagrams induced by seismic action, calculated by lateral forces method (a) and modal response spectrum analysis method (b)

To take into account that the seismic action can act in both the positive and the negative directions, and that the sign (positive/negative) results of a modal response spectrum analysis is unknown, the seismic action effects obtained from a modal response spectrum analysis must be considered in the analysis both with positive and negative sign. Thus, the seismic load combination given by equation (2.24) must be adapted, taking into account the seismic action with ± sign, as follows:

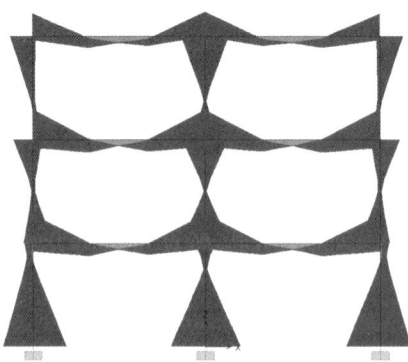

Figure 2.20 – Moment resisting frame – bending moment diagrams by combination of gravity and seismic effects using the modal response spectrum analysis

2.5 METHODS OF STRUCTURAL SEISMIC ANALYSIS

$$\sum_{j\geq 1} G_{k,j} + P \pm A_{Ed} + \sum_{i\geq 1} \psi_{2,i} Q_{k,i} \tag{2.43}$$

This procedure is implemented in most structural analysis software. For example, Figure 2.20 illustrates the bending moment diagram obtained by combining the effects due to gravity loads and the seismic effects determined with the modal response spectrum analysis method according to equation (2.43), where it can be recognized that the results represent the envelope corresponding to both the positive and the negative directions of the seismic action.

Another limitation of the modal response spectrum analysis method is that it provides only the response peak values, and not its time variation. Therefore, the seismic action effects obtained using this analysis method are not compatible. For example, a column from a three-dimensional moment resisting frame, with rigid beams on both directions, will be subjected to biaxial bending ($M_{y,Ed}$ and $M_{z,Ed}$) and axial force (N_{Ed}). The modal response spectrum analysis provides peak values of these forces and moments, which should be used for both strength and stability checks. However, the actual bending moment peak values on the two axes will not generally occur at the same time. It would be more correct to check the strength and stability of the structural element for each time step, with the corresponding moment values. Therefore, checking structural elements based on the modal response spectrum analysis method is conservative, even though it may sometimes be uneconomic because it may lead to unwanted overstrength in the structural members.

In order to obtain more accurate results it is necessary to directly determine the modal responses for all required response features, after which they are combined using either SRSS or CQC procedures. Indeed, it is not correct to determine some response features based on the total response. For example, the overall deformed shape of a structure obtained by using a modal response spectrum analysis is often meaningless, because all the displacements are positive. In order to check a structure at the serviceability limit state in a seismic load combination it is necessary to compute the relative interstorey displacements. Hence, the correct application of the modal response spectrum analysis method requires the computation of the relative interstorey displacements for all considered modes of vibration and subsequently

2. EN 1998-1: General and Material Independent Parts

combining their values using either the SRSS or the CQC procedures. This implies that if the structural analysis software does not provide this possibility, the user should extract the relative interstorey displacements based on the total structural displacements, and subsequently combine them by hand calculation.

2.5.4 Nonlinear static pushover analysis

The pushover analysis is a static non-linear analysis under permanent vertical loads and gradually increasing lateral loads, which should approximately represent the earthquake induced forces. This method of analysis was developed as a simpler and alternative method of analysis when compared to nonlinear dynamic analysis, but still accounting for the plastic behaviour of the structural elements (see Figure 2.21) and geometrical non-linearity occurring under seismic action.

The nonlinear static analysis is based on two hypotheses: (1) the dynamic structural response is governed by only one mode of vibration, which (2) is represented by a distribution of the lateral forces applied to the floor masses, as for the lateral force method, and kept constant during the seismic action.

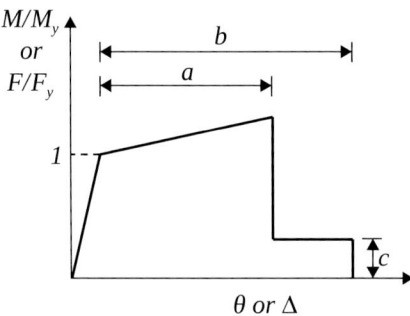

Figure 2.21 – Generalized force-deformation representation for the nonlinear behaviour of steel elements or components in case of nonlinear static analysis

The nonlinear static analysis principle is illustrated in Figure 2.22. The step "0" consists of applying the entire value of the gravitational loads in the seismic condition (see equation (2.24)), which will remain constant for all the subsequent steps. With this regard, it should be noted that structural analysis softwares allow to control the application of forces to the numerical model of the structure by means of two different strategies: 1) force control;

2.5 METHODS OF STRUCTURAL SEISMIC ANALYSIS

2) displacement control. The first case should be adopted when the magnitude of the load distribution that will be applied is known a-priori and the structure is expected to resist elastically these forces. Under force control, all loads are applied incrementally from zero to their maximum specified magnitude. This is the case to be adopted for the step "0" of the pushover analysis.

Displacement control should be used when it is known how far the structure moves (i.e. the lateral displacement demand monitored at the roof level), but the required corresponding lateral load is unknown. It is clear that this approach should be used when lateral forces are incrementally applied to the structural model (see steps 1 to 4 in Figure 2.22). To use displacement control, it is generally necessary to select a displacement component to be monitored, which can be a single degree of freedom chosen at a node of the structural model. In most cases, the horizontal displacement in the direction of the applied forces of a node belonging to the top storey is selected. Moreover, it is necessary to give the magnitude of the displacement that is the target for the analysis. Due to this assumption, it should be noted that the magnitude of the applied lateral forces is not important, only their distribution with the height of the structure.

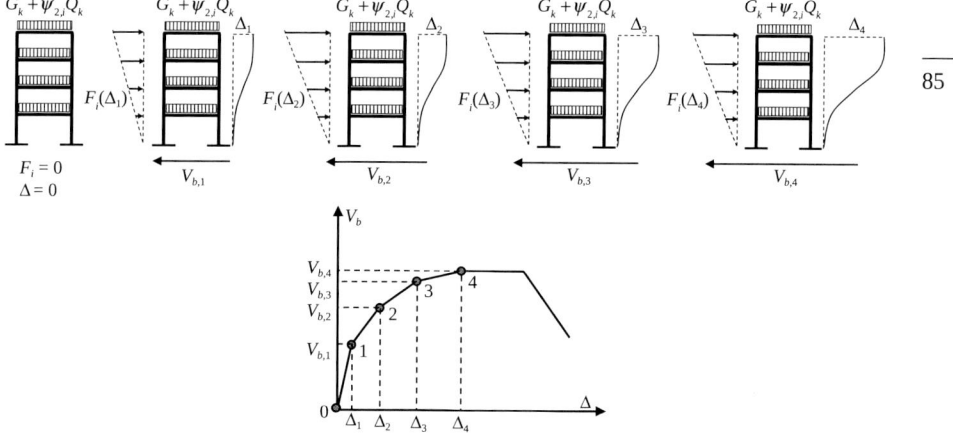

Figure 2.22 – Nonlinear static analysis principle

The graphical representation of the relationship between the base shear force "V_b" and the control displacement "Δ" (top lateral displacement of the structure) represents the capacity curve of the structure (see Figure 2.22). When running a first order global analysis, this curve linearly increases until

2. EN 1998-1: GENERAL AND MATERIAL INDEPENDENT PARTS

the first plastic hinge is formed (point 1), when the structural lateral stiffness decreases. Once the lateral forces increase, more and more plastic hinges are formed, until a plastic mechanism is reached.

It is important to highlight that structural stiffness changes as plastic hinges are formed. Therefore the fundamental mode of vibration changes and, consequently, the relevant distribution of lateral forces also changes. Figure 2.23 illustrates the variation of a lateral forces distribution corresponding to the evolution of plastic hinges pattern. To compensate for the inability of classical nonlinear static analysis to account for this phenomenon, clause 4.3.3.4.2.2(1) requires the use of two lateral force distributions:

- a "uniform" distribution, based on mass proportional lateral forces, regardless of height (uniform response acceleration), see Figure 2.24a;
- a "modal" distribution, where the lateral forces are proportional to the fundamental mode of vibration weighted with the masses at each storey. This distribution corresponds to lateral forces determined as in the lateral force method according to equation (2.33), see Figure 2.24b.

The response curves obtained using these two force distributions represent lower and upper bounds of the structural response. In general, the "uniform" distribution leads to larger demand estimates at the lower storeys, while the "modal" distribution overestimates the demand for the upper levels.

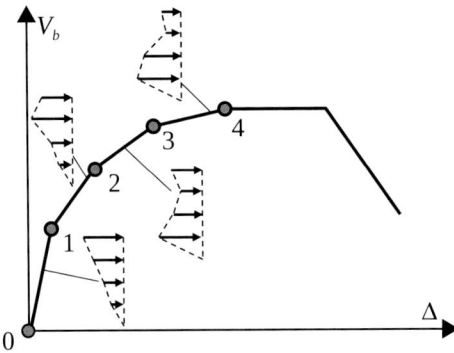

Figure 2.23 – Lateral forces distribution variation function of structural plastic deformations

2.5 METHODS OF STRUCTURAL SEISMIC ANALYSIS

The structural checks must be based on the most unfavourable effects resulting from the two lateral force distributions. Additionally, if the structure is not symmetrical, the lateral forces must be applied to both positive and negative directions.

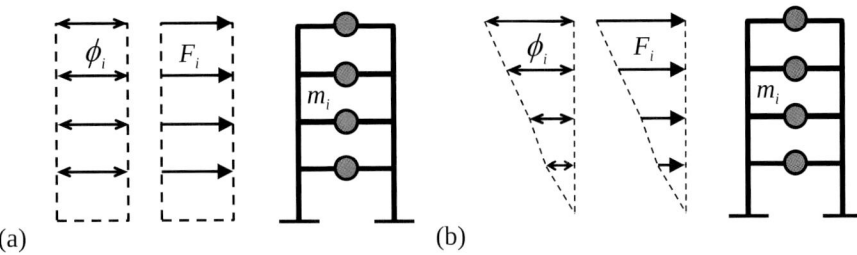

(a)   (b)

Figure 2.24 – Uniform (a) and modal (b) distribution of lateral forces for the nonlinear static analysis

According to EC8-1 (section 4.3.3.4.2.1), the nonlinear static analysis method may be used to verify the structural performance of newly designed and of existing buildings for the following purposes:

- to verify or revise the overstrength ratio α_u/α_1 factor (i.e. the structural redundancy), used to estimate the behaviour factor q, see Figure 2.25a;
- to estimate the expected plastic mechanisms and the distribution of damage, see Figure 2.25b;
- to assess the structural performance of existing or retroffited buildings;
- to design new structures, as an alternative to design based on linear elastic analysis using the behaviour factor q.

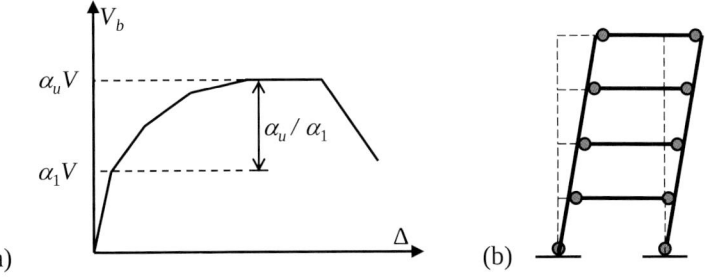

(a)   (b)

Figure 2.25 – α_u/α_1 factor – structural redundancy (a) and plastic mechanism (b) obtained using a nonlinear static analysis

2. EN 1998-1: GENERAL AND MATERIAL INDEPENDENT PARTS

The first two features can be directly obtained from the output of the pushover analysis. Indeed, the response curve expressed in terms of base shear vs. roof displacement provides the structural capacity and the behaviour of a structure in the plastic domain under the seismic loading action, but without representing or being directly related to any level of seismic action. This implies that pushover itself cannot be used for design or assessment of structure unless the seismic demand is explicitly accounted for. Therefore, clause 4.3.3.4.2.1 recommends to use as seismic demand the target displacement indicated in clause 4.3.3.4.2.6(1)P, which is defined as the seismic demand derived from the elastic response spectrum (see equation (2.20)) in terms of the displacement of an equivalent single-degree-of-freedom system.

Annex B of EC8-1 describes a procedure for the determination of the target displacement from the elastic response spectrum. Once the target displacement is estimated, the structure can be analysed using a nonlinear static analysis for this displacement and the structural performance is evaluated on the basis of comparison between displacement demand and corresponding plastic capacity provided by the structure. This procedure was developed by Fajfar (1999, 2000) and it is called the N2 method. It combines a nonlinear static analysis of the multi dynamic degree of freedom system (the analysed structure) with a spectral analysis of a single dynamic degree of freedom system. In order to be consistent, the basic assumption used in pushover-based methods is that the structure vibrates predominantly in a single mode. Hence, the overall structural response can be assimilated to that of an equivalent SDOF system having its dynamic properties (period, masses and stiffness) equal to that of the fundamental mode of the MDOF system.

Using inelastic response spectra, the displacement demand of the equivalent SDOF system is determined, and then, using a reverse transformation, it is possible to go back to the MDOF system and its displacement demand. The fundamental steps of N2 method implemented in Annex B of EC8-1 are the following:

1. Initial data are established:
    - the structure;
    - the (pseudo-) acceleration elastic response spectrum.

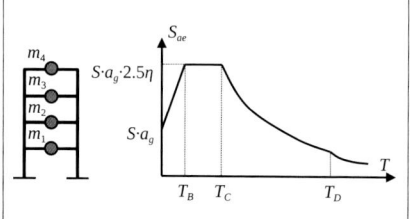

2.5 METHODS OF STRUCTURAL SEISMIC ANALYSIS

2a. The deformed shape (i.e. the fundamental mode of vibration) $\{\phi\}$ is chosen, normalized to the roof displacement d_n; 2b. The lateral forces distribution $\overline{F_i} = m_i \cdot \phi_i$ is determined; 2c. The capacity curve F_b-d_n is determined by performing a nonlinear static analysis on the MDOF system.	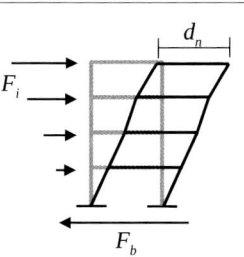
3a. The mass of the equivalent SDOF system is determined as: $m^* = \sum m_i \cdot \phi_i$; 3b. The MDOF system is converted in an equivalent SDOF system using the factor $$\Gamma = \frac{m^*}{\sum m_i \phi_i^2}$$	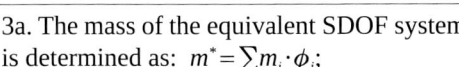 $F^* = \dfrac{F_b}{\Gamma} \quad d^* = \dfrac{d_n}{\Gamma}$
4. An equivalent bilinear force-displacement relationship is determined, with the yield force F_y^* and the yield displacement d_y^*.	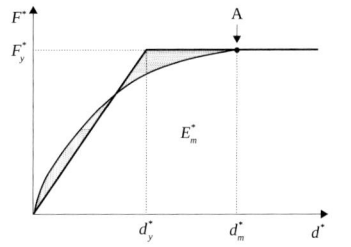
5a. The period of vibration of the equivalent SDOF system T^* is computed; 5b. The elastic displacement of the equivalent single dynamic degree of freedom system d_{et}^* is computed.	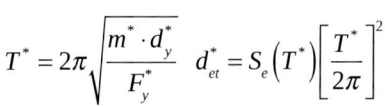 $T^* = 2\pi \sqrt{\dfrac{m^* \cdot d_y^*}{F_y^*}} \quad d_{et}^* = S_e(T^*)\left[\dfrac{T^*}{2\pi}\right]^2$
6. The displacement demand of the equivalent single dynamic degree of freedom system d_t^* is computed. For $T^* < T_C$: - If $F_y^*/m^* \geq S_e(T^*)$ the response is elastic and $d_t^* = d_{et}^*$ - If $F_y^*/m^* < S_e(T^*)$ the response is nonlinear and $$d_t^* = \dfrac{d_{et}^*}{q_u}\left(1 + (q_u - 1)\dfrac{T_C}{T^*}\right) \geq d_{et}^*$$ where $q_u \dfrac{S_e(T^*) \cdot m^*}{F_y^*}$	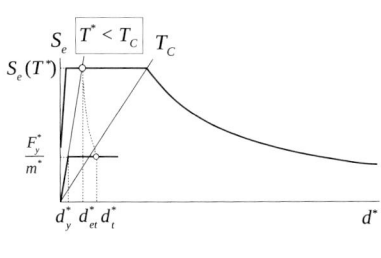

2. EN 1998-1: General and Material Independent Parts

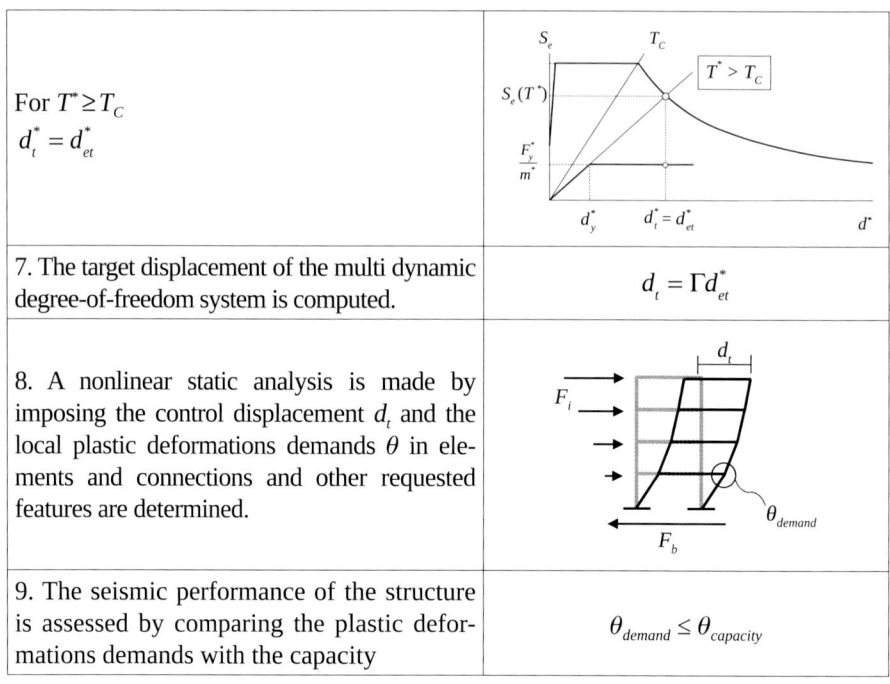

For $T^* \geq T_C$ $d_t^* = d_{et}^*$	
7. The target displacement of the multi dynamic degree-of-freedom system is computed.	$d_t = \Gamma d_{et}^*$
8. A nonlinear static analysis is made by imposing the control displacement d_t and the local plastic deformations demands θ in elements and connections and other requested features are determined.	
9. The seismic performance of the structure is assessed by comparing the plastic deformations demands with the capacity	$\theta_{demand} \leq \theta_{capacity}$

It is important to note that N2 method has important limitations: 1) it is applicable only to regular structures, whose dynamic response is mainly characterized by the fundamental mode of vibration; 2) assuming the lateral forces distribution constant with increasing plastic structural deformations is a severe approximation. However, the nonlinear static analysis has the advantage that is simpler than the nonlinear dynamic analysis, providing also valuable information on the plastic response of the structure under seismic actions.

2.5.5 Nonlinear time-history dynamic analysis

It is widely acknowledged that nonlinear dynamic time-history analysis is the most accurate method for simulating the seismic response of structures. The main advantage of this method is that the structural model accurately simulates the damage pattern of the structure. The type of plastic mechanism (either partial or global) can be checked and directly measured, namely by monitoring if non-dissipative structural elements experience some plastic deformation, and comparing plastic deformation demand with the capacity for the dissipative elements. Contrarily to the elastic (linear) analysis, this implies that the element ductility is directly verified and if non-dissipative elements remain in the elastic domain during the seismic action, their strength and stability are checked. The main disadvantages of nonlinear dynamic time history analysis are the large amount of time to develop the structural model and to perform the analysis, needing more sophisticated design software, the modelling of the plastic behaviour of the structural elements, the complexity of the nonlinear analysis and the large number of results that need to be processed. Therefore, this analysis method is not used in the design of standard buildings. Instead, nonlinear dynamic analysis is useful in the assessment of the seismic performance of new buildings with high importance. Another field of application for nonlinear dynamic analysis is the seismic performance assessment of seismic noncompliant existing buildings. These structures do not meet seismic compliance demands and structural details from modern design provisions that provide a ductile response. Therefore, elastic design methods are less reliable in this case.

As previously mentioned, nonlinear dynamic time-history analysis requires an inelastic structural model and the seismic response is obtained by direct numerical integration of the equations of motion (see equation (2.1)). It is clear that the accuracy of this analysis firstly depends on the effectiveness of the structural model to mimic the inelastic hysteretic response of the plastic zones, which differs with the type of element subjected to cyclic actions (e.g. see Figure 2.26). Hence, comparisons and calibration with experimental results from sub-assemblage and large-scale testing of structures are still needed, in order to ensure satisfactory accuracy. In the following sub-chapter 2.6, the main features related to different modelling strategies are described.

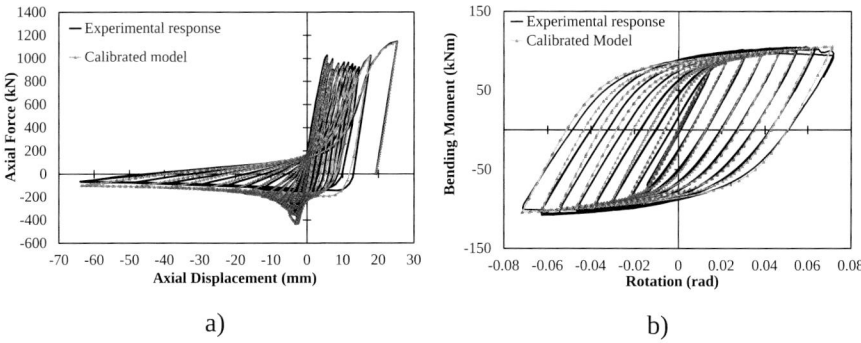

Figure 2.26 – Non-linear hysteretic modelling of structural elements: a) bracing member under cyclic axial loading (D'Aniello *et al*, 2013); b) beam under cyclic bending (Tenchini *et al*, 2014)

2. EN 1998-1: General and Material Independent Parts

The seismic action in nonlinear dynamic analyses is directly represented by accelerograms. Due to the probabilistic nature of the seismic action, characterizing structural response based only on one accelerogram is inappropriate. Therefore, EC8-1 (section 3.2.3.1.2) imposes the use of a minimum of 3 accelerograms. From a statistical point of view, three records represent a very small number of cases. For this reason, under this assumption, clause 4.3.3.4.3(3) recommends to consider the most unfavourable value of the structural response quantity to be used as design effect E_d in order to verify the structural capacity. Alternatively, provided that a larger number of accelerograms is adopted for dynamic time history analyses (minimum of seven different accelerograms), it is possible to use the average of the structural response quantities from all of these analyses as the design value of the action effect E_d in the relevant verifications.

There are several numerical methods that can be used for direct integration of the equations of motion (2.1). The objective of these methods is to satisfy dynamic equilibrium at discrete time intervals, once the initial conditions at $t=0$ are known. Most of these methods use equal time intervals: Δt, $2\Delta t$, $3\Delta t$, … $N\Delta t$ and can be classified in two large categories: explicit methods and implicit methods (Wilson, CSI Technical Report).

The explicit methods do not solve step by step the set of linear equations, but they basically use the differential equation at time "t" in order to predict a solution at time "$t + \Delta t$". For most structures that contain very rigid elements, obtaining a stable solution implies adopting a very small time step. Therefore, all explicit methods are stable related to the size of the integration step.

Conversely, implicit methods try to check the differential equation at time "t" after the solution in step "$t + \Delta t$" has been found. These methods solve the set of linear equations at every step, but they have the advantage that they allow to use larger time steps. Implicit methods can be conditionally or unconditionally stable.

Both methods assume that the solution is a smooth function with continuous higher derivatives. Nevertheless, the "exact" nonlinear solution requires that the accelerations (i.e. the second derivative of the displacements) are not smooth functions. Indeed, it should be noted that acceleration discontinuities are inevitably present in nonlinear dynamic analyses, due to different sources of nonlinearities (e.g. material behaviour, connections and buckling of members). This situation leads to choose implicit integration methods with simple step and unconditionally stable for most practical cases.

Several direct integration methods can be used (Wilson, CSI Technical Report), and among the most common the following are worth to be mentioned: Newmark method, constant acceleration method, θ factor Wilson method, Hilber method, α (HHT) factor Hughes and Taylor method, etc. Due to problems related to nonlinearities (both mechanical and geometrical, but even non-constant boundary conditions, and force nonlinearity as well), iterative-incremental procedures are generally adopted. These techniques imply the application of loads in predefined time increments, and the dynamic equilibrium is solved using an iterative scheme, which is applied until convergence or an imposed limit on the number of iterations is reached. After each incremental step, the stiffness matrix of the structure is updated, in order to reflect the modification of structural stiffness due to nonlinear behaviour.

Different types of convergence criteria can be used, such as displacement/rotation, force/moment or energy based (Seismosoft, 2014). The displacement/rotation based convergence criterion verifies at each individual degree-of-freedom of the structure that the current iterative displacement/rotation is less than or equal to a user-specified tolerance, thus guaranteeing to control directly the degree of precision or, inversely, the approximation adopted in the solution of the problem. This criterion is also sufficient to guarantee the overall accuracy of the solution obtained for the large majority of analyses. The force/moment based convergence criterion is recommended when the displacement based convergence criterion is not sufficient to guarantee equilibrium of the internal forces of the structural elements. The maximum precision will be obtained if the displacement and the force convergence criteria are combined; instead, the maximum stability of the analysis procedure is obtained if a single convergence criterion is adopted.

Usually, the integration time step size is set in the range of 0.005 - 0.02 s. Algorithms that automatically adjust the size of the integration step between predefined limits can be used, increasing the step of integration when the solution easily converges and decreasing it when the convergence is more difficult to achieve.

Finally, it is worth mentioning IDA (Incremental Dynamic Analysis), which is a computational analysis method that it is becoming very popular to evaluate of the seismic performance of structures (Vamvatsikos and Cornell, 2002). In particular, IDAs are mostly used to estimate the mean annual rate of exceeding certain levels of structural demand (e.g., maximum peak interstorey drift ratio, ductility demand, residual interstorey drift ratios, etc.) or specified limit-states (e.g., global dynamic instability, rotation capacity, local instability, etc.).

This method of analysis consists in performing multiple nonlinear dynamic time history analyses under a suite of ground motion records, whereby each of them is scaled to several levels of seismic intensity. It is important to highlight that the scaling levels should be selected in order to examine the structural behaviour from the initial elastic response to the inelastic response and finally to overall dynamic instability, which corresponds to collapse. The results from this type of analysis are generally post-processed in terms of IDA curves (Vamvatsikos and Cornell, 2002), one for each ground motion record, reporting the seismic intensity (i.e. a scalar Intensity Measure "IM", as the scaled peak ground acceleration) versus the structural response parameter (i.e. a measured Engineering Demand Parameter "EDP", as the interstorey drift ratio).

2.6 STRUCTURAL MODELLING

2.6.1 Introduction

The structural model for seismic analysis must be capable of simulating the (1) masses, (2) damping and (3) the distribution of structural mechanical properties in the structure. The level of refinement and complexity for the modelling of these features basically depends on the two aspects: 1) the regularity and complexity of the structural configuration; 2) the method of structural analysis.

Generally speaking, the first aspect is related to the conception of structural scheme. In case of steel buildings, the structure is made of two macro-systems linked at the floor levels by horizontal diaphragms, namely a part devoted to sustain gravitational forces and another conceived to resist seismic actions. Decoupling the structural resistance from gravity and seismic forces is really profitable from a practical point of view, because it allows simplifying and better controlling the resisting mechanisms, and leads to more economical design solutions.

Typical lateral forces resisting systems are moment resisting frames structures (with rigid joints), braced frames (centrically or eccentrically) and structural walls (see Figure 2.27). As it can be easily recognized, except moment resisting frames with rigid joints, the other lateral force resisting systems impose architectural restrictions and, consequently, limitations in terms of their arrangement in the structure.

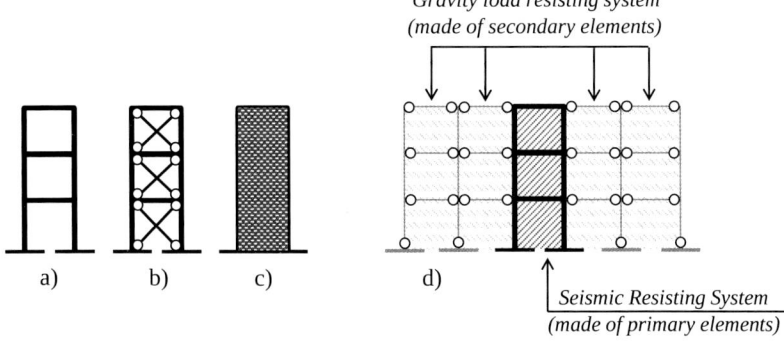

Figure 2.27 – Lateral forces resisting systems: moment resisting frames with rigid joints (a), centrically braced frames (b), structural walls (c); combined gravitational and seismic resisting system (d)

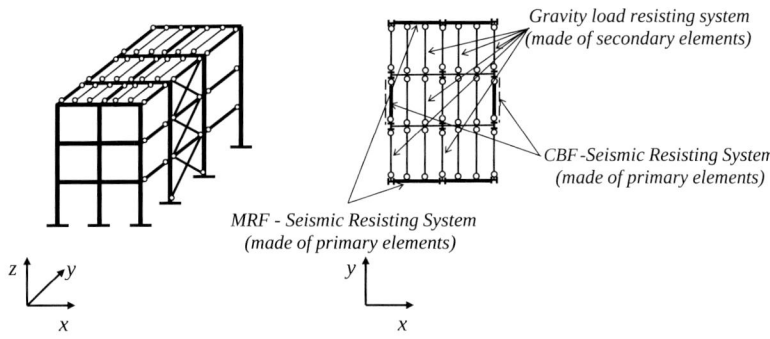

Figure 2.28 – Example of gravitational and lateral loads resisting systems

Figure 2.28 illustrates the layout of a steel structure with two spans in the x direction, three bays in the y direction and three storeys. Gravitational loads are taken by pinned secondary beams and internal frames. The lateral loads resisting systems are located along the perimeter of the building in order to provide good strength and torsional stiffness. In particular, moment resisting frames (MRFs) with rigid beam-to-column joints are disposed in the x direction and concentrically braced frames in the y direction. Steel-concrete composite slabs (not represented in the sketch) fulfil the role of horizontal diaphragms, thus transferring the seismic action induced by the inertial forces to all seismic resisting systems.

2.6.2 Modelling of masses

The correct modelling of the structural mass is crucial in seismic analysis, because the seismic forces are inertial actions that appear due to accelerations induced to the structural masses.

The structural analyst should clearly distinguish between mass and weight, which must be never confused. Indeed, mass refers loosely to the amount of "matter" in a body, whereas weight refers to the force experienced by an object due to acceleration of gravity (e.g. a body with a mass equal to 1.0 Kilogram has a weight approximately equal to 9.81 Newton, being 9.81 m/s² the acceleration of gravity).

Generally, the mass of a body can be concentrated in its centre of gravity and it is characterized by three translational and three rotational components. The mass of a structure is distributed in its whole volume (see Figure 2.29a). However, to simplify the structural analysis, in most cases, the mass can be considered lumped in the nodes of the structural model (see Figure 2.29b). Generally, the rotational components of the masses have a minor influence on the structural dynamic response and are neglected. In case of planar structural models, the masses have components in the two translational directions (i.e. horizontal and vertical). Considering the presence of horizontal rigid diaphragms and neglecting the vertical component of the seismic motion, the resultant of the horizontal component of structural masses can be considered lumped at each floor (see Figure 2.29c). However, when it is necessary to consider the vertical component of the seismic action, the vertical components of the mass must be modelled.

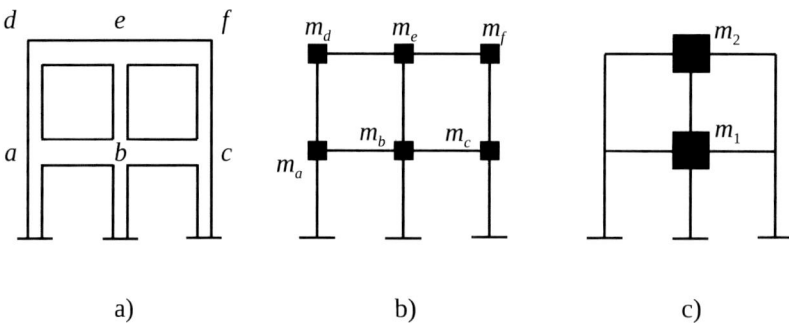

Figure 2.29 – Distributed mass in a planar frame (a), its model with masses lumped into structural nodes (b) and with masses lumped at each floor (c)

2.6 STRUCTURAL MODELLING

In case of three-dimensional structural models, if the slabs are deformable in-plane (i.e. they do not behave as a rigid diaphragm, such as timber slabs or reinforced concrete slabs with large openings), the masses should be assigned to each structural node and their values correspond to the tributary area associated to each node (Figure 2.30a). In such cases, rotational components of masses may be neglected, while only the translational components may be considered in the structural analysis. If the vertical component of the seismic excitation is neglected, in three-dimensional models only the translational components of the mass in the two horizontal directions may be considered.

If the slabs can be considered infinitely rigid in their plane, the overall mass belonging to each floor can be lumped at the centre of gravity at that level (clause 4.3.1(4)). This assumption is very important from the point of view of the computational effort, because reducing the number of lumped masses reduces the number of dynamic degrees of freedom, hence the size of the structural model. The masses lumped at the centre of mass of each floor have three components, namely two horizontal translations and one rotation around the vertical axis (see Figure 2.30b). The mass translational components are determined by summing all the masses associated to that level, as follows:

$$M_x = M_y = \sum m_i \qquad (2.44)$$

The rotational component of mass at storey level M_{zz} is the moment of inertia of the masses and is given by:

$$M_{zz} = \sum m_i d_i^2 \qquad (2.45)$$

where d_i is the distance from centre of mass to the discrete mass m_i (see Figure 2.30c). In case of masses that are uniformly distributed on the floor level, the moment of inertia of masses is computed as the product between the polar moment of inertia and the value of the uniformly distributed mass on the planar surface.

In structural analysis software the rigid diaphragm effect is usually modelled by assigning a "rigid diaphragm" constraint to the nodes of the floor, namely one internal constraint per storey level. If the rigid diaphragm constraint is not directly implemented in the software, it can be modelled using equivalent horizontal bracing that connect the nodes of the floor with those of the main frames.

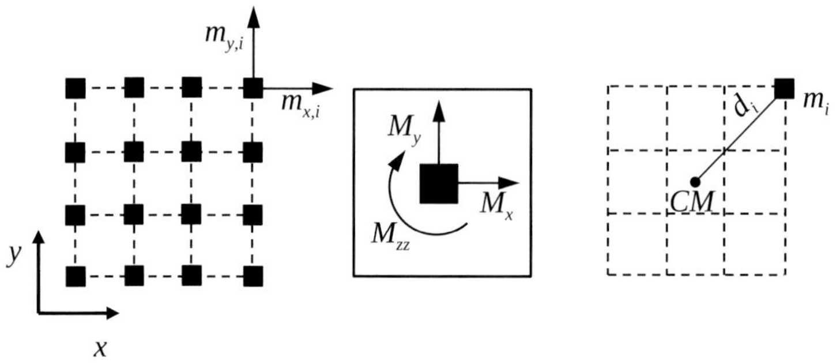

Figure 2.30 – Masses lumped into the structural nodes in case of flexible slab (a); mass concentrated into the centre of masses in case of rigid diaphragms (b); mass m_i and distance d_i to compute the moment of inertia of masses (c)

From the design point of view, it is important to establish when it is reasonable to assume that the slab is rigid in-plane. Clause 5.10(1) states that solid reinforced concrete slabs provide a rigid diaphragm effect if they are at least 70 mm thick and both horizontal directions have the minimum reinforcement according to EN 1992:1-1 (CEN, 2005a), also provided that the slab has regular shape and small gaps. If the concrete slab has a very irregular plan shape, large gaps, or irregular mass and stiffness distributions, it must be explicitly checked to in-plane internal forces generated by the seismic action. As a general remark, clause 4.3.1(4) clarifies that the slab can be considered as a rigid diaphragm, if, when it is modelled with its actual in-plane flexibility, its horizontal displacements do not exceed by more than 10 % those resulting from the analysis with a rigid diaphragm.

2.6.3 Modelling of damping

The modelling of damping is not required for lateral force method, modal response spectrum analysis and nonlinear static analysis. Indeed, in these types of analysis, the damping properties of the system are directly included in response spectra by calculating the damping correction factor η (see equation 2.19) that modifies the ordinates of the elastic spectrum (equation (2.20)), and characterises the seismic demand.

However, the explicit modelling of damping is necessary for linear and nonlinear dynamic analyses. Indeed, the numerical integration of the equations of motion (2.1) requires the characterization of the damping matrix $[c]$, which

2.6 Structural Modelling

depends on both structural geometric features and component materials properties. Unfortunately, it is unfeasible to determine the matrix [c] on the basis of those characteristics. Hence, for SDOF systems it is possible to express the damping matrix as proportional to either the mass or the stiffness matrix, as follows:

$$[c] = a_0 [m] \qquad (2.46)$$

$$[c] = a_1 [k] \qquad (2.47)$$

When damping is proportional to the mass, the fraction of the damping ratio corresponding to mode "i" is given by:

$$\xi_i = \frac{a_0}{2} \frac{1}{\omega_i}. \qquad (2.48)$$

In this case, damping is inversely proportional to the circular frequency (see Figure 2.31a). The coefficient a_0 can be determined so as to represent the critical damping fraction from any mode, for example, ξ_i in mode i. Equation (2.48) gives:

$$a_0 = 2\xi_i \omega_i \qquad (2.49)$$

Once the coefficient a_0 is known, the damping matrix can be determined from equation (2.46), and the damping ratio for any other mode "j" is given by equation (2.48).

Similarly, the damping ratio can be related to the coefficient a_1 in case of stiffness proportional damping. In this case:

$$\xi_i = \frac{a_1}{2} \omega_i \qquad (2.50)$$

The damping ratio increases linearly with the circular frequency (see Figure 2.31a). The coefficient a_1 can be determined so as to represent the damping ratio ξ_j for any mode "j", as follows:

$$a_1 = \frac{2\xi_j}{\omega_j} \qquad (2.51)$$

Similarly, once the coefficient a_1 is obtained, the damping matrix can be determined from equation (2.47) and the damping ratio from any other mode "j", is given by equation (2.50).

It should be noted that neither type of damping is suitable for MDOF systems, because experimental tests show similar values of damping ratios for several modes of vibration. Therefore, in case of building structures it is more convenient to use the "Rayleigh damping", which is obtained by combining both mass and stiffness proportional damping models, as follows:

$$[c] = a_0[m] + a_1[k] \qquad (2.52)$$

Hence, the Rayleigh damping ratio for mode "i" can be expressed as:

$$\xi_i = \frac{a_0}{2}\frac{1}{\omega_i} + \frac{a_1}{2}\omega_i \qquad (2.53)$$

where the coefficients a_0 and a_1 can be determined so as to represent the damping ratios for any couple of modes, for example ξ_i and ξ_j in modes "i" and "j". Thus, rewriting equation (2.53) for the two modes "i" and "j" in matrix format yields:

$$\frac{1}{2}\begin{bmatrix} 1/\omega_i & \omega_i \\ 1/\omega_j & \omega_j \end{bmatrix} \begin{Bmatrix} a_0 \\ a_1 \end{Bmatrix} = \begin{Bmatrix} \xi_i \\ \xi_j \end{Bmatrix} \qquad (2.54)$$

If the same damping ratio ξ is considered for both modes, equation (2.54) can be solved as follows:

$$a_0 = \xi\frac{2\omega_i\omega_j}{\omega_i + \omega_j} \qquad a_1 = \xi\frac{2}{\omega_i + \omega_j} \qquad (2.55)$$

Having determined the coefficients a_0 and a_1, the damping matrix can be determined using equation (2.52), and the damping ratio for any mode "k" is given by equation (2.53) and graphically represented in Figure 2.31b.

In practical situations, modes "i" and "j" are chosen so that the damping ratios have reasonable values for all modes of vibration that significantly contribute to the structural response. For example, if the damping ratio ξ is assigned to modes 1 and 4, modes 2 and 3 will have a damping ratio slightly smaller than ξ, while from the fifth to the higher modes the damping is larger than ξ. This implies that all modes higher than the fifth will have damping ratios that increase with the circular frequency and their response becomes practically ineffective because it has been eliminated by excessive damping.

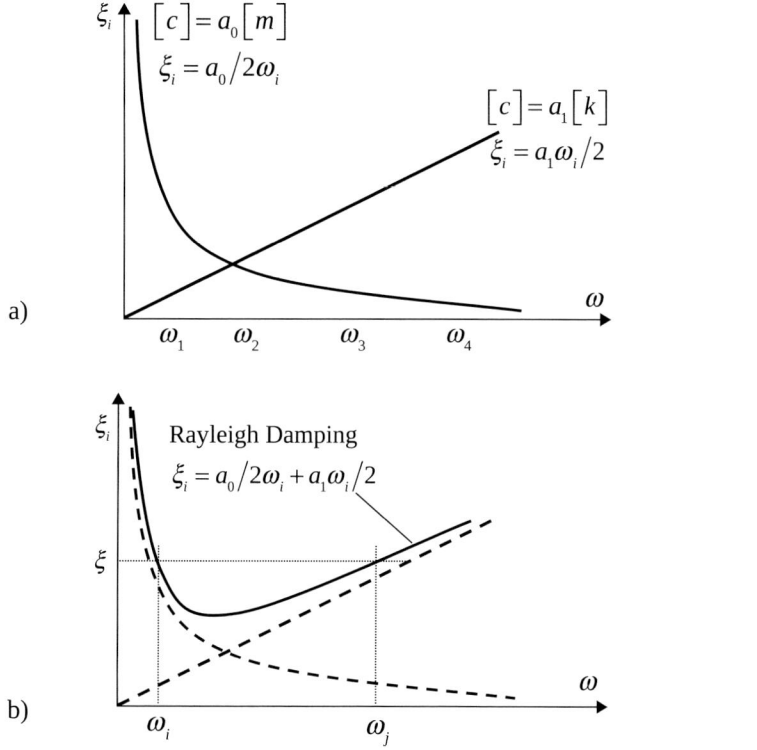

Figure 2.31 – Variation of modal damping with circular frequency: mass and stiffness proportional damping (a); Rayleigh damping (b)

2.6.4 Modelling of structural mechanical properties

The assumptions and the level of refinement required for structural models differ with the type of analysis. For elastic analyses (i.e. lateral force method, modal response spectrum and linear dynamic analysis) the structural model must adequately represent the distribution of stiffness of members into the building (see clause 4.3.1(1)P). In addition, the elastic structural model of a steel building should include the connections, the web of the columns (for moment resisting frames) and the stiffness of foundation joints (e.g. soil-to-footings interaction) if those components are characterized by significant deformability that potentially modifies the overall dynamic behaviour and the internal distribution of forces. Additionally, the structural model shall include non-structural elements that may significantly influence the response of the main seismic structure.

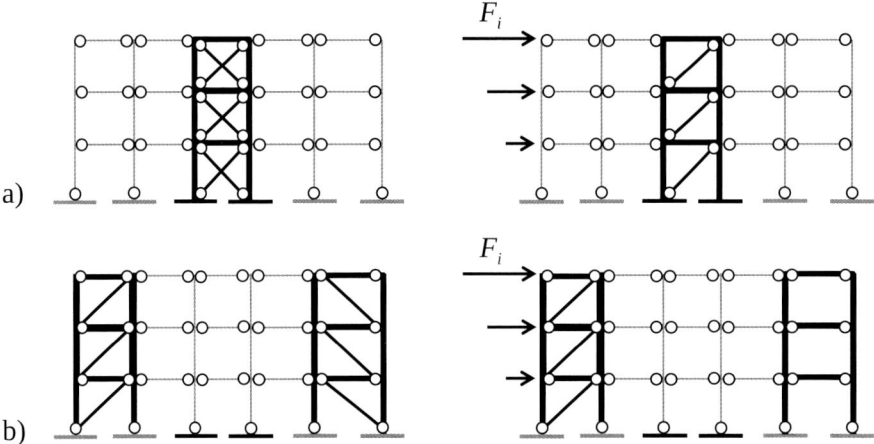

Figure 2.32 – Structural model for elastic analysis of either X concentrically braced frame (a) and diagonal braced frame (b)

In most cases, the elastic structural model requires the definition of the geometry of the members and their relevant elastic properties, namely elastic modulus of material and characteristics of the cross sections (i.e. area, second moment of area, etc.). These assumptions are generally very simple to be implemented in the most structural schemes used for steel buildings. However, some specific problems may arise when elastic structural models of both X and single diagonal concentrically braced steel frames (see Figure 2.32a) should be developed. Indeed, clause 6.7.2(2)P specifies that in elastic analysis (i.e. using either the lateral force method or modal response spectrum analysis) only tensioned braces are considered, the compressed ones being neglected due to buckling. For the lateral force method, the compressed braces can be easily identified and eliminated from the structural model (see Figure 2.32b). On the contrary, in modal response spectrum analysis, the compressed braces cannot be identified, because the results from this type of analysis represent the envelope of seismic effects considered in both directions. However, one

2.6 STRUCTURAL MODELLING

of each two braces in every braced span can be eliminated (see Figure 2.32b), considering the force developed in the braces as tensile only, irrespective of the sign obtained in the analysis. It is clear that whichever type of elastic analysis (i.e. lateral force method or modal response spectrum analysis) if the structure is not symmetrical, it is necessary to carry out the analysis considering two different models, the first with bracing members tilted in a direction and the second with braces tilted in the opposite direction.

When a plastic analysis method (nonlinear static or dynamic analysis methods) is used, the structural model must include both tensioned and compressed braces (see Figure 2.32a), which should be simulated accounting for the behaviour before and after buckling. In case of non-linear static and dynamic analyses (i.e. pushover and time history analyses) the structural model must include both geometric non-linearities and the post-elastic behaviour of the structure (see clause 4.3.3.4.1(2)).

Geometric non-linearity is very important in seismic analysis. Indeed, the seismic displacement demand at each storey generally overcomes the range of displacements and deformation usually accepted for first-order small displacement and small deformation theory. In this case, secondary effects are induced into the structure at both local and global level, which are generally known as *P*-delta effects. There are two distinct types of *P*-delta effects: P-δ (sometimes referred to as "small *P*-delta" or "*P*-small delta"), and P-Δ (sometimes referred to as "large *P*-delta" or "*P*-large delta"). The former corresponds to individual members within the frame, thus influencing the structural response at the local level, being "*P*" the source of external action and "δ" the distance between one end of the member and its offset from the other end. Indeed, P-δ effects (i.e. secondary bending moments, shear and axial forces) are induced into the structural members by the large deflections that those elements experience under seismic excitation due to the fact that the ends of the member are no longer co-linear in the deflected position. This effect can be accounted for updating the matrix of stiffness and that of forces as respect to the deformed shape of the structure.

At global level, the resultant "*P*" of vertical loads acting at a generic floor will produce overturning action and secondary internal forces into the members due to the horizontal interstorey drift "Δ". If 3D structural models are developed, the large displacement option directly allows accounting for this effect. In case of 2D (i.e. planar) models the gravity loads applied are those given by the tributary area of loads acting directly on the planar frame (indicated as A_L in Figure 2.33), which do not reflect the actual amount of vertical forces producing overall overturning effects. Hence, it is necessary to account for the influence given by the complement of vertical loads acting on the tributary area of masses (indicated as A_M in Figure 2.33) that is attributed to that planar frame. In order to account for this phenomenon, it is possible to introduce the so-called leaning column into the nonlinear numerical model (see Figure 2.33). This type of column is a fictitious zero-stiffness vertical element supporting the amount of vertical loads that are not directly applied to the 2D frame model. The leaning column shall also be connected at each storey level by a rigid diaphragm.

It should be noted that also in elastic analyses it is necessary to account for P-Delta effects. The common method is based on a simplified approach that consists in magnifying the effects due to seismic action (namely the internal forces) by means of a factor depending on the deformability of the structure and the gravity and horizontal loads. This is the method explicitly provided by EC8 for structural design and it will be shown and discussed in sub-chapter 2.10.

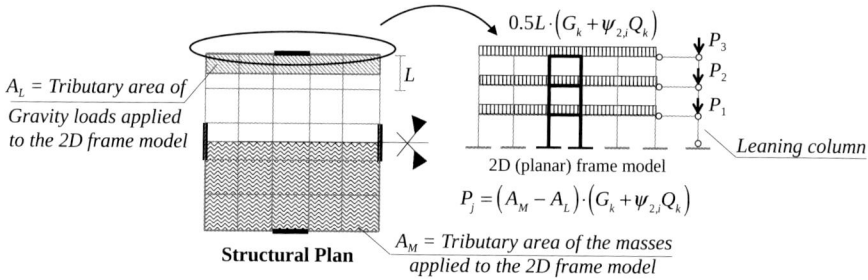

Figure 2.33 – The concept of leaning column to account for P-Δ effects in 2D models

Another key aspect that should be taken into account when non-linear structural models have to be developed is the simulation of the plastic response of structural members. The accuracy and the reliability of the analysis strictly depend on the capability of reproducing the intrinsic inelastic hysteretic behaviour of structures. Generally speaking, the hysteretic models used to simulate the nonlinear response introduce significant simplifications when compared to the experimental behaviour. These simplifications could lead to a incorrect prediction of the structural response (e.g. peak and residual distribution of interstorey drifts, local ductility demand) or even behaviour modes (i.e. damage pattern).

2.6 STRUCTURAL MODELLING

Hence, comparisons with experimental results from either sub-assemblage or large-scale testing of structures and their components can be still necessary in several cases, in order to ensure adequate levels of confidence in this numerical methodology. However, the need for validating against experimental data strictly depends on the numerical approaches used to simulate the plastic behaviour of structural elements. For most steel building structures, two different categories of modelling strategies are used to simulate the inelastic behaviour of the structural elements for the analysis of the seismic action:

i) lumped plasticity models;
ii) distributed plasticity models.

A structural member modelled using lumped plasticity is made of an elastic part and plastic hinges fixed a-priori in one or more sections along the element length, where all plastic deformations are concentrated. For members belonging to moment-resisting frames the plastic hinges are generally located at both ends of each member (see Figure 2.34), where the maximum bending moments generally occur. This plasticity model has the advantage of being numerically efficient and allows adopting different constitutive relationships (i.e. hysteretic curves) that can model a wide range of phenomena like: strain hardening, strength and stiffness degradation, slips, etc., as shown in Figure 2.35. The plastic hinge model may be derived to reproduce the pure bending moment-rotation response (e.g. for modelling plasticity of beams or long links), the axial force-axial displacement response (e.g. for modelling plasticity of braces) or the shear force-shear distortion response (e.g. for modelling plasticity of shear links). In addition they can be extended in order to account for the interaction between bending moment and axial force (e.g. for modelling plasticity of a column). An advantage of this approach is the simplicity and the immediacy to interpret the results, because structural analysis provides plastic rotations in hinges that can be directly compared to the relevant capacity. However, the concentrated plasticity model has several disadvantages. The most severe is that the hysteretic response calibration requires skilled experience and the accuracy of the model should often be validated against experimental results, because the hysteretic properties are defined using a set of empirical rules for the shape of hysteretic loops without directly representing the physical phenomena. Another weak aspect is the predefined position of plastic hinges, which can lead to unrealistic and unsafe prediction of the plastic distribution and the relevant demand (e.g. when the maximum moment is not at the end of the element).

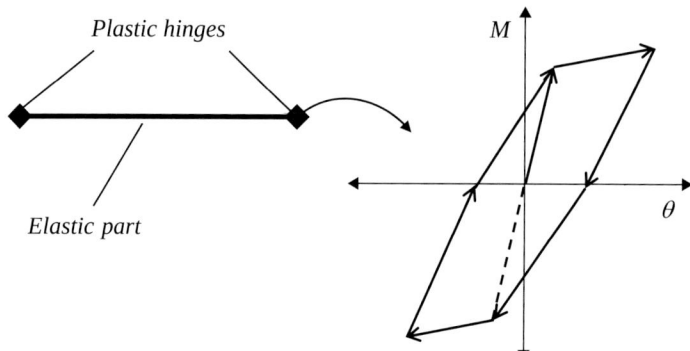

Figure 2.34 – Modelling of a structural member with lumped plastic hinges at both ends

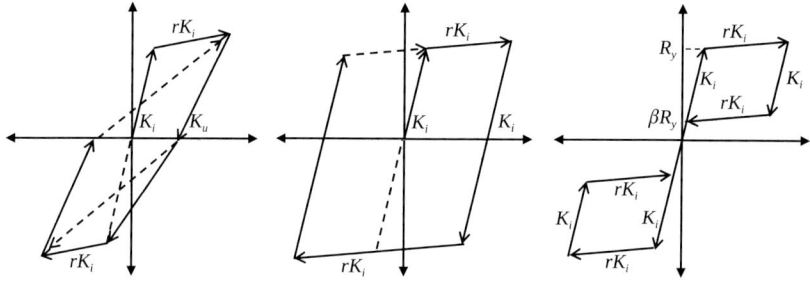

Figure 2.35 – Some examples of constitutive relations for modelling the cyclic behaviour of structural components by means of plastic hinges

Distributed plasticity models (see Figure 2.36) account for distributed inelasticity through the integration of the material response over the cross section and integration of the section response along the length of the element, that is discretized into a number of integration nodes. The cross section behaviour is reproduced by means of the fibre approach, assigning a uniaxial stress-strain relationship to each fibre. This type of modelling approach overcomes most of the limitations of the lumped plasticity approach. First, they require a smaller number of experimental parameters to be specified, namely the basic input data to be implemented are the material properties, the element geometry and the distribution of the fibers in the cross sections. In addition, by subdividing each element in a series of segments, the model can accurately simulate the evolution of plasticity in the cross section and its progression along the member. Another advantage is the direct modelling of the moment-axial force interaction, which does not require the definition of the interaction domain. The main disadvantages of distributed plasticity models are the higher computational effort (i.e. time and requirements of the computer hardware) and the more difficult interpretation of results.

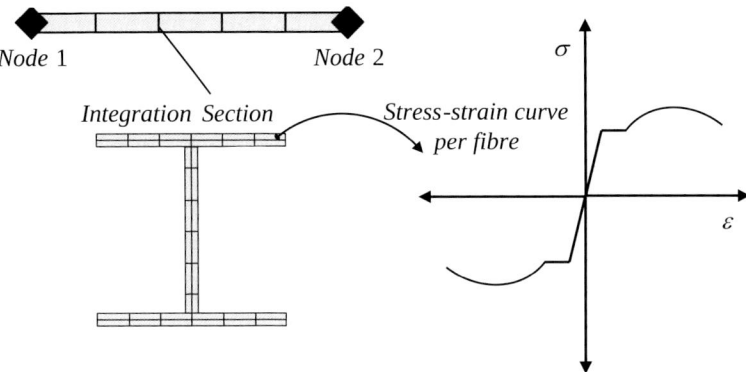

Figure 2.36 – Modelling of a structural member with distributed plasticity approach

2.7 ACCIDENTAL TORSIONAL EFFECTS

2.7.1 Accidental eccentricity

Although EC8-1 encourages conceiving regular structures (both in plan and elevation), many buildings cannot satisfy this optimal condition due to aesthetical and/or functional needs. It can be easily recognized that structures with irregular distribution of mass and/or stiffness at a storey level should exhibit an important torsional contribution to the seismic response. Therefore, the most accurate modelling of the torsional effects will be achieved if a three-dimensional model of the structure is used, especially when a modal response spectrum analysis or a nonlinear or linear dynamic time history analysis is adopted. However, even for perfect regular structures in plan, the "actual" distribution of both mass and stiffness may differ from that nominally estimated at design stage (e.g. assumption of uniformly distributed mass upon each floor), and the seismic motion may have torsional components around the vertical axis of the building, thus leading to supplementary torsional effects in the structure. Moreover, the seismic motion may present some spatial variation at the base of the building, inducing some

2. EN 1998-1: GENERAL AND MATERIAL INDEPENDENT PARTS

torsional acceleration and consequently torsional effects. In order to account for all these effects, Eurocode 8 introduces the concept of accidental torsional effects. Therefore, for buildings with slabs acting as rigid diaphragms in their plane, the effects generated by the uncertainties associated to masses at each storey and/or the spatial variation of the ground seismic motion are considered by introducing an additional accidental eccentricity e_{ai} to shift the position of the centroid of masses at the *i*-th storey of the building from its nominal location. This eccentricity has to be accounted for in each main direction of the structure and for each floor. According to clause 4.3.2(1)P, the accidental eccentricity is given by:

$$e_{ai} = \pm 0.05 L_i \qquad (2.56)$$

where L_i is the floor dimension perpendicular to the direction of the seismic action of the i-th floor.

In order to apply this provision, it is necessary to determine both translational and rotational components of the mass that should be concentrated into the relevant centroid at each storey and then shifting it from its nominal position in each direction by the additional eccentricity e_{ai} calculated according to equation (2.56), as shown in Figure 2.37. This procedure can be used for all structural analysis methods. However, in order to establish the most unfavourable sense of the eccentricity on the two directions, four structural analyses are necessary, namely one for each combination of the eccentricities in the two directions ($\pm e_{ax}$, $\pm e_{ay}$). In what concerns the distribution of eccentricities along the height of the structure, the most unfavourable response is generally obtained when the eccentricities are oriented at the same direction at each storey.

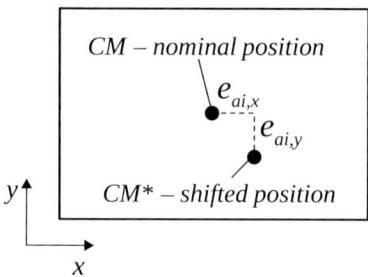

Figure 2.37 – Torsional effects simulated by means of accidental eccentricity e_{ai} in 3D structural models

Since the need to carry out at least four analyses can result in a large number of load combinations and relatively large calculation effort, EC8-1 provides alternative simplified approaches to account for torsional effects that differ for each adopted method of structural analysis. Hereinafter, a concise overview of those methods is described.

2.7.2 Accidental torsional effects in the lateral force method of analysis

The lateral force method is the simplest method for seismic analysis and a simplified approach of accidental torsional effects is consequently justified.

Clause 4.3.3.2.4(1) requires for a lateral force analysis on a three-dimensional structural model that the accidental torsional effects can be taken into account by multiplying the action effects in the individual load resisting elements by a magnification factor δ, as shown in Figure 2.38 and given by:

$$\delta = 1 + 0.6 \frac{x}{L_e} \quad (2.57)$$

where:

- x is the in-plane distance between the structural element under consideration (i.e. the seismic resisting frame) and the centre of mass of that storey, measured perpendicularly to the direction of seismic action considered (see Figure 2.38);
- L_e is the in-plane distance between the two outermost lateral load resisting systems, measured perpendicularly to the direction of seismic action considered (see Figure 2.38).

The factor δ given by equation (2.56) is derived assuming the following hypotheses:

- torsional effects are fully resisted by the seismic resisting systems located on the considered direction, neglecting any contribution by the seismic resisting systems placed on the orthogonal direction;
- both strength and stiffness of the seismic resisting systems are uniformly distributed in plan.

2. EN 1998-1: GENERAL AND MATERIAL INDEPENDENT PARTS

In case the lateral force analysis is carried out on 2D structural models (clause 4.3.3.2.4(2)), one for each main horizontal direction, the accidental torsional effects may be taken into account by multiplying the action effects in the individual load resisting elements by a larger δ factor given as follows:

$$\delta = 1 + 1.2 \frac{x}{L_e} \qquad (2.58)$$

This approach is equivalent to use a 10 % accidental eccentricity, instead of the 5 % standard one, to account for possible static eccentricities between mass and stiffness centres that are not otherwise considered when planar models are used.

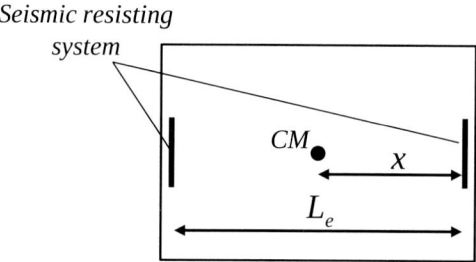

Figure 2.38 – Torsional effects simulated by means of the magnification factor δ

2.7.3 Accidental torsional effects in modal response spectrum analysis

Clause 4.3.3.3.3(1) accounts for accidental torsional effects when the seismic response is estimated by means of modal response spectrum analysis using 3D structural models by introducing torsional moments applied at the centre of mass of each storey (see Figure 2.39). These supplementary torsional moments should be calculated as follows:

$$M_{ai} = e_{ai} F_i \qquad (2.59)$$

where:
- M_{ai} is the torsional moment applied at storey i about its vertical axis;
- e_{ai} is the accidental eccentricity of the centre of masses of storey i, evaluated according to equation (2.56);
- F_i is the horizontal seismic force applied to storey i, per relevant direction of the seismic action.

Torsional moments are computed for all directions and orientations of the actions considered in the analysis. It should be noted that for computing torsional moments, the forces F_i are obtained using the lateral force method (see section 2.5.2), even if the structure is not regular in elevation. In particular, those equivalent static forces are distributed according to equation (2.33), thus being solely representative of the fundamental mode of vibration, which differ from those given by superposition of modal responses.

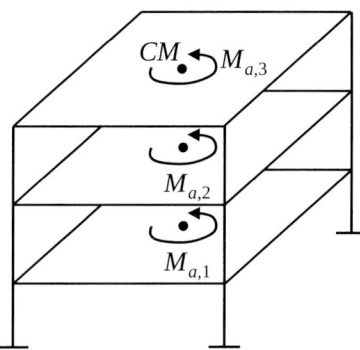

Figure 2.39 – Accidental eccentricity effects modelled using torsional moments applied to the centroid of mass at each storey

When the modal response spectrum analysis is carried out using 2D planar structural models (clause 4.3.3.3.3(3)), accidental torsional effects can be considered by multiplying the effects of the actions obtained by the modal superposition by the magnification δ factor given by equation (2.58).

2.7.4 Accidental torsional effects in nonlinear static pushover analysis

Nonlinear static analysis was developed for plane structural models, subjected to a single seismic action component. Therefore, the application of this method of analysis in case of structures affected by accidental or primary torsional effects, or affected by higher modes of vibration, raises a large number of open issues, for which there is yet no unique solution.

It is known (clause 4.3.3.4.2.7(1)P) that a nonlinear static pushover analysis carried out using unidirectional distributions of the lateral forces may significantly underestimate deformations (i.e. the seismic demand) at the stiff/strong in-plane side of a torsional flexible structure, which is characterized by a pure torsional or torsionally coupled fundamental mode of vibration. The same applies for the stiff/strong side deformations in one direction of a structure with a predominantly torsional second mode of vibration. The stiff/strong side in plan is the one that develops smaller horizontal displacements than the opposite side when the structure is subjected to lateral static forces parallel to it. For torsional flexible structures, dynamic inelastic displacements on the stiff/strong side may significantly increase due to the influence of the predominant torsional mode.

As it can be easily understood, the prediction of the seismic response of a torsional flexible structure is a complex issue. However, in order to preserve the simplicity that characterizes the nonlinear static analysis (when compared to nonlinear dynamic time history analyses), EC8-1 requires correcting the displacement demand obtained using a nonlinear static analysis in order to account for the amplifications of displacements due to torsional effects. This approach is based on the assumption that the torsional effects influence in the same manner both the elastic and the inelastic displacement demands. Therefore, according to clause 4.3.3.4.2.7(2), when a 3D model is used to perform a nonlinear static analysis, the torsional effect induced by accidental eccentricity can be obtained by magnifying the displacements on the stiff/strong side with a factor determined using a modal response spectrum analysis on the elastic three-dimensional model. Implementing these provisions involves the following steps:

1. A nonlinear static analysis is made on the three-dimensional structural model for each horizontal direction and orientation (positive or negative), using codified (i.e. "modal" or "uniform", see section 2.5.4) unidirectional distributions of lateral forces. The target displacement is determined referring to the centre of mass of the last storey of the structure (see Annex B of EC8-1) for each main horizontal direction of the structure, considering the largest of the values for the positive and negative orientation of the lateral forces;
2. A modal response spectrum analysis is carried out on the 3D elastic structural model, taking into account both seismic action directions (see section 2.5.3);
3. Correction factors defined by the ratio between displacements normalized to the top of the structure obtained from the modal response spectrum analysis (step 2) and nonlinear static analysis (step 1) are determined. With this regard, it should be clarified that, for both pushover and modal analyses, the profile of normalized displacements of the points belonging

to the top storey of the structure is given as the displacement at the roof storey at an arbitrary point divided by the displacement experienced by the centre of mass of the roof of the structure. Values smaller than 1 of the displacements normalized to the top of the structure obtained from the modal response spectrum analysis are considered equal to 1.0. The correction factors are computed for each horizontal direction and correction factors must be obtained for all alignments of the frames;
4. Nonlinear static analysis results from step 1 are adjusted, multiplying the relevant response features with the corresponding correction factor. For example, in an end frame parallel to the x axis, all response features are multiplied by the correction factor determined for that frame using nonlinear static analysis in the x direction. The relevant response features are, for example, plastic deformations for ductile elements and internal forces for brittle elements.

It should be noted that the last step of this procedure is physically correct, because the response features from a nonlinear analysis (i.e. plastic deformations in ductile components, and forces in brittle components) are not proportional to the structural global deformation, especially if the models of the structural components include strength degradation. However, the procedure adopted by EC8 can be considered sufficiently accurate if the corrected displacements at the top of the structure applied to each frame alignment do not modify the plastic mechanism of the structure and do not substantially modify the state of structural components (e.g. yielding or collapse).

When 2D planar models are used to perform a pushover analysis, clause 4.3.3.4.2.7(3) allows to account for accidental torsional effects by amplifying the imposed displacement demand by a factor δ given by equation (2.57). Alternatively, the amplification factors can be determined using an elastic analysis on a 3D structural model, modelling accidental torsional effects using static torsional moments determined as described in section 2.6.3. Implementing this procedure involves the following steps:

1. A nonlinear static analysis is made on a 2D planar structural model and the target displacement of the structure is determined (see annex B of EN 1998-1);
2. The correction factor is determined using equation (2.58). Alternatively, the correction factor can be determined by computing the ratio between

the top displacement of the structure on an elastic 3D model that includes torsional moments (see section 2.6.3) and the top displacement of the structure on the same model but using only translational components of the seismic action;
3. Nonlinear static analysis results from step 1 are adjusted by multiplying the displacement demand obtained in step 1 by the correction factor determined in step 2.

2.7.5 Accidental torsional effects in linear and nonlinear dynamic time history analysis

In case of linear or nonlinear dynamic time history analysis, EN 1998-1 does not provide any simplified method to account for accidental torsional effects. In case of a 3D model of the structure, the principle described in section 2.6.1 should be directly applied, namely shifting the centre of mass from its nominal position in each direction with the additional eccentricity estimated according to equation (2.56).

2.8 COMBINATION OF EFFECTS INDUCED BY DIFFERENT COMPONENTS OF THE SEISMIC ACTION

The seismic action is actually characterized by three orthogonal translational components (two horizontal and one vertical) that act simultaneously (the torsional components being mostly negligible). All components of the seismic action may have a different contribution (i.e. smaller or larger) to the effects induced into the structure (i.e. lateral displacements of the structural joints, internal forces and ductility in the structural elements, etc.).

Figure 2.40 illustrates this aspect with reference to the axial force in a corner structural column that is induced by both horizontal components of the seismic action (F_x and F_y). The seismic action component F_x generates the axial force N_{1x}, and the seismic action component F_y generates the axial force N_{1y}. Total axial force, as seismic action effect on both horizontal directions is N_1.

The peak values of the ground acceleration for the horizontal components of the seismic action do not occur at the same time. Moreover, a structure generally has different periods of vibration in the two main directions because

2.8 COMBINATION OF EFFECTS INDUCED BY DIFFERENT COMPONENTS OF THE SEISMIC ACTION

of different horizontal stiffness in those directions. Hence, since the seismic response is mainly determined by the period of vibration, the seismic effects from different components occur at different time. These aspects should be accounted for when the seismic effects are considered and, obviously, different approaches should be used for the different methods of structural analysis.

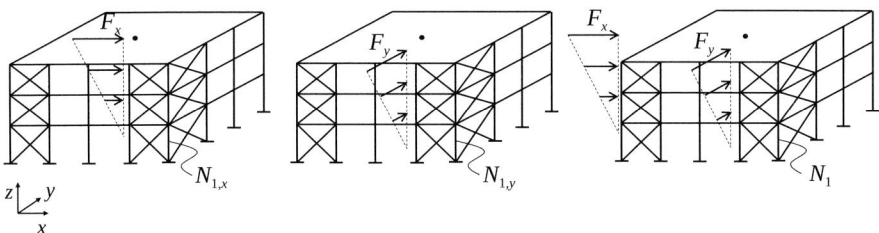

Figure 2.40 – Example of effects induced by two horizontal components of seismic action

In case of the lateral force method or the modal response spectrum analysis, the peak values of seismic induced effects are directly determined. Using the example depicted in Figure 2.40, the structure is consecutively analysed for the two horizontal components F_x and F_y, thus separately obtaining the peak values of the structural response (N_{1x} and N_{1y}). However, the total effect (N_1) of both horizontal components of seismic action is not equal to the algebraic sum of seismic effects separately considered on the two directions (N_{1x} and N_{1y}), because these methods of analysis directly estimate the peak values of the response, but not when they occur. Therefore, it is necessary to use a procedure for combining seismic induced effects that reflects this. According to clause 4.3.3.5(2), the peak value of the seismic induced effect, given by simultaneous action of two orthogonal horizontal components, may be obtained by using the square root of sum of squares combination method for each horizontal component:

$$E_{Ed} = \sqrt{E_{Edx}^2 + E_{Edy}^2} \qquad (2.60)$$

where:
- E_{Edx} represents the action effects due to seismic action on the horizontal x axis;
- E_{Edy} represents the action effects due to seismic action on the horizontal y axis, which is perpendicular to x axis of the structure.

2. EN 1998-1: General and Material Independent Parts

As an alternative, the seismic effects due to the two horizontal components can be computed using the following combinations (clause 4.3.3.5.1(3)):

$$E_{Edx} \text{"+"} 0.3E_{Edy} \quad (2.61)$$

$$0.3E_{Edx} \text{"+"} E_{Edy} \quad (2.62)$$

where "+" means "to be combined with". The sign of each component in the above combinations is chosen so that the effect of the considered action would be unfavourable (clause 4.3.3.5.1(5)P). This combination rule is based on the assumption that when one component of the seismic action achieves the peak value, the other component achieves 30 % of its peak value.

From the two combination methods of the effects induced by the components of the seismic action, the square root of sum of squares method is recommended, because it is simpler and it is independent of the chosen system of axes. This last aspect is very important for structures with complex geometry, for those irregular in-plan and for those that do not have a clearly defined system of two main orthogonal directions (for example buildings with circular shape plan).

The previous rules of combinations can be extended to those cases when the vertical component of the seismic action should be accounted for (clause 4.3.3.5.2(4)). Hence, equations. (2.60), (2.61) and (2.62) become:

$$E_{Ed} = \sqrt{E_{Edx}^2 + E_{Edy}^2 + E_{Edz}^2} \quad (2.63)$$

$$0.3E_{Edx} \text{"+"} 0.3E_{Edy} \text{"+"} E_{Edz} \quad (2.64)$$

$$E_{Edx} \text{"+"} 0.3E_{Edy} \text{"+"} 0.3E_{Edz} \quad (2.65)$$

$$0.3E_{Edx} \text{"+"} E_{Edy} \text{"+"} 0.3E_{Edz} \quad (2.66)$$

where E_{Edz} represents the seismic effects due to the vertical component of the seismic action.

Although the combination rules of seismic effects given by equations (2.60), (2.61) and (2.62) are substantiated only for elastic analysis methods, clause 4.3.3.5.1(6) extends these rules also for nonlinear static analysis. In this case, E_{Edx} and E_{Edy} represent the forces and deformations due to the application of the target displacement in the x and y directions.

In case of a linear or nonlinear dynamic time history analysis employing 3D structural models, there are no needs for special combination rules of components of the seismic action, because the total response is directly obtained by simultaneously applying accelerograms on two or three main directions of the building (clause 4.3.3.5.1(7)P).

2.9 CALCULATION OF STRUCTURAL DISPLACEMENTS

Clause 4.3.4(1)P recommends calculating the structural displacements under the assumption of the so-called equality displacement rule, which is based on the observation that the displacement demand of a perfectly elastic SDOF system is the same as an elastic-perfectly plastic SDOF system with the same initial stiffness. This principle is clarified by Figure 2.41 that illustrates the relationship between the base shear force and top lateral displacement of a ductile structure. Following an elastic analysis under design forces F_{Ed}, displacement d_e is obtained. This displacement is determined based on the design spectrum, whose ordinates are obtained reducing those of elastic spectrum by the behaviour factor q. If the structure would have an infinitely elastic behaviour, the non-reduced seismic action would correspond to the elastic force $q \cdot F_{Ed}$ and displacement $q \cdot d_e$.

This approach is rigorously correct only for structures with a fundamental period of vibration larger than T_C. (i.e. the corner period of the acceleration response spectrum, see section 2.3.2). For simplicity sake, EC8-1 does not account for the relationship between the characteristics of seismic motion (i.e. control period T_C), the period of vibration of the structure and the inelastic displacements of the structure.

In light of the above consideration, the displacements induced by the design seismic action shall be calculated on the basis of the elastic deformations of the structural system by means of the following:

$$d_s = q_d \cdot d_e \qquad (2.67)$$

2. EN 1998-1: GENERAL AND MATERIAL INDEPENDENT PARTS

where:
- d_s is the displacement of a point of the structural system induced by the design seismic action;
- q_d is the displacement behaviour factor, considered equal to q unless otherwise specified;
- d_e is the displacement of the same point of the structural system, as determined by a linear analysis under the design seismic loading.

It should be noted that the torsional effects should be accounted for in the calculation of the displacements d_e (clause 4.3.4(2)P). If the lateral force method is adopted, the displacement d_e can be obtained by magnifying the values given by the elastic analysis by the factor δ (see section 2.6.2).

If nonlinear analyses (both static and dynamic) are adopted to evaluate the seismic response of the structure, the displacements are directly obtained from the analysis without any further modification (clause 4.3.4(3)), provided that second order effects are accounted for.

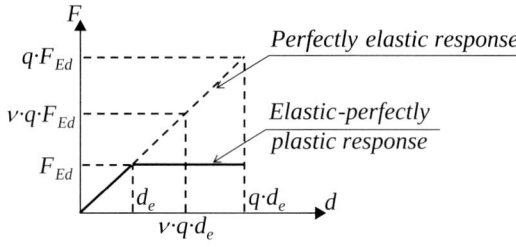

Figure 2.41 – Calculation of structural displacements according to EN 1998-1

2.10 SECOND ORDER EFFECTS IN SEISMIC LINEAR ELASTIC ANALYSIS

As previously discussed in section 2.6.4, another important aspect to be taken into account is the influence of second order (P-Δ) effects on frame stability. Indeed, in case of large lateral deformation the vertical gravity loads can act on the deformed configuration of the structure and increase the overall deformation and forces in the structure thus leading to potential collapse in a sidesway mode under seismic conditions.

2.10 SECOND ORDER EFFECTS IN SEISMIC LINEAR ELASTIC ANALYSIS

In general, the majority of structural analysis software can automatically account for these effects in the analysis. However, it is convenient to address directly the second order effects in order to control and optimise the structural design.

If linear elastic (both static and dynamic) analyses are used to estimate the design seismic response, according to clause 4.4.2.2(2), second-order (P-Δ) effects are specified through an interstorey drift sensitivity coefficient (θ) given as:

$$\theta = \frac{P_{tot} \cdot d_r}{V_{tot} \cdot h} \tag{2.68}$$

where:

- P_{tot} is the total gravity load, at and above the storey considered in the seismic design situation;
- V_{tot} is seismic shear at the storey under consideration;
- h is the interstorey height;
- d_r is the design inter-storey drift, given by the product of elastic inter-storey drift from the analysis and the behaviour factor q (i.e. $d_e \times q$).

In case of multi-storey frames, the interstorey drift sensitivity coefficient θ should be evaluated at each storey and the maximum should be considered to verify the building stability. Frame instability is assumed for $\theta \geq 0.3$ (clause 4.4.2.2(4)P). If $\theta \leq 0.1$, second-order effects may be neglected, whilst for $0.1 < \theta \leq 0.2$, P-Δ effects may be approximately taken into account in the seismic action effects by the Merchant-Rankine multiplier, given as follows (clause 4.4.2.2(3)):

$$\alpha = \frac{1}{(1-\theta)} \tag{2.69}$$

The multiplier α is used to magnifiy all seismic induced effects formerly calculated by the first-order linear elastic analysis.

The discussion about P-Δ effects deserves some additional remarks. It should be noted that equation (2.68) has been directly derived from EN 1993:1-1 clause 5.2.1(4)B (CEN, 2005b), which is based on the Horne method (1949). According to this method, the critical buckling load of 2D

2. EN 1998-1: GENERAL AND MATERIAL INDEPENDENT PARTS

multi-storey frames depends on their initial elastic lateral stiffness K_e. Hence, there is a contradiction between EN 1993:1-1 and EN 1998-1 approaches as EC3 assumes the elastic stiffness K_e, while EC8 (see equation (2.68)) considers the secant stiffness K_{sec}, as shown in Figure 2.42, since the critical seismic load is obtained as the ratio between the overturning moment of gravity loads under the plastic displacements induced by the earthquake (i.e. $d_e \times q$) and that given by the design seismic shear (i.e. V_E/q, which is an elastic force) at each storey.

The EC8 approach adopts the secant stiffness K_{sec} because it aims at considering the stability condition of the frame when the failure mechanism is fully developed and moderate to severe damage is uniformly distributed into the structure. This explains the dependency of the EC8 compliant stability coefficient with the behaviour factor considered in the design. This condition is very severe, and it is also not consistent with the elastic theory at the basis of Horne method. It is worth highlighting that American codes like NEHRP provisions (FEMA 750, 2009) assume the elastic lateral stiffness K_e to calculate the stability coefficient θ, consistently with the theoretical assumptions at the basis of the Horne method, thus being independent from the energy dissipation capacity of the structure.

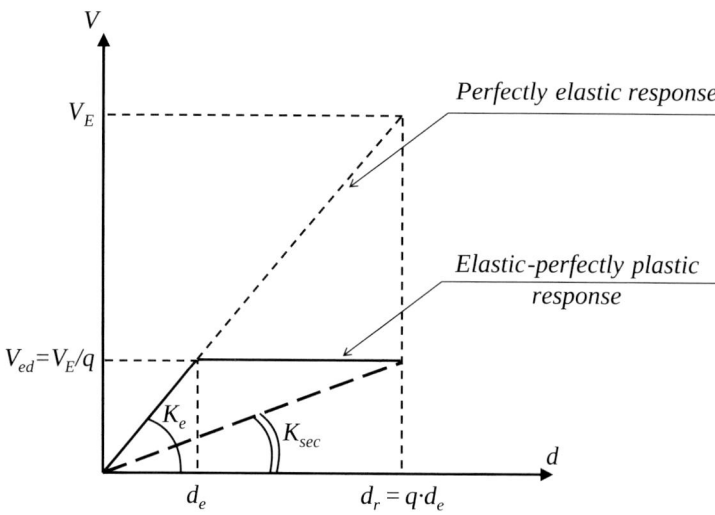

Figure 2.42 – Elastic vs secant lateral stiffness for calculation of stability coefficients according to Eurocodes

It is trivial to observe that for structural systems sensitive to lateral displacement EC8 and FEMA approaches lead to noticeable differences.

Steel moment resisting frames (MRFs) are typically drift sensitive structures, which are often characterized by EC8 storey stability coefficients exceeding 0.1, owing to the lateral flexibility of this type of structural scheme. Hence, the stability criterion typically influences the design of MRFs. As a direct consequence, for MRFs the need to reduce interstorey drift ratios and P-Δ effects could lead to oversize the structural members in order to increase the lateral stiffness (Tenchini *et al*, 2014). On the contrary braced structures are generally stiffer and less influenced by the stability criterion.

Finally, it is worth also noting that the linear modal response spectrum analysis may induce some additional problems to calculate and verify the EC8 stability criterion. Indeed, the storey shear forces obtained by the linear modal response spectrum analysis are generally smaller than those obtained by the lateral force method. Since the seismic storey shear force is the denominator of equation (2.68), larger stability coefficients are commonly obtained with modal response spectrum analysis.

In light of the above considerations, it is recognized that the issues related to the overall stability of building structures under seismic excitation are still far from being appropriately covered.

2.11 DESIGN VERIFICATIONS

2.11.1 Safety verifications

The no-collapse requirement under the seismic design situation at ULS is deemed to be satisfied if the following conditions are met (clause 4.4.2.1(1)P):

- Resistance;
- Global and local ductility;
- Equilibrium;
- Resistance of horizontal diaphragms (flooring system);
- Resistance of foundations;
- Seismic resistant joints.

2. EN 1998-1: General and Material Independent Parts

A brief description of those design conditions is provided hereinafter, with reference to the sections and clauses of EC8-1 where specific requirements are clearly recommended.

2.11.1.1 Resistance condition

The following resistance condition for all structural and non-structural elements shall be satisfied (clause 4.4.2.2.(1)P):

$$E_d \leq R_d \qquad (2.70)$$

where:
- E_d is the design value of the action effect, due to the seismic design situation, including, if necessary, second order effects (see sub-chapter 2.10);
- R_d is the corresponding design resistance of the element, calculated in accordance with the rules specific to the material used and in accordance with the mechanical models which relate to the specific type of structural system.

2.11.1.2 Global and local ductility conditions

The verification check given by equation (2.70) represents a general criterion that is widely applied in case of structural design. In order to guarantee adequate overall and local ductility it is necessary to assume a number of criteria that characterize judiciously the design effects E_d in order to enforce that the structure develops a complete or global plastic mechanism, which enables the use of larger behaviour factors (i.e. those for DCM or DCH concept). In addition, another important feature is the proper conception and execution of local detailing and proper choice of material properties, which allows to prevent local brittle failures.

Partial global plastic failure mechanisms (e.g. limited number of plastic hinges are formed or premature failure due to buckling of columns) or soft story mechanisms (e.g. due to concentration of plastic hinges in columns) must be avoided by imposing a rational hierarchy of resistance between dissipative and non-dissipative components. With this aim, sections 5 to 9 of EC8-1 provide specific rules for structures made of each type of building materials (namely concrete, steel, composite steel-concrete, timber and masonry).

However, to guarantee overall ductile behaviour, clause 4.4.2.3(4) recommends a general principle that was conceived taking as reference moment resisting frames, namely the "strong columns-weak beams" concept. This criterion aims at enforcing a global plastic failure mode with plastic hinges formed at the ends of the beams only, while the columns should remain elastic along the building height except for their bases and their top (roof level) where plastic hinges are accepted. This principle is deemed to be met if following condition is satisfied:

$$\sum M_{Rc} \geq 1.3 \cdot \sum M_{Rb} \tag{2.71}$$

where:

$\sum M_{Rc}$ is the sum of the design values of the bending strength of the columns framing the joint. For each column, EC8 recommends to consider the minimum value of flexural resistance of the column accounting for the presence of axial forces produced by the seismic design situation;

$\sum M_{Rb}$ is the sum of the design values of the bending strength of the beams framing to the joint. When partial strength connections are used, the moments of resistance of these connections are taken into account in the calculation of $\sum M_{Rb}$.

2.11.1.3 Equilibrium condition

Since seismic design actions are basically characterized by lateral horizontal forces, it is necessary to prevent the loss of overall stability due to the overturning and/or sliding effects induced by those actions. Therefore, clause 4.4.2.4(1)P recommends to conceive and design the structures to ensure the overall stability of the building against these global instability problems in the seismic design situation specified in clause 6.4.3.4 of EN 1990 (CEN, 2002a).

In special cases (e.g. highly irregular structures), the equilibrium may be verified by means of energy balance methods, or by geometrically non-linear methods (clause 4.4.2.4(2)).

2. EN 1998-1: General and Material Independent Parts

2.11.1.4 Resistance of horizontal diaphragms

Clause 4.4.2.5(1)P recommends designing both floor and roofing slabs as in-plan diaphragms in order to be able to transmit, with adequate overstrength, the effects of the design seismic action to the lateral load resisting systems to which they are connected.

This requirement is considered to be satisfied if for the relevant resistance verifications the seismic action effects in the diaphragm obtained from the analysis are multiplied by an overstrength factor γ_d (clause 4.4.2.5(2)). EC8 recommends for non-ductile diaphragms such as concrete slabs to apply on overstrength factor $\gamma_d = 1.3$, while $\gamma_d = 1.1$ for those having ductile failure modes. However, the National Annexes may specify different values, but always larger than 1.0.

In case the flooring and roofing systems cannot provide the rigid diaphragmatic behaviour, horizontal braces must be designed to guarantee both in-plane strength and stiffness. To that purpose the horizontal bracings can be calculated as a lattice systems subjected to horizontal forces magnified by $\gamma_d = 1.3$.

2.11.1.5 Resistance of foundations

The foundation system should be designed to comply with both EN 1998-5 (CEN, 2004c) and EN 1997-1 (clause 4.4.2.6(1)P) (CEN, 2004b).

Consistently with the capacity design principles, the action effects on the foundation elements should be derived accounting for the development of possible overstrength for the dissipative zones of the structural elements connected to the foundations. As a general principle, all foundations are considered non-dissipative, thus they should remain elastic ($q = 1.0$) during design earthquake (clause 4.4.2.6(2)P). For foundations of individual vertical elements (walls or columns), this condition is satisfied if the design values of the action effects E_{Fd} on the foundations are derived as follows (clause 4.4.2.6(4)):

$$E_{Fd} = E_{F,G} + \gamma_{Rd} \Omega E_{F,E} \qquad (2.72)$$

where:

$E_{F,G}$ is the action effect due to the non-seismic actions included in the combination of actions for the seismic design situation;

$E_{F,E}$ is the action effect from the analysis of the design seismic action;

γ_{Rd} is the overstrength factor, taken as being equal to 1.0 for $q \leq 3$, or as being equal to 1.2 otherwise;

Ω is the value of $(R_{di}/E_{di}) \leq q$ of the dissipative zone or element "i" of the structure which has the highest influence on the effect E_F under consideration;

R_{di} is the design resistance (capacity) of the element "i";

E_{di} is the design value of the action effect on the zone or element "i" in the seismic design situation.

For foundations of structural walls or of columns of moment-resisting frames, Ω is the minimum value of the ratio V_{Rd}/V_{Ed} for all dissipative zones in shear or the ratio M_{Rd}/M_{Ed} for all plastic hinges in the vertical elements, in the seismic design situation.

For common foundations of several vertical elements, equation (2.72) is applied for the bases of all vertical elements.

2.11.1.6 Seismic joint condition

Clause 4.4.2.7(1)P highlights the need to protect buildings from any potential earthquake-induced pounding from adjacent structures or between structurally independent units of the same building. This requirement is deemed to be satisfied if the following features are guaranteed (clause 4.4.2.7(2)):

a) for buildings, or structurally independent units, that do not belong to the same property, if the distance from the property line to the potential points of impact is not less than the maximum horizontal displacement of the building at the corresponding level, calculated in accordance with equation (2.67) (see sub-chapter 2.9);

b) for buildings, or structurally independent units, belonging to the same property, if the distance between them is larger than the square root of the sum of the squares (SRSS) of the maximum horizontal displacements of the two buildings or units at the corresponding level, calculated in accordance with equation (2.66) (see sub-chapter 2.9).

2.11.2 Damage limitation

In the modern performance based seismic design philosophy, the damage control performance level is a fundamental aspect in order to guarantee the full structural operativeness under frequent earthquakes (having a larger probability of occurrence than the design seismic action) and to reduce the cost of repairing under major seismic events. In clause 4.4.3.2(1), the damage limitation requirement for buildings results in an upper limit on the interstorey drift ratio demand d_r/h under the serviceability limit state (SLS), where d_r is the interstorey demand and h the storey height. In general, member sizes will be controlled by the limit on interstorey drift ratio. For this reason, it is recommended that in practice compliance with the damage limitation requirement should be established, before proceeding with dimensioning and detailing of members to satisfy the no-collapse requirement.

The interstorey drift ratio limit α is set equal to:

- 0.5 %, if there are brittle non-structural elements attached to the structure so that they are forced to follow structural deformations (normally partitions);
- 0.75 %, if non-structural elements (partitions) attached to the structure as above are ductile;
- 1 %, if no non-structural elements are attached to the structure.

The interstorey drift demand d_r for a generic storey is evaluated as the difference between the average lateral displacements d_s at the top and bottom of the storey under consideration. It should be determined under the frequent seismic action, which is defined by multiplying the entire elastic response spectrum of the design seismic action for 5 % damping by the same factor ν that reflects the effect of the mean return periods of these two seismic actions. Indeed, the corresponding mean return period for the design earthquake for SLS is smaller than the one for ULS. Hence, the seismic forces and the corresponding structural effects at SLS are ν times smaller than those at ULS. Thus, as shown in Figure 2.41, the value of seismic forces at SLS in the

2.11 Design Verifications

infinitely elastic system will be $\nu\, q \cdot F_{Ed}$, and the corresponding displacements will be assumed equal to $\nu\, d_s = \nu\, q \cdot d_e$, where d_s is the displacement of the structural system induced by the design seismic action, q is the behaviour factor and d_e is the displacement of the structural system, as determined by a linear elastic analysis under the design seismic forces.

If the analysis is non-linear, the interstorey drift ratio should be determined for a seismic action (acceleration time-history for time-history analysis, acceleration-displacement composite spectrum for pushover analysis) derived from the elastic spectrum (with 5 % damping) of the design seismic action times ν.

Once calculated the lateral displacements, the EC8 damage limitation requirement is expressed by the following equation:

$$d_r \nu \leq \alpha h \qquad (2.73)$$

where:

- α is the limit related to the typology of non-structural elements;
- d_r is the design interstorey drift;
- h is the storey height;
- ν is a displacement reduction factor depending on the importance class of the building, whose values are specified in the National Annex. Clause 4.4.3.2(2) recommends $\nu = 0.5$ for importance classes I and II and $\nu = 0.4$ for importance classes III and IV.

Chapter 3

EN 1998-1:
DESIGN PROVISIONS FOR STEEL STRUCTURES

3.1 DESIGN CONCEPTS FOR STEEL BUILDINGS

As discussed in detail in chapter 2 (see section 2.3.5), the seismic design forces are smaller than those obtained from a linear elastic response, because EC8-1 takes into account the capacity of structures to dissipate energy through inelastic deformations. In fact, the design forces are obtained using the "design spectrum", whose ordinates are given by those of the elastic spectrum reduced by the behaviour factor q (EC8-1 clause 3.2.2.5(2)).

For steel structures, the values of the behaviour factor q depend on the structural system according to the relevant ductility classes considered by EC8-1, section 6.1.2. As summarized in Figure 3.1, seismic resistant steel buildings may be designed in accordance with one of the following concepts (clause 6.1.2(1)P):

- Concept a): Low-dissipative structural behaviour;
- Concept b): Dissipative structural behaviour.

In concept a), the action effects may be calculated on the basis of an elastic global analysis neglecting the non-linear behaviour, because a low dissipative structure does not fulfil the conditions to apply a plastic design for predominantly static loading. In this case, the behaviour factor assumed in the calculation must be less than 2. Structures designed in accordance with concept a) belong to the low dissipative structural class "DCL" (Ductility Class Low). Hence, the resistance of members and connections should be evaluated in accordance

3. EN 1998-1: Design Provisions for Steel Structures

with EN 1993-1-1 (CEN, 2005b) without any additional requirement (clause 6.1.2(4)). The classification as "low dissipative" should be more properly stated as "not ductile enough", meaning that those structures may provide only poor, or even no post-elastic deformation capacity (e.g. beyond their elastic limit capacity). It is worth noting that when the conceptual design rules are correctly applied, and the design of a low dissipative structure is properly carried out according to EN 1993-1-1, the structure should possess a minimum of design overstrength and static redundancy, enabling the reduction of the seismic action by applying a behaviour factor $q \leq 1.5-2.0$. For non-base-isolated structures, this simplified design is recommended only for low seismicity regions. Although the designation of low seismicity zone should be established by the competent National Authorities, a threshold value of design ground acceleration for the specific soil type is recommended as $\gamma_I \cdot S \cdot a_{gR} = 0.1g$. It should be noted that in case of very low seismicity (namely for the cases where $\gamma_I \cdot S \cdot a_{gR} < 0.05g$) EC8 allows to neglect the seismic action in the design of buildings.

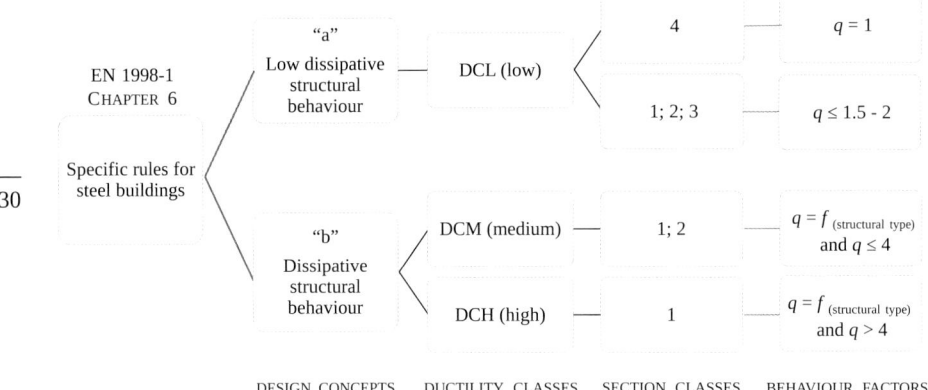

Figure 3.1 – Design Concepts according to EN 1998-1 (CEN, 2004a)

In concept b), the capability of parts of the structure (dissipative zones) to undergo plastic deformations in case of an earthquake is taken into account. The behaviour factor q assumed in the calculation may be larger than 2 and it depends on the type of seismic resistant structural scheme. Structures designed in accordance with concept b) may belong to a medium structural ductility class "DCM" (Ductility Class Medium) or to a high ductility class "DCH" (Ductility Class High). These classes directly correspond to an increased ability of the structure to dissipate energy through inelastic behaviour. Depending on the ductility class, specific design requirements are provided for both local and global structural aspects. It should be noted that the use of large values for q-factors (e.g. $q = 6.5$ for moment resisting frames) corresponds to a significant reduction of the seismic design forces. Consequently, this leads to reduced constructional costs but also lower stiffness and resistance than those designed according to concept a). The value of the q-factor differs with each structural typology owing to their specific capability to dissipate the seismically induced energy at both global and local levels, as it will be shown further.

A well designed seismic resistant structure should be characterized by a good balance between strength, stiffness and ductility between its members, connections and supports (Bertero, 1997). As general rule, a dissipative structure has ductile components in selected zones designed as dissipative, which undergo plastic deformations without premature failure, while the structure stands-up (e.g. collapse prevention), due to its redundancy based on over strengthened components designed to behave predominantly in the elastic range. A dissipative structure will suffer damage during severe earthquakes (such as the design seismic event at ULS with a return period of 475 years). However, plastic deformation may occur even at SLS for rigid structural systems like concentrically braced frames, where the buckling of braces in compression can be activated for low levels of lateral actions. Thus, a dissipative structure will survive the ULS seismic action and protect human lives, but will need to be repaired after the event.

The reparability of a structure following a seismic event depends on the level of permanent damage that it exhibits following the event. The residual or permanent drift gives a good indication of the reparability of a structure and it is generally accepted as a criterion for the assessment of a building's condition as it provides an indication of the structural damage during inspections after an earthquake. Based on the residual drift, the structural damage may be estimated as low, moderate or high. Although it is difficult to prescribe precise limit values, some figures are proposed and applied in practice (Ohi and Takanashi, 1998). For a residual drift exceeding roughly 3 %, it is supposed that the structural damage is so heavy that the building must be demolished. For values of the residual drift over 1 % (also supported by Vayas and Dinu, 2000), the damage is considered to be moderate to heavy so that it is possible for the building to be repaired. This criterion is characterized as a strength criterion due to the fact that if the structural strength is not sufficient, large inelastic deformations will occur that will potentially lead to unacceptably high residual drifts.

It should be noted that the possibility to repair the building is directly related to its level of damage. With this regard, residual or permanent interstorey drift ratios are easy and rational indexes providing effective data. Ohi and Takanashi (1998) correlated the actual levels of damage observed in steel structures with residual interstorey drift ratios. In particular, "moderate damage" is associated to permanent interstorey drift ratios ranging from 0.6 % to 1 % and it is associated to easiness of repair. "Heavy damage" is associated to permanent interstorey drift ratio roughly exceeding 3 %, which corresponds to difficult rehabilitation due to both technical and economic reasons. When the permanent drift ratios exceed 5 %, those authors recommend the demolition of the structure without further investigation.

Section 2.1 of EN 1998-3 accounts for permanent drifts as performance index characterizing the state of damage in the structure per limit states. However, only a qualitative description is provided, namely either large or moderate permanent drifts at Near Collapse and Significant Damage (i.e. ULS) limit state, respectively.

In line with Ohi and Takanashi (1998) and SEAOC (1995), see sub-chapter 1.3, Grecea *et al* (2004) proposed 1 % residual interstorey drift ratios as performance index at ULS for EC8 compliant steel frames. The damage associated to this index generally ranges from moderate to heavy, thus allowing repair of the building.

More detailed recommendations are provided by FEMA 356, which specifies the thresholds for both transient and permanent interstorey drift ratios for moment and braced frames at each limit states for the immediate occupancy (IO), life safety (LS) and collapse prevention (CP) performance levels, as summarized in Table 3.1. It should be noted that the FEMA 356 limit states defined as Collapse Prevention, Life Safety and Immediate Occupancy basically correspond to EC8 Near Collapse, Significant Damage (i.e. ULS) and Damage Limitation (i.e. SLS).

Table 3.1 – Acceptance criteria for interstorey drift ratios according to FEMA 356

Structural system	Limit State		
	Collapse Prevention	Life Safety	Immediate occupancy
Moment Frames	5 % transient or permanent	2.5 % transient; 1 % permanent	0.7 % transient; negligible permanent
Braced Frames	2 % transient or permanent	1.5 % transient or 0.5 % permanent	0.5 % transient; negligible permanent

As a general remark, it is worth noting that the design of a structure located in a seismic area should comply with both seismic and non-seismic requirements, such as the limitation of floor vertical deflections under gravity loading and the control of inter storey drift and lateral strength against wind actions. This aspect is meaningful, because the design resistance for ULS under the design earthquake might be less critical than the requirements for non-seismic design actions. Whenever the non-seismic design actions are predominant, the complete design process may lead to structures stronger than strictly needed to resist the design earthquake (Fardis *et al*, 2005). Designing those structures as dissipative will lead to largely uneconomic and unfeasible structures. This is generally true in the following cases:

- Buildings located in low seismicity areas;
- Single storey large span buildings (e.g. industrial halls) built in moderate-to-high snow areas. In these cases, the ULS design is dominated by snow loading; the seismic combination will only be relevant for the design of the bracing system;
- Flexible structures, in which the requirements for SLS control the design. This is the case of high rise multi-storey moment resisting frames, where the magnitude of the lateral loads due to the wind action are generally lower than those induced by the earthquake. However, since the lateral displacements of the building induced by the wind action are more severely limited to guarantee adequate comfort to the occupants, this condition may be more restrictive than the ULS criteria imposed by the seismic action.

3.2 REQUIREMENTS FOR STEEL MECHANICAL PROPERTIES

3.2.1 Strength and ductility

It is well known that the nominal yield stress f_y is a minimum guaranteed value, which is generally lower than the actual steel strength. Indeed, the various product standards for steel (EN 10025–1 (CEN, 2004d); EN 10210 (CEN, 2006a); EN 10219 (CEN, 2006b)) impose a guaranteed minimum value for yield stress. There is currently no obligation to comply

3. EN 1998-1: Design Provisions for Steel Structures

with a maximum value for the yield stress f_y. Consequently, steel products are not usually supplied with an upper bound yield stress. However, according to the capacity design philosophy, it is important to know the maximum yield stress of the dissipative parts. Hence, clause 6.2(2)P states that *"the distribution of material properties, such as yield strength and toughness, in the structure shall be such that dissipative zones form where they are intended to in the design"*. To ensure this principle, clause 6.2(3) stipulates the three following alternative options:

a) the actual maximum yield strength $f_{y,max}$ of the steel of dissipative zones satisfies the following expression:

$$f_{y,max} \leq 1.1 \gamma_{ov} f_y \qquad (3.1)$$

where:
  f_y is the nominal yield strength specified for the steel grade;
  γ_{ov} is an overstrength factor based on a statistical characterization of steel products. The recommended value is 1.25 (clause 6.2(3)a), but the designer should use the value provided by the relevant National Annex.

b) the design of the structure is made on the basis of a single grade and nominal yield strength f_y for the steels both in dissipative and non dissipative zones; an upper value $f_{y,max}$ is specified for the steel of dissipative zones; the nominal value f_y of the steels specified for non dissipative zones and connections exceeds the upper value of the yield strength $f_{y,max}$ of dissipative zones. In this case there is no need for γ_{ov} which can be set equal to 1 (clause 6.2(4));

c) the actual yield strength $f_{y,act}$ of the steel of each dissipative zone is determined from measurements and the overstrength factor is computed for each dissipative zone as $\gamma_{ov,act} = f_{y,act}/f_y$, f_y being the nominal yield strength of the steel of dissipative zones.

Of the three options mentioned in EC8-1, the most common case is condition a), whereby the designer does not have information about the actual value of the yield strength in dissipative members/connections and a unique value of the material overstrength factor is recommended by EC8-1 as $\gamma_{ov} = 1.25$. The second condition corresponds to the situation (impractical in most cases) whereby the steel producers supply a so-called "seismic" steel, with a guaranteed upper limit of the yield strength (Fardis *et al*, 2005). The third condition is possible in a limited number of situations, for example when assessing existing constructions.

The material overstrength factor γ_{ov} deserves some additional remarks. Indeed, both practical experience and experimental data (Dubina *et al*, 2008; Simões da Silva *et al*, 2009; da Silva *et al*, 2017) show that the overstrength factor decreases with increasing steel grade. Lower grades of Mild Carbon Steel, such as S 235 and S 275, are characterized by a higher yield overstrength factor (e.g. even larger than 1.6, as shown in Landolfo (2013)) than the upper bound value $\gamma_{ov} = 1.25$ recommended by EC8. On the contrary, S 355 is generally characterized by the smaller difference between nominal and real yield strength, γ_{ov} being close to the EC8 recommended value. Hence, due to this better agreement with the recommended value, S 355 can be reasonably considered as a "good seismic" steel.

Regarding the bolts in bolted connections of primary seismic members of a building, clause 6.2(9) recommends to use high strength bolts of grade 8.8 or 10.9.

Another important property of steel used in dissipative components is ductility. EC8-1 does not directly recommend specific requirements for ductility. However, it is suggested for steels in dissipative zones to satisfy the following conditions:

- A ratio between ultimate and yield stress $f_u/f_y > 1.20$ (as indicated by the Italian (NTC, 2008) and the Romanian (MRDPA, 2013) codes);
- A tensile elongation A % > 20 % (as indicated by the Italian (NTC, 2008) and the Romanian (MRDPA, 2013) codes).

3.2.2 Toughness

Another important property of steel for dissipative zones and welds of seismic resistant structures is toughness, which represents a measure of the amount of energy required to fracture the material. In general, the tougher the material, the more is the energy required to cause a crack growing to fracture. Also, whatever the steel grade in terms of yield strength, a low toughness grade always corresponds to small ductility.

3. EN 1998-1: Design Provisions for Steel Structures

Under seismic actions, ductile components are expected to experience large inelastic cyclic deformation and high strain rates, which can reduce their available ductility and may induce failure by low cycle fatigue. In order to guarantee adequate toughness to prevent brittle failure, clause 6.2(7) states: "*The toughness of the steels and the welds should satisfy the requirements for the seismic action at the quasi-permanent value of the service temperature* (see EN 1993-1-10 (CEN, 2005e))"

EN 1993-1-10 (CEN, 2005e) provides maximum permissible values of element thickness in its Table 2.1 that depends on steel grade and sub-grade (toughness), reference temperature (T_{Ed}) and stress level (σ_{Ed}). The stress level in the structural members is determined from the internal forces E_d, calculated according to the following load combination (clause 2.2(4) of EN 1993-1-10):

$$E_d = E\left\{A\left[T_{Ed}\right] \text{"+"} \Sigma G_K \text{"+"} \psi_1 Q_{K1} \text{"+"} \Sigma \psi_{2,i} Q_{Ki}\right\} \tag{3.2}$$

It is unclear from the information provided in EC8-1 and EN 1993-1-10 how to determine material toughness requirements in the seismic design situation. Firstly, no load combination is provided to be used for the determination of the stress level σ_{Ed} in case of seismic actions. It should be considered that in seismic design situations the stress level in the dissipative zones reaches the yield strength ($\sigma_{Ed} = f_y(t)$). In contrast, EN 1993-1-10 refers to stress levels up to $\sigma_{Ed} = 0.75 f_y(t)$, and no extrapolation is allowed. Detailed information on toughness oriented rules in EN 1993-1-10 is available in Nussbaumer et al (2011). According to this design manual, the $\sigma_{Ed} = 0.75 f_y(t)$ value corresponds to the maximum possible "frequent stress", where for the ultimate limit state verification yielding of the extreme fibre of the elastic cross section has been assumed ($\sigma_{Ed} = f_y(t) / 1.35 = 0.75 f_y(t)$). Consequently, the value $\sigma_{Ed} = 0.75 f_y(t)$ given by EN 1993-1-10 would correspond to the case of yielded cross section, and it can be presumably used for selection of material toughness and thickness in the seismic design situation.

Regarding the strain rate values in the seismic design situation, no guidance is given in EN 1993-1-10 or EC8-1. Anyway, some indications on the values of strain rate for seismic actions can be found in the European standard for anti-seismic devices (CEN, 2009), which clarifies that devices made of steel are mostly insensitive to strain rate in the usual range of seismic effects.

3.3 STRUCTURAL TYPOLOGIES AND BEHAVIOUR FACTORS

3.3.1 Structural types

The structural typologies for steel buildings addressed by EC8-1 are illustrated in Figure 3.2 and are classified according to the behaviour of their primary seismic resisting systems, as follows (clause 6.3.1(1)P):

a) <u>Moment resisting frames (MRF)</u> - the horizontal forces are mainly resisted by members acting in an essentially flexural manner (see Figure 3.2a). In these structures, the dissipative zones should be located in plastic hinges located at ends of beams or in the connections of the beams to the columns (clause 6.6.1(1)P), so that energy is dissipated by means of cyclic bending (clause 6.3.1(2)). The dissipative zones may also be located in columns:

- at the base of the frame;
- at the top of the columns in the upper floor (i.e. the roof) of multi-storey buildings;
- at the top and bottom of columns in single storey buildings in which the design axial force N_{Ed} in columns is less than the 30 % of the column axial plastic strength $N_{pl,Rd}$.

The design provisions and detailing rules for MRFs are described and discussed in sub-chapter 3.5.

b) <u>Frames with concentric bracings (CBF)</u> - the horizontal forces are mainly resisted by members subjected to axial forces. In these structures, the dissipative zones are located in the diagonals in tension. Bracings may be designed in two configurations (clause 6.3.1(3)):

- bracings with active diagonals in tension (e.g. either cross or single diagonal shape, as shown in Figure 3.2b), where the horizontal forces are solely resisted by the diagonals in tension, neglecting the contribution of diagonals in compression. In addition, a single diagonal scheme is allowed provided that at least two braced spans

are used together with their respective mirror image along the same line of bracing;
- bracings with diagonals in V or inverted V (e.g. chevron) shape, where the horizontal forces are resisted by both diagonals (see Figure 3.2c).

It is important to note that K-braced frames (see Figure 3.2e) are not allowed in seismic resistant systems. Indeed, this structural scheme is highly dangerous, because if one of the diagonal members buckles, the increased force that the tension brace might develop may result in a transverse unbalanced force applied at mid-height of the column. In this case, a plastic hinge can form in the column at the intersection with the braces, leading to undesirable brittle failure of the column.

The design provisions and detailing rules for CBFs are described and discussed in sub-chapter 3.6.

c) <u>Frames with eccentric bracings (EBF)</u> - the horizontal forces are mainly resisted by axially loaded members, but the eccentricity of the layout of the bracings (see Figure 3.2d) is such that energy can be dissipated in seismic links by means of either cyclic bending or cyclic shear. EBFs should be designed to enforce plastic engagement of all dissipative links (clause 6.3.1(4)).

The design provisions and detailing rules for EBFs are described and discussed in sub-chapter 3.7.

d) <u>Inverted pendulum structures</u> – either more than 50 % of the mass is placed in the upper third of the height of the structure (see Figure 3.2i), or structures in which the dissipation of energy takes place mainly at the base of a single element of the building (see Figure 3.2j). An example of inverted pendulum structure is a single storey building with plastic hinges located at both columns ends, provided that the design column axial force N_{Ed}, is smaller than 30 % of column cross section axial resistance $N_{pl,Rd}$. If the axial load exceeds this threshold, it is not possible to design the structure as an inverted pendulum scheme, owing to the lack of ductility at the base of columns (clause 6.3.1(5)).

If the structural scheme of an inverted pendulum structure is assimilable to a portal frame, the capacity design rules are equal to those for MRFs. However, section 6.9 of EC8-1 imposes the following specific requirements: (i) the non-dimensional slenderness of the columns should be limited to 1.5 (clause 6.9(3); (ii) the interstorey drift sensitivity coefficient θ as defined by equation (2.61) (see sub-chapter 2.10) should be limited to 0.20 (clause 6.9(4)).

e) <u>Steel structures associated to concrete cores or concrete walls</u> - horizontal forces are mainly resisted by the concrete cores or walls (see Figure 3.2f). In this case, the steel structure shall resist gravity loadings only. Therefore, this structural typology is not discussed in this design manual, since the seismic design requirements are those for reinforced concrete structures.

f) <u>Dual Frames made of Moment resisting frames combined with Braced frames</u> – the horizontal forces are carried-out by both sub-systems, proportionally to their stiffness (see Figure 3.2g). To achieve an effective "dual response", although not specified in EC8-1, it is recommended that the MRF (i.e. the less stiff system) contributes with at least 25 % to the total strength and stiffness of the dual frame (as recommended by AISC 341-10, 2010). Anyway, clause 6.10.2(1) imposes that dual structures with both moment resisting frames and braced frames acting in the same direction should be designed using a single q factor. In addition, the MRF and the BF should conform the relevant detailing rules and the requirements provided for those structural typologies, respectively (clause 6.10.2(2)). In light of these considerations, this chapter details the design provisions for the main subsystems (e.g. MRFs, CBFs and EBFs), without addressing dual frames in a specific section.

g) <u>Moment resisting frames combined with reinforced concrete infills</u> – the horizontal forces are carried by this hybrid system (see Figure 3.2h), where both MRF and concrete infills work together. This system should be designed according to chapter 7 of EC8-1 (i.e. "Specific rules for composite steel – concrete buildings") and it is outside the scope of this design manual.

3. EN 1998-1: Design Provisions for Steel Structures

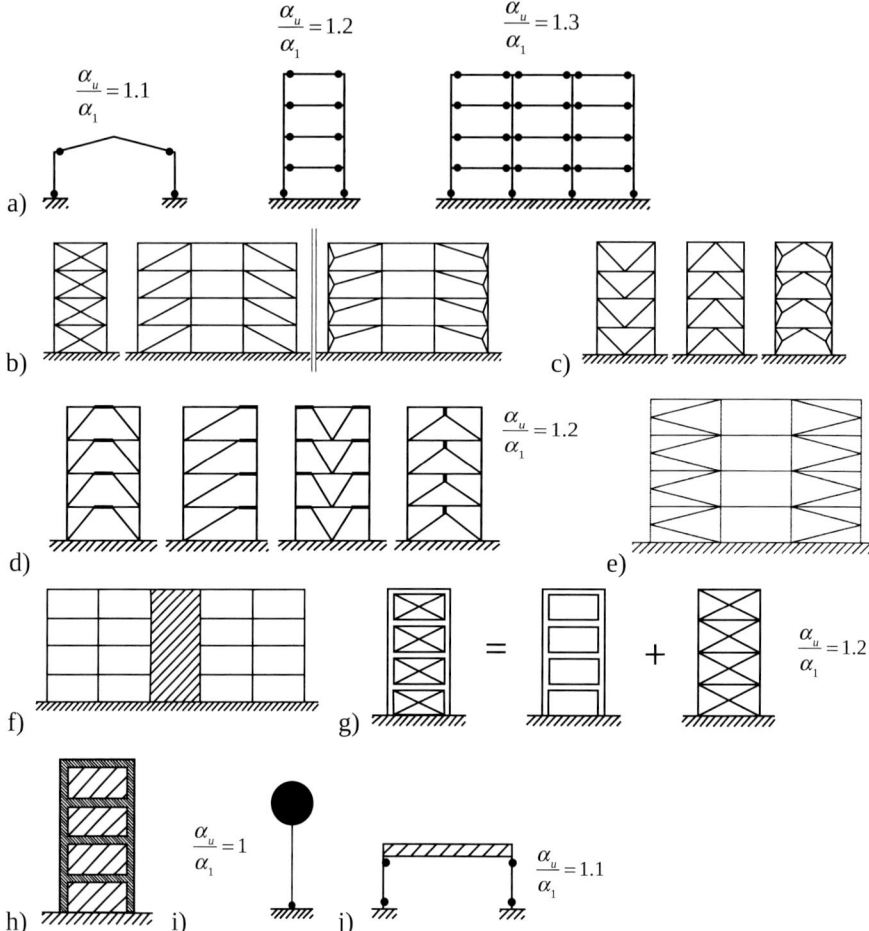

Figure 3.2 – The structural schemes considered in EC8: a) Moment resisting frames (dissipative zones in the beams and at the bottom of the columns); b) Frames with diagonal concentric bracings (dissipative zones in tension diagonals only); c) Frame with V bracings (dissipative zones in both tension and compression diagonals); d) Frames with eccentric bracings (dissipative zones in bending or shear links); e) K-braced frames (not allowed); f) Structures with concrete cores or concrete walls; g) Moment resisting frame combined with concentric bracing (dissipative zone in moment frame and in tension diagonals); h) Moment resisting frames combined with infills; i) Inverted pendulum with dissipative zones at the column base; j) Inverted pendulum with dissipative zones in columns

In practice, several strategies may be adopted to optimize the design of steel structures in seismic areas. Recently, as a practical way of ensuring that dissipative zones form where they are intended to in design (see section 3.2.1) irrespective of the variability of the yield stress of steel, the authors (Dubina *et al*, 2015) have developed the dual-steel concept for buildings in seismic areas. According to this concept, mild carbon steel (MCS) is used in dissipative members while high strength steel (HSS) is used in non-dissipative "*predominantly elastic*" members, leading to more reliable and cost efficient structures, when compared to conventional *homogeneous steel* solutions. *Dual-Steel* structures enable to fulfil by design the three critical tasks of a robust structure: (i) to secure plastic deformation capacity in structural members, targeted as dissipative, which are key members in any seismic-resistant structure; (ii) to prepare multiple routes for transfer of forces and ensure their redistribution through yielding of other members; (iii) to provide sufficient over-strength to structural members that cannot be allowed to collapse at any cost. In dual-steel structures, the role of lower-yield steel is to work like a fuse, dissipating the seismic energy through plastic deformations, while the rest of the structure remains elastic or undamaged. Further detailed information on guidance and recommendations for the application of this concept may be found in the freely available final report of the European project HSS-SERF - High Strength Steel in Seismic Resistant Building Frames (Dubina *et al*, 2015).

As a final note, it is emphasized that the combination of dual frame systems and dual steel materials is in line with the design principles of Eurocode 8, giving freedom to the designers to optimize their structures in an efficient way.

3.3.2 Behaviour factors

The general design approach recommended by EC8 aims to control the inelastic structural behaviour by avoiding the formation of soft storey mechanisms and brittle failure modes. In order to achieve this purpose, the design rules are based on the capacity design of members and on detailing the dissipative zones parts with specific rules to improve their ductility and deformation capacity. This philosophy is specifically intended for buildings designed for ductility classes *M* and *H*. For these cases, the design forces calculated by means of linear elastic analysis (namely obtained from either the lateral force method or a modal response spectrum analysis) may be obtained by reducing the elastic spectrum by the behaviour factor q, which accounts for ductility and the dissipative capacity of the structural system.

3. EN 1998-1: Design Provisions for Steel Structures

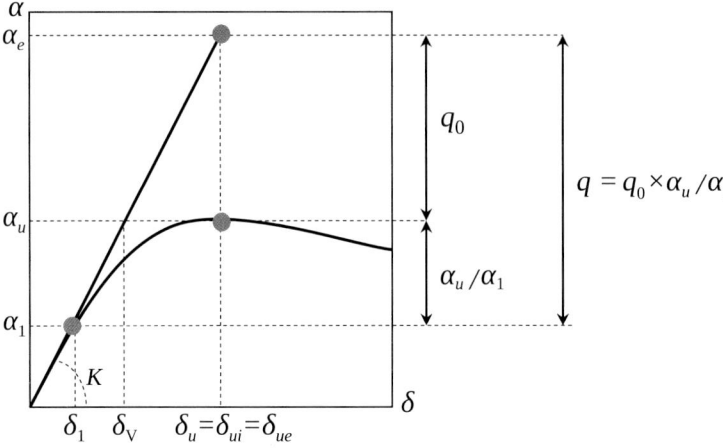

Figure 3.3 – The behaviour factor q according to EC8-1

As described in Figure 3.3, the behaviour factor q for regular structural systems is given by:

$$q = \frac{\alpha_u}{\alpha_1} \cdot q_0 \qquad (3.3)$$

where q_0 is the reference value of the behaviour factor and α_u/α_1 is the plastic redistribution parameter accounting for the system overstrength due to redundancy. The parameter α_1 is the multiplier of the horizontal seismic design action to first reach the plastic resistance in any member in the structure and α_u is the multiplier of the horizontal seismic design action necessary to form a global mechanism. The ratio α_u/α_1 may be obtained from a nonlinear static 'pushover' global analysis according to EC8-1, but it is limited to 1.6 (clause 6.3.2(6)). Clause 6.3.2(3) proposes the following reference values for α_u/α_1 (see also Figure 3.2):

- 1 for inverted pendulum structures;
- 1.1 for one-storey frames;
- 1.2 for one-bay multi-storey frames, eccentric bracing or dual systems with moment resisting frames and concentrically braced frames;
- 1.3 for multi-storey multi-bay moment-resisting frames.

Table 3.2 (Table 6.2 of EC8-1) also provides upper limits for the behaviour factors q for the structural schemes depicted in Figure 3.2.

Table 3.2 – Upper limits of q factors for systems regular in elevation

Structural type	Ductility Class	
	DCM	DCH
a) Moment resisting frame (MRF)	4	5 (α_u/α_1)
b) Frame with concentric bracings		
– Diagonal bracings	4	4
– V-bracings	2	2.5
c) Frame with eccentric bracings	4	5 (α_u/α_1)
d) Inverted pendulum	2	2 (α_u/α_1)
e) Structures with concrete cores or concrete walls	(see reinforced concrete frames)	
f) Moment resisting frame with concentric bracings	4	4 (α_u/α_1)
g) Moment resisting frames with infills:		
– Unconnected concrete or masonry infills, in contact with the frame	2	2
– Connected reinforced concrete infills	(see steel – concrete frames)	
– Infills isolated from MRF (see MRF)	4	5 (α_u/α_1)

However, in most design cases, the building structure combines MRF with braced frames (either CBF or EBF), resulting in Dual Frame systems, whereby both components are able to dissipate seismic energy. EC8-1 gives roughly some requirements for dual frames made of MRFs with X-CBFs, but it does not provide recommendations for other cases. Specific requirements for these types of structures are given by the Romanian Seismic Design Code (MRDPA, 2013). Table 3.3 summarizes the recommended upper limits for the behaviour factors of Dual Frames regular in elevation, provided that the MRF contributes at least with 25 % to the capacity of the dual system. It is noted that the latter requirement is consistent with AISC 341-10.

3. EN 1998-1: Design Provisions for Steel Structures

Table 3.3 – Upper limits of q factors for dual systems regular in elevation

Dual Frames	Ductility Class	
	DCM	DCH
MRF + CBF (X and alternative); $\frac{\alpha_u}{\alpha_1} = 1.2$ *(diagram)* − Dissipative zones located in MRF and in CBF tension braces	4	
MRF + CBF (inverted V); $\frac{\alpha_u}{\alpha_1} = 1.2$ *(diagram)* − Dissipative zones located in MRF and CBF braces	2	$2.5 \frac{\alpha_u}{\alpha_1}$
MRF + EBF ; $\frac{\alpha_u}{\alpha_1} = 1.2$ *(diagram)* − Dissipative zones located in MRF and CBF links	4	$5 \frac{\alpha_u}{\alpha_1}$

3.4 DESIGN CRITERIA AND DETAILING RULES FOR DISSIPATIVE STRUCTURAL BEHAVIOUR COMMON TO ALL STRUCTURAL TYPES

3.4.1 Introduction

As described in sub-chapter 3.1, three ductility classes (i.e. DCL, DCM and DCH) are defined in EC8-1, depending on the accepted level of the plastic engagement in the dissipative parts. Indeed, in case of DCL, poor plastic deformation is expected and the code allows to perform global elastic analysis using a q factor within the range 1.5-2.0 and to check the strength of elements (both members and connections) according to EN 1993:1-1 without the verification of the capacity design rules. DCL design should be applied only in low seismicity areas. On the contrary, for DCM and DCH, it is expected to have moderate and large plastic engagement in the dissipative zones, respectively. Therefore, EC8-1 prescribes specific design rules both at global and local levels in order to guarantee sufficient ductility in the dissipative elements (clause 6.5.1(1)); there are rules common to all structural schemes and other rules specifically conceived for each typology.

In general, structures shall be designed such that yielding or local buckling or other phenomena due to hysteretic behaviour do not affect the overall stability of the structure (clause 6.5.2(1)P). The dissipative zones shall have adequate ductility and resistance. The resistance shall be verified in accordance with EN 1993-1-1 (clause 6.5.2(2)P). The dissipative zones may be located in the structural members or in the connections (clause 6.5.2(3)). If the dissipative zones are located in the structural members, the non-dissipative parts and the connections of the dissipative parts to the rest of the structure shall have sufficient overstrength to allow the development of cyclic yielding in the dissipative parts (clause 6.5.2(4)P), whereas when dissipative zones are located in the connections, the connected members shall have sufficient overstrength to allow the development of cyclic yielding in the connections (clause 6.5.2(5)P). This means that in the first case, non-dissipative connections shall be used (discussed in section 3.4.3) while in the second case dissipative joints are required (see section 3.4.4).

3.4.2 Design rules for cross sections in dissipative members

For dissipative structures (DCM or DCH), EC8-1 recommends to use a q factor larger than 2 (see clause 6.1.2(5)P and Table 6.1 of EC8-1).

3. EN 1998-1: Design Provisions for Steel Structures

This assumption can be applied provided that the dissipative elements in compression or bending under seismic loading satisfy a set of cross section requirements, namely by restricting the local slenderness ratios to limit local buckling phenomena under large deformation demand (clause 6.5.3(1)P). EC8-1 adopts the classification given by EN 1993-1-1 for cross sections, relating the restrictions on cross-sectional class of the dissipative elements to the value of the q factor for each Ductility Class, as summarized in Table 3.4 (clause 6.5.3(2) and Table 6.3 of EC8-1). It should be noted that the use of class 3 cross sections in dissipative members limits the reference value of the behaviour factor q to a maximum of 2, which can be justified by overstrength and structural redundancy. This requirement results from the poor structural performance of class 3 members, which cannot exploit their full sectional capacity. Indeed, EN 1993-1-1 recommends considering the elastic strength for this category of steel elements. However, EC3 provisions are too severe because the actual transition from full plastic to elastic capacity gradually varies with the local slenderness as shown by the results of the Semi-Comp project (Greiner et al, 2009), which updated the resistance model for class 3 profiles introducing partial plasticity strength by linear interpolation from class 2 to class 4.

Table 3.4 – Requirements on cross-sectional class of dissipative elements depending on Ductility Class and reference behaviour factor

Ductility class	Reference value of behaviour factor q	Required cross-sectional class (see EN 1993:1-1)
DCM	$1.5 < q \leq 2.0$	class 1, 2 or 3
	$2.0 < q \leq 4.0$	class 1 or 2
DCH	$q > 4.0$	class 1

Currently, EC8-1 does not account for structures with class 4 members. However, several more recent studies (Dubina et al, 2006; Dubina et al, 2010; Fiorino et al, 2012, 2014) demonstrated that light gauge steel structures, made of class 4 sections, fabricated by cold forming or thin plate welding, can be effectively used in seismic resistant structures. This category of structures is generally known as cold formed steel housing (CFS), and it is becoming very popular all over the world, because it represents a suitable solution to the demand for low-cost high performance houses (Landolfo, 2011). In the framework of EN 1998-1 (CEN, 2004a), CFS structures can be designed in seismic areas as non-dissipative structure using $q=1$ in accordance with EN 1993 provisions, provided that the general rules related to layout of the structure are accounted for, namely homogeneity and regularity in order to have satisfactory 3D behaviour, adequate details, redundancy etc.

It is noted that class 4 cross sections may also be used in seismic regions. The design practice of class 4 cold-formed cross sections the reader is explained in the corresponding ECCS design manual (Dubina *et al*, 2012) where this issue is addressed in detail.

3.4.3 Design rules for non-dissipative connections

Section 6.5.5 of EC8-1 provides general rules for all types of non-dissipative connections in dissipative zones of the structure, in order to ensure sufficient overstrength and to avoid localization of plastic strains. In particular, the following criterion to ensure connection overstrength must be applied (clause 6.5.5(3)):

$$R_d \geq 1.1 \cdot \gamma_{ov} \cdot R_{fy} \tag{3.4}$$

where:
- R_d is the resistance of the connection;
- R_{fy} is the plastic resistance of the connected dissipative member based on the design yield stress of the material;
- γ_{ov} is the overstrength factor (see section 3.2.1).

Apart from the splices connecting dissipative elements, there are other important non-dissipative connections that should be rationally conceived, namely gravity load resisting bolted beam-to-column joints (e.g. single or double web cleat joints or flush end-plate joints, as shown in Figure 3.4). In order to ensure adequate rotation capacity, clause 6.5.5(5) requires that the design shear resistance of the bolts should be higher than 1.2 times the design bearing resistance. This provision aims at avoiding a brittle failure mode in those joints that despite being non-dissipative are nevertheless required to provide adequate rotational capacity for the interstorey drift demand under seismic actions (the rotation demand for these types of joints may be approximately assumed equal to the interstorey drift normalized to the interstorey height).

More details about the design criteria and seismic detailing of beam-to-column and brace-to-structure connections are provided in chapter 4.

3. EN 1998-1: Design Provisions for Steel Structures

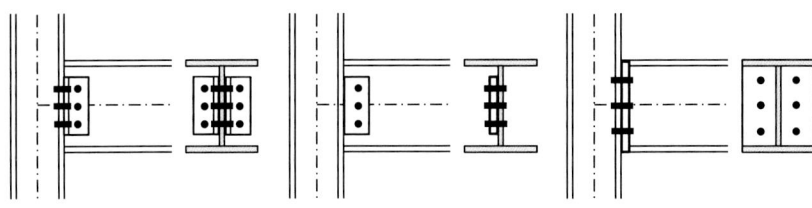

Figure 3.4 – Examples of gravity load resisting bolted beam-to-column connections

3.4.4 Design rules and requirements for dissipative connections

EC8-1 clause 6.5.5(6) allows the use of partial strength and dissipative connections. It is further emphasized that EC8-1 explicitly allows the use of dissipative semi-rigid and/or partial strength joints for MRF (clause 6.6.4(2)), EBF (clause 6.7.3(9)) and EBF (clause 6.8.4(2)) structural system. However, the adequacy of the design of the joints (both full and partial strength) should be supported by experimental evidence, in order to ensure that their behaviour is in line with the specific requirements defined for all structural types and structural ductility classes. This requirement is wisely justified by the lack of provisions and qualification procedures in EC8-1. It is clear that this approach is unfeasible for most design cases, because it requires experimental tests on sub-assemblages of full scale beam-to-column joints. This is the reason why experimental evidence may be based on existing data from the technical literature (clause 6.5.5(7)). However, this possibility does not offer much guidance to designers. Hence, current design practice is oriented toward full strength non-dissipative joints that can be designed according to EN 1993-1-8, even though also this category of joints deserves some remarks. More details about this issue are provided in chapter 4, including guidance on the appropriate selection of joints for dissipative zones of the structure that satisfy clauses 6.5.5(6) and 6.5.5(7).

3.4.5 Design rules and requirements for non-dissipative members

As described in sub-section 2.11.1.2, EC8-1 imposes the "strong columns-weak beams" concept for moment resisting frames (see equation 2.64). However, Eurocode 8 mandates a more general capacity design rule that should be applied to all non-dissipative members for all structural schemes (MRFs included), which can be expressed for the generic element "i" by the following equation:

$$E_{Ed,i} = E_{Ed,G,i} + 1.1 \cdot \gamma_{ov} \cdot \Omega \cdot E_{Ed,E,i} \qquad (3.5)$$

where $E_{Ed,G,i}$ is the action effect (e.g. axial force, bending moment or shear force) in member "i" induced by gravity loads; $E_{Ed,E,i}$ is the design seismic action effect in member "i"; $E_{Ed,i}$ is the combined design effect in the considered member to be adopted for its structural verification; γ_{ov} is the material overstrength; and Ω represents the minimum design overstrength of the dissipative elements, expressed as follows:

$$\Omega = \min\left(\frac{R_{d,E,i}}{E_{Ed,E,i}}\right) \qquad (3.6)$$

where $R_{d,E,i}$ is the strength (i.e. either bending, shear or axial resistance) of the i-th dissipative member with respect to its relevant peak effect $E_{Ed,i}$ (i.e. either bending moment, shear or axial force) due to the seismic combination.

Equation (3.5) aims at increasing the size of non-dissipative elements accounting for the resistance of the dissipative members. Hence, the over-sizing of the dissipative zones should be avoided in order to achieve rational and sufficiently economic structures. Unfortunately, there are some cases where the over-sizing of dissipative members is unavoidable, like structures prone to lateral deformability and P-Delta effects (e.g. MRFs). Indeed, stability and drift criteria (see sub-chapter 2.10 and section 2.11.2, respectively) require enlarging the size of the members to increase the lateral stiffness. At the end of the design process, the need to fulfill those requirements leads to structures with a large amount of strength and ductility, often excessively higher than effectively required by the seismic loads (Elghazouli, 2009).

3.5 DESIGN CRITERIA AND DETAILING RULES FOR MOMENT RESISTING FRAMES

3.5.1 Code requirements for beams

In MRFs with full strength rigid joints, beams are the dissipative elements of the structure, as stated in clause 6.6.1(1)P. Therefore, according

3. EN 1998-1: Design Provisions for Steel Structures

to section 6.5.3 of EC8-1, depending on the ductility class and the behaviour factor q used in the design, beams have to satisfy the requirements regarding the cross-sectional classes (e.g. class 1 for DCH concept and class 1 or 2 for DCM concept).

As previously discussed, in order to achieve a ductile global collapse mechanism, MRFs should be designed to form plastic hinges either at the ends of beams or in the beam-to-column connections, but avoiding the plastification of the columns with the exception of the base of the frame, at the top level of multi-storey buildings and for single storey buildings (i.e. "weak beam/strong column" behaviour). This type of failure mode is the most favourable performance, exploiting the beneficial dissipative capacity of the steel beams. On the contrary, plastic hinges developed in the columns (i.e. "strong beams/weak columns" behaviour, which typically characterizes non-seismic design) may lead to premature storey collapse mechanisms, because of the possibly poor and limited rotation capacity of the columns.

Aside from the requirements regarding the cross-sectional classes (previously described in section 3.4.2), clause 6.6.2(2) imposes the following conditions for MRFs in order to avoid that compression and shear forces acting on beams could impair the full plastic moment resistance and the rotation capacity at the location where the formation of hinges is expected:

$$\frac{M_{Ed}}{M_{pl,Rd}} \leq 1 \tag{3.7}$$

$$\frac{N_{Ed}}{N_{pl,Rd}} \leq 0.15 \tag{3.8}$$

$$\frac{V_{Ed}}{V_{pl,Rd}} \leq 0.5 \tag{3.9}$$

where M_{Ed}, N_{Ed} and V_{Ed} are the design forces, while $M_{pl,Rd}$, $N_{pl,Rd}$ and $V_{pl,Rd}$ are design resistances in accordance with EN 1993-1-1.

In general, owing to the presence of floor diaphragms, axial forces in the beams of MRFs are negligible. Instead, the shear forces could be significant and should be limited to avoid flexural-shear interaction in plastic hinge zones. Hence, as depicted in Figure 3.5, the shear force demand at both ends of each beam of MRF spans should be calculated using capacity design principles as follows:

$$V_{Ed} = V_{Ed,G} + V_{Ed,M} \tag{3.10}$$

where $V_{Ed,G}$ is the shear force due to the gravity forces in the seismic design situation and $V_{Ed,M}$ is the shear force corresponding to plastic hinges formed at the beam ends (namely $V_{Ed,M} = (M_{pl,A}+M_{pl,B})/L_h$, where $M_{pl,A}$ and $M_{pl,B}$ are the beam plastic moments with opposite signs at the end sections A and B and L_h is the length between those sections. L_h should accounts for the actual dimension of the beam-to-column joints and be defined as the net distance between any protruding parts of both end connections as shown in Figure 3.5.

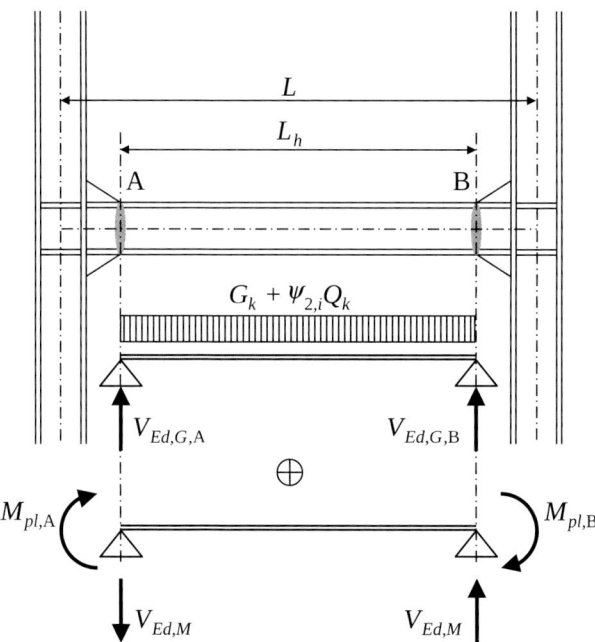

Figure 3.5 – Design shear forces for girders of MRF span: combination of gravity loads with plastic bending moments at both ends of the beam

There are situations, like long-span beams and large gravity forces, in which plastic hinges can form away from the beam ends (Mazzolani and Piluso, 1996, 1997). The EC8-1 procedure may be adjusted to account for these effects, by considering the probable positions of the plastic hinges from the bending moment diagrams obtained combining gravity and seismic effects.

3. EN 1998-1: Design Provisions for Steel Structures

3.5.2 Code requirements for columns

The columns should be designed according to the general capacity design rule given in section 3.4.5 (see equation (3.5)). Hence, in order to obtain the "weak beams/strong columns" behaviour, the forces acting on the columns calculated by an elastic model have to be amplified by the magnification coefficient Ω. For MRFs, equation (3.6) may be written as follows:

$$\Omega = \min\left(\frac{M_{pl,Rd,i}}{M_{Ed,i}}\right) \quad (3.11)$$

where $M_{Ed,i}$ is the design value of the bending moment in beam "i" in the seismic design situation and $M_{pl,Rd,i}$ is the corresponding plastic moment. It is important to highlight that this ratio should be calculated for all beams in which dissipative zones are located.

Once Ω has been calculated, the columns should be verified for all resistance checks including those for element stability, according to the provisions of chapter 6 of EN 1993-1-1 for the most unfavourable combination of bending moments M_{Ed}, shear force V_{Ed} and axial force N_{Ed}, based on (clause 6.6.3(1)P):

$$M_{Ed} = M_{Ed,G} + 1.1 \cdot \gamma_{ov} \cdot \Omega \cdot M_{Ed,E} \quad (3.12)$$

$$V_{Ed} = V_{Ed,G} + 1.1 \cdot \gamma_{ov} \cdot \Omega \cdot V_{Ed,E} \quad (3.13)$$

$$N_{Ed} = N_{Ed,G} + 1.1 \cdot \gamma_{ov} \cdot \Omega \cdot N_{Ed,E} \quad (3.14)$$

where:

$M_{Ed,G}$, $V_{Ed,G}$ and $N_{Ed,G}$, are the forces in the column due to the non-seismic actions included in the combination of actions for the seismic design situation;

$M_{Ed,E}$, $V_{Ed,E}$ and $N_{Ed,E}$ are the forces in the column due to the design seismic action;

γ_{ov} is the material overstrength factor (see section 3.2.1).

In addition to the member checks based on the Ω criterion, clause 4.4.2.3(4) requires verifying the local hierarchy at every joint as discussed in sub-section 2.11.1.2 (see equation (2.71)).

3.5.3 Code requirements for beam-to-column joints

According to section 1.4.4 of EN 1993-1-8 (CEN, 2005d), *"a beam-to-column joint consists of a web panel and either one connection (single sided joint configuration) or two connections (double sided joint configuration)"*. This definition is crucial because it clarifies the distinction between the term "connection" (which is the ensemble of mechanical elements directly connecting the members, as welded plates, bolts, etc.), the "web panel" of the column (which is portion of web column bounded by the two flanges of the column profile together with doubler plates, if used), and the joint (that includes both the connection and the web panel of the column).

Beam-to-column joints are key elements characterizing the seismic performance of MRFs. If the steel beams are designed to behave as dissipative parts of MRFs, the beam-to-column joints should be designed as full-strength, which could be either rigid or semi-rigid. Rigid joints are generally made of fully welded, hybrid welded/bolted or stiffened bolted configurations, while semi-rigid joints are usually made of bolted end plated configurations.

As discussed in section 3.4.3, section 6.5.5 of EC8-1 provides general rules for all types of non-dissipative connections belonging to the dissipative members of the structure, in order to ensure sufficient overstrength to avoid localization of plastic strains. In particular, the overstrength criterion given by equation (3.4) must be applied. Further details are described and discussed in chapter 4.

The design of beam-to-column joints is a complex and wide topic, which is beyond the scope of this design manual. The fundamental rules and the framework for the calculation of joints properties are given in EN 1993-1-8 (CEN, 2005d) and detailed guidance is given in the ECCS Eurocode design manual dedicated to joints in steel and composite structures (Jaspart and Weynand, 2016). However, it is important to highlight that EN 1993-1-8 is not a seismic code. Hence, a number of requirements are missing for the design of seismic resisting joints. In addition, EC8-1 does not provide normative design rules to achieve the required seismic performance for specific types of joints. In contrast, the US approach is oriented at codified design guidelines based on prequalified connections (AISC 358, 2010) or on prototype tests. Indeed, prequalified joints simplify the design and manufacturing process of beam-to-column joints. To this end, there is clearly a need to develop European guidance on appropriate prequalified connections, which are currently missing. It is noted that current prequalification work being carried out in Europe in the framework of the EQUALJOINTS project (Landolfo, 2016) should provide detailed guidance on appropriate dissipative and non-dissipative beam-to-column joints for steel MRFs.

As previously mentioned in section 3.4.4, EC8-1 allows to use semi-rigid partial-strength connections in the seismic design of MRFs, provided that they can be shown to have stable and ductile behaviour under cyclic loading. In order to achieve this requirement, clause 6.6.4(2) stipulates that: (i) dissipative semi rigid and/or partial strength joints should have a rotation capacity consistent with the global deformations; (ii) the members framing into the connections must be stable at the ultimate limit state; and (iii) the effect of connection deformation on global drift must be accounted for using non-linear analysis.

Moreover (clause 6.6.4(3)), the rotation capacity θ_p of the dissipative joints, given by:

$$\theta_p = \delta/0.5L \tag{3.15}$$

where δ is the beam deflection at midspan (excluding the elastic rotation of the column (clause 6.6.4(5)) and L is the beam span (see Figure 3.6), should be larger than 35 mrad for structures of ductility class DCH and 25 mrad for structures of ductility class DCM with $q > 2$. The rotation capacity of the plastic hinge region θ_p should be ensured under cyclic loading without degradation of strength and stiffness greater than 20 %. This requirement is valid independently of the intended location of the dissipative zones.

Owing to missing codified rules, it is clear that these requirements can only be verified by means of qualification tests to be carried out on joint sub-assemblages.

As an additional requirement to qualify the performance of the joint by means of testing, clause 6.6.4(4) states that the column web panel shear deformation should not contribute for more than 30 % of the plastic rotation capability θ_p.

This approach enables the use of a wide variety of bolted connections already studied in scientific literature that could have economical and practical benefits. More information about the most convenient and commonly used typologies of connections is given in chapter 4.

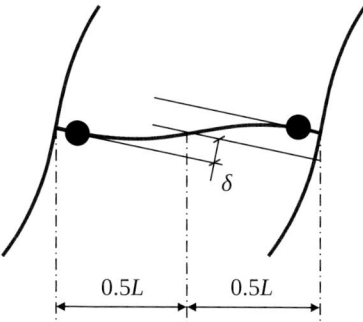

Figure 3.6 – Beam deflection for the calculation of θ_p

Another important aspect to be accounted for is the design of the web panel of beam-to-column joints. Indeed, due to the transfer of bending moments from the beams to the columns, large shear forces could be developed in the panel zone, as clarified in Figure 3.7, where the bending moment and shear diagrams produced by seismic loading are shown for a column. The ideal dimensionless structural node generally adopted in the structural models is not suitable to account for the actual behaviour of the structural node, which has finite dimensions. Indeed, the bending moments transferred by the steel beams with I or H profiles are basically characterized by a couple of forces applied at the centre of each beam flange (i.e internal resultant of tension and compression stresses). This couple of forces modifies the bending moment distribution as illustrated in Figure 3.7 and it is responsible of the concentration of shear forces in the column web panel.

Experimental tests have shown that shear yielding in panel zones is a very ductile mode of deformation. Panel zones can undergo many cycles of large inelastic distortions without deterioration in strength, while exhibiting cyclic hardening. However, if welded full strength and rigid joints are used, large inelastic shear distortion of the panel zone could trigger the initiation and propagation of fracture from the welds to the column flange. Hence, for non-dissipative joints the code is oriented to give an equal chance to start yielding in the panel zone and in the beam end. This is why the design expression (3.16) for the panel zone is similar to (3.7) for dissipative zones of beams. Excessive deformation of the panel zone is prevented by the following considerations: it

3. EN 1998-1: Design Provisions for Steel Structures

is known that yielding in shear raises the shear resistance of a panel in term of stress from $f_y/\sqrt{3}$ to f_y so that if the first yield is in the panel in shear, the latter soon develops overstrength with respect to the beam plastic mechanism and the beam end starts to yield. Clause 6.6.3(6) requires that the shear strength of web panels in beam-to-column connections should satisfy the following expression:

$$\frac{V_{wp,Ed}}{V_{wp,Rd}} \leq 1.0 \tag{3.16}$$

where $V_{wp,Ed}$ is the design shear force in the web panel due to the plastic resistance of the adjacent dissipative zones (namely beams or connections) and $V_{wp,Rd}$ is the shear resistance of the web panel in accordance with clause 6.2.4.1 of EN 1993-1-8, calculated neglecting the effect of the stresses due to axial force and bending moment on the plastic resistance in shear.

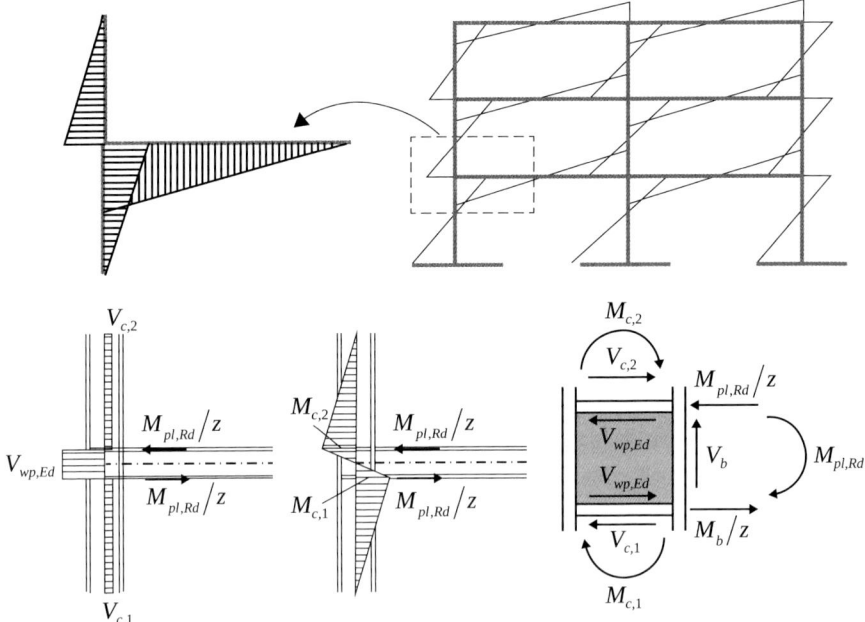

Figure 3.7 – Shear forces acting on the column web panel

This implies that the shear force demand can be determined from the summation of moments at the column faces, as determined by projecting the expected moments at the plastic hinge points to the column faces, as follows:

3.5 Design Criteria and Detailing Rules for Moment Resisting Frames

$$V_{wp,Ed} = \frac{\sum M_{pl,Rd,i}}{z} - \left(\frac{V_{c1,Ed} - V_{c2,Ed}}{2}\right) \qquad (3.17)$$

where $M_{pl,Rd,i}$ is the plastic bending moment of beam "i" (or its projected value at column face if some stiffeners are used to strengthen the connection) and z is the beam lever arm, which can be calculated as $z = d_b - t_f$; d_b is the beam depth and t_f the relevant flange thickness; $V_{c1,Ed}$ and $V_{c2,Ed}$ are the shear forces in the lower and upper portions of the column, respectively. According to clause 6.2.4.1 of EN 1993-1-8, the shear strength of the web panel in the presence of transverse web stiffeners is given by:

$$V_{wp,Rd} = V_{wc,Rd} + \Delta V_{wp,add,Rd} = \frac{0.9 f_{y,wc} \cdot A_{vc}}{\sqrt{3} \cdot \gamma_{M0}} + \frac{4 M_{pl,fc,Rd}}{d_s} \qquad (3.18)$$

where $V_{wc,Rd}$ is the design plastic shear resistance of the unstiffened column web panel and $\Delta V_{wc,Rd}$ the overstrength contribution due to a mechanism involving the plastic moment capacity of the column flanges $M_{pl,fc,Rd}$.

Studies carried out by Krawinkler (1978) showed that the overstrength $\Delta V_{wc,add,Rd}$ can be achieved for shear distortions larger than $4\gamma_y$, where γ_y is the shear distortion at yielding. This implies that shear yielding occurs if this contribution to shear strength of web panel is considered. It is clear that allowing plastic deformations in the panel reduces the plastic rotation demand in the beams. However, plastic shear distortion of panel zone can induce excessive strain demands into welds that are prone to brittle failure. In addition, plastic engagement of panel zone may result in residual distortion with may compromise the possibility to repair the structure after an earthquake with potentially high residual drifts. Therefore, although continuity plates may be used to strengthen the column flanges in order to resist tension and compression and to increase $M_{pl,fc,Rd}$ due to the bending transfer by beams, it is advisable, in order to avoid plastic deformation in the web panel, to verify the panel shear strength considering only the contribution given by $V_{wc,Rd}$. If strengthening is required to the web panel, additional plates can be welded to the column panel zone, as illustrated in clause 6.2.6.1 of EN 1993-1-8.

3. EN 1998-1: Design Provisions for Steel Structures

3.6 DESIGN CRITERIA AND DETAILING RULES FOR CONCENTRICALLY BRACED FRAMES

3.6.1 Code requirements for braces

CBFs are characterized by a truss behaviour due to the axial forces developed in the bracing members. According to clause 6.7.1(1)P, the diagonal bracings in tension are the dissipative zones which have to yield, thus preserving the connected elements from damage. Therefore, the response of a CBF is basically influenced by the behaviour of its bracing elements. However, it should be noted that the role of the bracing members differs with the configurations of the bracings. For X and Diagonal CBFs, the energy dissipation capacity of the braces in compression is neglected and the lateral forces are assigned to the tension braces only. On the contrary, in frames where the braces intersect the beams, e.g. with V and inverted-V bracings, both the tension and compression diagonals should be taken into account.

Sub-chapter 4.4 discusses and details appropriate solutions for the braces and their connections with respect to the potential failure modes.

Provided that capacity design requirements are fulfilled, the damage pattern is mostly concentrated in the braces and in their end connections. Indeed, besides the yielding in tension, flexural hinges may form at mid-length of the brace or in the gusset plates owing to the brace buckling in compression. The *M-N* interaction characterizes the strength deterioration of the braces in the post-buckling range. In addition, this failure mode can occur solely if the gusset plate connections are able to develop a linear hinge region (Astaneh-Asl *et al*, 1982) that allows brace end rotations caused by brace buckling. Under these assumptions, the braces may be modelled as pinned. If flexural full strength connections are adopted, the plastic hinges form only in the brace member, namely at both ends and in the middle, and fixed ends should be considered for the braces. More details about the design criteria and recommendations for brace connections are given in chapter 4.

Depending on the ductility class and the behaviour factor q used in the design, tension bracings have to satisfy the requirements regarding the cross-sectional classes reported in Table 3.4. Moreover, the diagonal braces have to be designed and arranged in such a way that, under seismic action reversals, the structure exhibits similar lateral force-displacement responses in opposite

3.6 DESIGN CRITERIA AND DETAILING RULES FOR CONCENTRICALLY BRACED FRAMES

directions at each storey (clause 6.7.1(2)P). This performance requirement is deemed to be satisfied if the following rule is met at every storey (clause 6.7.1(3)):

$$\frac{|A^+ - A^-|}{A^+ + A^-} \leq 0.05 \qquad (3.19)$$

where A^+ and A^- are the areas of the vertical projections of the cross sections of the tension diagonals (see Figure 3.8) when the horizontal seismic actions have a positive or negative direction, respectively.

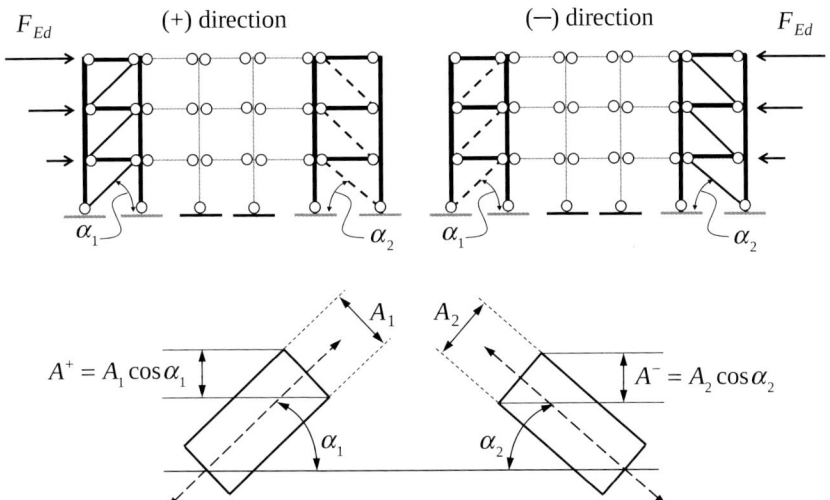

Figure 3.8 – Example of application of requirement given by equation (3.19)

As discussed in section 2.6.4 (see also Figure 2.30), in an elastic seismic analysis, for frames with diagonal bracings (e.g. X and Diagonal CBFs), only the tension diagonals should be taken into account (clause 6.7.2(2)P). Consequently, the tensioned bracing members have to be designed in order to guarantee that the yield resistance $N_{pl,Rd}$ of their gross cross section satisfies the following inequality (clause 6.7.3(5)):

$$N_{pl,Rd} \geq N_{Ed} \qquad (3.20)$$

where N_{Ed} is calculated from the elastic model ideally composed by a single brace (i.e. the diagonal in tension). In general, in order to make tension

3. EN 1998-1: Design Provisions for Steel Structures

alternatively developing in all the braces at any storey, two models should in principle be developed, one with the braces tilted in one direction and another with the braces tilted in the opposite direction, as shown in Figure 3.9, unless the structure is symmetrical.

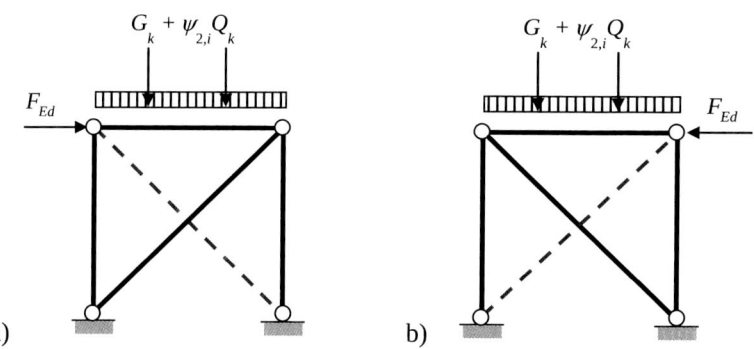

Figure 3.9 – Calculation models of X and Diagonal CBFs

For bracing members in a X configuration, the brace non-dimensional slenderness (EN 1993-1-1) must fall in the range $1.3 \leq \overline{\lambda} \leq 2.0$ (clause 6.7.3(1)). The lower bound value is imposed in order to limit the maximum compressive axial force transmitted to the columns. Indeed, in the simplified calculation model (see Figure 3.9) the relevant force distribution is not realistic both for the pre-buckling and the post-buckling stages. In the first case, the lateral shear capacity of a X-CBF is given by $2\chi \cdot N_{pl,Rd} \cdot \cos \alpha$, where α is the slope of the bracing and χ is the buckling reduction factor. If stocky bracings are adopted (i.e. $\overline{\lambda} < 1.3$ and $\chi > 0.5$), $2\chi \cdot N_{pl,Rd} \cdot \cos \alpha > N_{pl,Rd} \cdot \cos \alpha$, so that the lateral strength and the relevant distribution of internal forces is larger in the pre-buckling stage than in the post-buckling, where only one diagonal in tension is considered. It is clear that in such conditions, the simplified model with only the diagonal in tension is non-conservative, because it would lead to smaller forces in the columns. On the contrary, if slender bracings are assumed (i.e. $\overline{\lambda} \geq 1.3$ and $\chi \leq 0.5$), the lateral pre-buckling strength $2\chi \cdot N_{pl,Rd} \cdot \cos \alpha$ will be smaller than the yield strength $N_{pl,Rd} \cdot \cos \alpha$ when only the diagonal in tension is considered, and the simplified modelling is conservative. In addition, the simplified approach disregards the residual post-buckling strength of the compressed bracing. In reality, the brace post-buckling strength is far from being negligible. Hence, the compressed bracing in the post-buckling range may transfer force to the

3.6 Design Criteria and Detailing Rules for Concentrically Braced Frames

columns. This effect becomes negligible by increasing the slenderness of the braces. In light of these considerations, EC8-1 prevents the use of stocky braces that are characterized by significant post-buckling capacity, and enforces the use of slender braces, which are characterized by small to negligible residual axial strength after buckling. The upper bound value of slenderness is given in order to limit excessive vibrations and shocks at load reversal and undesired buckling under service loads.

In case of a single diagonal bracing configuration (i.e. all layouts in which the diagonals are not positioned as X bracings), no lower bound limit is considered for the brace slenderness and only the condition $\lambda \leq 2$ remains (clause 6.7.3(2) 2.0).

Clause 6.7.3(4) allows to disregard any limitation on brace slenderness for low-rise structures up to two storeys, thus enabling to use rods or cables, only for this category of buildings.

In case of frames with V and inverted-V bracing the calculation model should account for both diagonal members and those under compression should be designed to resist the applied compressive forces, according to the following:

$$\chi N_{pl,Rd} \geq N_{Ed} \qquad (3.21)$$

where χ is the buckling reduction factor calculated according to clause 6.3.1.2 (1) of EN 1993-1-1, and N_{Ed} is the required strength. Differently from the case of X-CBFs, the code does not impose a lower bound limit for the non-dimensional slenderness $\overline{\lambda}$, while the upper bound limit ($\overline{\lambda} \leq 2$) is retained (clause 6.7.3(3)).

Whichever configuration of bracing members is adopted, EC8-1 aims at enforcing the formation of a global mechanism, which means maximizing the number of diagonals that yield, by controlling the variation of the brace design overstrength at each storey of the structure.

The minimum design overstrength of the bracing elements (in line with the general criterion expressed by equation (3.5)) may be written as follows (clause 6.7.4(1)):

$$\Omega = \min\left(\frac{N_{pl,Rd,i}}{N_{Ed,i}}\right) \qquad (3.22)$$

3. EN 1998-1: Design Provisions for Steel Structures

Hence, once the ratio Ω is calculated, clause 6.7.3(8) mandates that the values of all other Ω_i should be in the range Ω to 1.25Ω. Consequently, this criterion requires to fit the distribution of the diagonal strengths $N_{pl,Rd,i}$ to the distribution of the computed action effect $N_{Ed,i}$, inevitably leading to changes of the cross section of the diagonals over the height of the building.

3.6.2 Code requirements for beams and columns

For all types of CBFs, in order to guarantee the formation of a global mechanism, the strength of the non-dissipative members (i.e. beams and columns) should be designed in accordance to the general hierarchy criterion expressed by equation (3.5) (see section 3.4.5). Considering that the main internal force regime is truss action, equation (3.5) should be rewritten as follows (clause 6.7.4(1)):

$$N_{pl.Rd}(M_{Ed}) \geq N_{Ed} = N_{Ed,G} + 1.1 \cdot \gamma_{ov} \cdot \Omega \cdot N_{Ed,E} \qquad (3.23)$$

where:

- $N_{pl,Rd}(M_{Ed})$ is the design axial force resistance of the beam or the column calculated in accordance with section 6.2.9 of EN 1993-1-1, taking into account the interaction with the design value of bending moment, M_{Ed}, in the seismic design situation;
- $N_{Ed,G}$ is the axial force in the beam or in the column due to the non-seismic actions included in the combination of actions for the seismic design situation;
- $N_{Ed,E}$ is the axial force in the beam or in the column due to the design seismic action;
- γ_{ov} is the material overstrength factor (see section 3.2.1);
- Ω is the minimum overstrength ratio given by equation (3.20).

In addition, Eurocode 8 provides specific requirements for the brace-intercepted beams belonging to both V and inverted V CBFs. Indeed, for those types of structural schemes, the seismic response is significantly influenced by the beam behaviour. After the compression brace buckles, the resulting axial forces in the compression and in the tension braces has a vertical component acting on the connected beam, which may induce significant bending moment at the brace-intercepted section. In this situation, the formation of a plastic hinge at mid-span of the beam must be avoided, because it would result in a drop of storey lateral resistance with the consequent inelastic concentration of drift at the storey with yielded beams. In order to prevent this undesirable behaviour, clause 6.7.4(2) requires that: i) the beams of the braced span must be able to carry all non-seismic actions without considering the intermediate support given by the diagonals; ii) the beams have to be designed to carry also the vertical component of the force transmitted by the tensioned and compressed braces. As clarified in Figure 3.10, this vertical component is calculated assuming that the tensioned brace transfers a force equal to its yield resistance ($N_{pl,Rd}$) and the compressed brace transfers a force equal to a percentage of its original buckling strength ($N_{b,Rd}$) to take into account the strength degradation under cyclic loading. The reduced compression strength is given by $\gamma_{pb} N_{pl,Rd}$ where γ_{pb} is found in the National Annexes. The recommended value is 0.30.

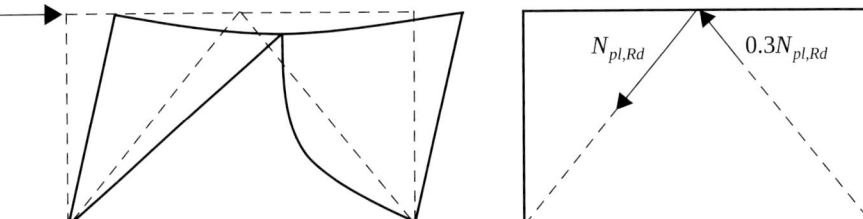

Figure 3.10 – Capacity design criterion for brace-intercepted beams in chevron CBFs

Regarding the design of the columns, clause 6.7.4(3)P provides additional requirements only for CBFs with diagonal bracings in which the tensioned and compressed diagonals are not intersecting. In this case, the strength of columns adjacent to compressed braces should also be verified for the axial forces $N_{Ed,E}$ transmitted by the buckled diagonals assuming their design buckling resistance (i.e. $\chi N_{pl,Rd}$) without any further reduction, as shown in Figure 3.11. This requirement aims at overcoming the limits of elastic analysis to predict the complex nonlinear behaviour of CBFs. Obviously, pushover analyses may be carried out (clause 6.7.2(3a)) in order to predict the force distribution in the columns better than those derived by equation (3.23).

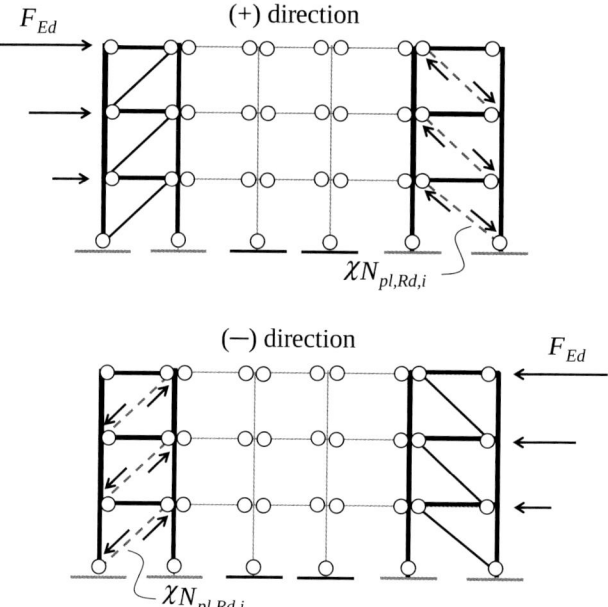

Figure 3.11 – Capacity design criterion for columns in Diagonal CBFs

3.7 DESIGN CRITERIA AND DETAILING RULES FOR ECCENTRICALLY BRACED FRAMES

3.7.1 Code requirements for seismic links

The link is the key distinguishing feature of an EBF, which is a portion of the beam set out between the braces or between one brace and the column. According to clause 6.8.1(1)P, the links are the components devoted to dissipate seismic energy. As demonstrated by the results of the DUAREM project (Sabau *et al*, 2014), links may be designed to be detachable with bolted end connections, thus allowing to be easily replaceable even though EC8-1 does not provide any such requirements. Further details on this issue are given in sub-chapter 4.5.

Usual EBF arrangements are shown in Figure 3.2d where the links are identified by a bold segment for each structural configuration. The four EBF arrangements presented in Figure 3.2d are named (i) split-K-braced frame, (ii) D-braced frame, (iii) V-braced and (iv) inverted-Y-braced frame.

3.7 Design Criteria and Detailing Rules for Eccentrically Braced Frames

On the basis of the type of plastic mechanism to be developed, links are classified into three categories: short (which dissipate energy by yielding essentially in shear), intermediate (in which the plastic mechanism involves bending and shear) and long (which dissipate energy by yielding essentially in bending), depending on the structural and geometric properties of the links (clause 6.8.2(2)).

The mechanical parameter influencing the plastic mechanism is the link length "e", which is related to the ratio between the plastic bending moment and the plastic shear of the cross section of the link. In order to clarify this concept it is useful to refer to a link subjected to symmetric flexural actions equal to the plastic bending moment $M_{p,link}$ and a shear force equal to the plastic shear strength $V_{p,link}$, calculated according to clause 6.8.2(3) as follows:

$$M_{p,link} = f_y \cdot b \cdot t_f \cdot (d - t_f) \qquad (3.24)$$

$$V_{p,link} = \left(f_y / \sqrt{3}\right) \cdot t_w \cdot (d - t_f) \qquad (3.25)$$

where f_y is the value of the yield stress of steel, d is the depth of the cross section, t_f is the flange thickness and t_w is the web thickness.

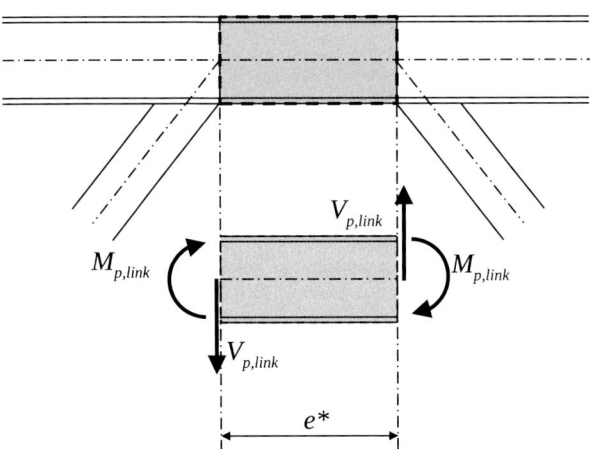

Figure 3.12 – Theoretical limit for link length (perfectly plastic behaviour no bending-shear interaction)

Assuming perfect plasticity, no flexural-shear interaction and equal link end moments, the theoretical length dividing the behaviour of short links

3. EN 1998-1: Design Provisions for Steel Structures

(governed by shear) and long links (governed by flexure) is shown in the free-body scheme of Figure 3.12 and it can be expressed as:

$$e^* = 2M_{p,link}/V_{p,link} \tag{3.26}$$

A large number of experimental and analytical activities (such as Kasai and Popov, 1985, 1986a,b; Hjelmstad and Popov, 1983, 1984; Okazaki et al, 2005, 2006 and 2009) indicated that the assumption of no M-V interaction is reasonable, but an assumption of perfect plasticity is not correct. In fact, substantial strain hardening occurs in shear links. According to tests performed on American wide-flange steel profiles, EC8-1 assumes that the average ultimate link shear forces reach the value of $1.5V_{p,link}$. Kasai and Popov (1986) estimated that the maximum link end moments are $1.2M_p$, in order to avoid concentration of plastic bending strains that may lead to severe flange buckling, to fracture from key area to the link web or to failure of link flange-to-column welds. Thus, from free-body equilibrium of forces acting on the link, if the end moments are limited to $1.2M_{p,link}$ and the link shear is assumed to reach $1.5V_{p,link}$, the limiting link length is

$$e_s = \frac{2 \cdot (1.2M_{p,link})}{1.5 \cdot V_{p,link}} = 1.6\frac{M_{p,link}}{V_{p,link}}.$$

Consequently, in the cases where equal moments (i.e. antisymmetric conditions) could form simultaneously at both ends of the link (e.g. the split-K configuration) the following equations (clause 6.8.2(8)) can be used to classify the link mechanical response:

Short links:
$$e \le e_s = 1.6\frac{M_{p,link}}{V_{p,link}} \tag{3.27}$$

Long links:
$$e \ge e_L = 3\frac{M_{p,link}}{V_{p,link}} \tag{3.28}$$

Intermediate links:
$$e_s < e < e_L \tag{3.29}$$

Based on equations (3.27), (3.28) and (3.29), Figure 3.13 explains the classification criteria for seismic links, relating the link length to the bending-

3.7 Design Criteria and Detailing Rules for Eccentrically Braced Frames

shear interaction domain. Short links develop only shear yielding, long links develop only flexural yielding, while intermediate links develop both shear and flexural yielding.

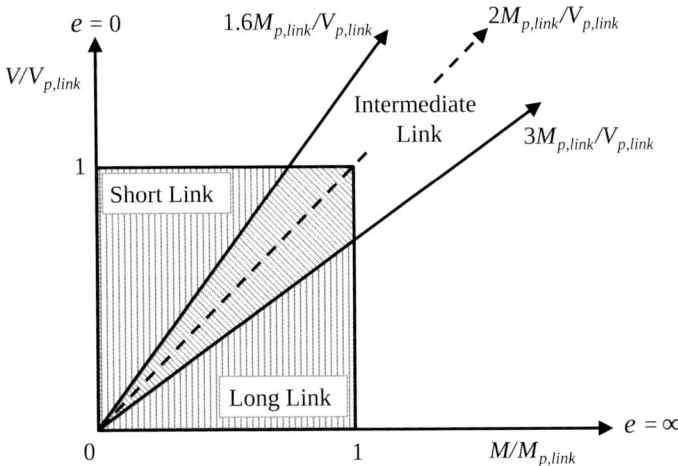

Figure 3.13 – M-V interaction domain: classification of seismic links

Equations (3.26), (3.27) and (3.28) can be generalized to the design cases where only one plastic hinge would form at one end of the link (e.g. the case of inverted-Y configuration, where the link end connected to the beam develops the plastic flexural hinge, while the opposite end connected to the braces is in the elastic range) as follows (clause 6.8.2(8)):

Short links:
$$e \leq e_s = 0.8 \cdot (1+\alpha) \frac{M_{p,link}}{V_{p,link}} \qquad (3.30)$$

Long links:
$$e \geq e_L = 1.5 \cdot (1+\alpha) \frac{M_{p,link}}{V_{p,link}} \qquad (3.31)$$

Intermediate links:
$$e_s < e < e_L \qquad (3.32)$$

where α is the ratio of the smaller bending moments $M_{Ed,A}$ at one end of the link in the seismic design situation, to the greater bending moments $M_{Ed,B}$ at the end where the plastic hinge would form, both moments being taken as absolute values.

3. EN 1998-1: Design Provisions for Steel Structures

It is interesting to note that the type of link plastic mechanism is directly related to the ductility capacity, namely plastic rotation, provided by the link in order to withstand the seismic ductility demand. The link rotation is defined as the rotation angle θ_p between the link and the element outside of the link. Figure 3.14 clarifies this definition and simple equations are given for estimating the link rotation, which should be consistent with global deformation, for the EBF configurations reported in Figure 3.2d (i.e. split-K-braced frame, V-braced, D-braced frame and inverted-Y-braced frame). It can be observed that the shorter is the link length and the greater is the ductility demand.

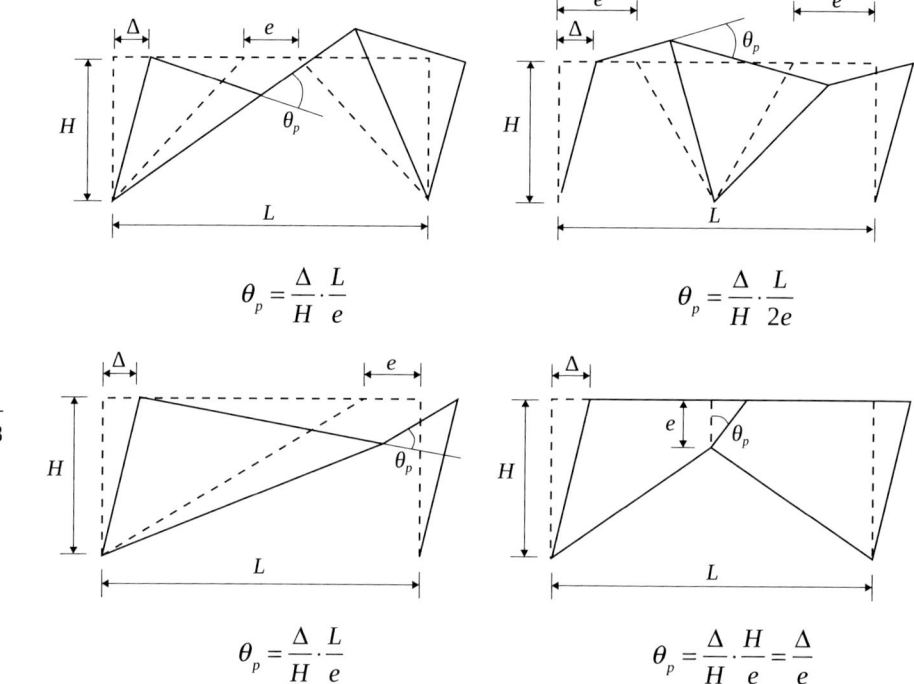

Figure 3.14 – The link rotation angle θ_p for the EBF configurations given in Figure 3.2

The allowable link deformation capacity θ_p depends on the link length, as depicted in Figure 3.15. With this regard, clause 6.8.2(10) states that the link rotation should not exceed the following values:

Short links $\qquad\qquad \theta_p \leq \theta_{pR} = 0.08$ radians $\qquad\qquad$ (3.33)

3.7 Design Criteria and Detailing Rules for Eccentrically Braced Frames

Long links $\quad \theta_p \leq \theta_{pR} = 0.02$ radians $\quad$ (3.34)

Intermediate links $\quad \theta_p \leq \theta_{pR} =$ the value determined by linear interpolation between the above values. $\quad$ (3.35)

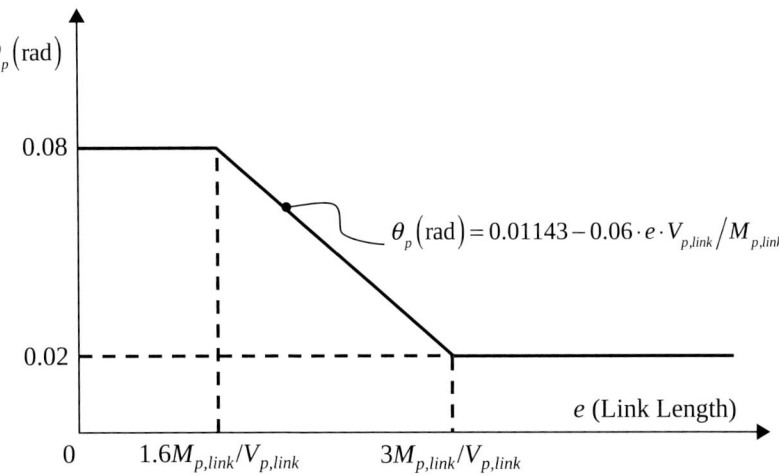

Figure 3.15 – Allowable link deformation capacity θ_p

In order to guarantee a sufficient link rotation capacity, link web stiffeners should be designed. Stiffeners are fundamental details that allow to prevent inelastic web buckling, which impairs the link performance in the range of the expected ductility demand. Hjelmstad and Popov (1983) developed several cyclic tests in order to relate the web stiffeners spacing to link energy dissipation. Subsequently, Kasai and Popov (1986) developed simple rules to relate stiffeners spacing and maximum link inelastic rotation θ_p up to web buckling. Assuming that the link web buckling modes are similar to those of a plate under shear loading, they applied plastic plate shear buckling theory to relate the stiffeners spacing to the maximum deformation angle of a shear link, thus deriving simple expressions for each required link deformation capacity. Accordingly, clause 6.8.2(12)a proposes to design the stiffener spacing for short links by means of the following equations:

$$a = 30t_w - \frac{d}{5} \quad \text{for } \theta_p = 0.08 \text{ radians} \quad (3.36)$$

3. EN 1998-1: Design Provisions for Steel Structures

$$a = 52t_w - \frac{d}{5} \quad \text{for } \theta_p < 0.02 \text{ radians} \qquad (3.37)$$

where a is the distance between equally spaced stiffeners, d is the link depth and t_w is the web thickness. Linear interpolation should be used for values of θ_p between 0.08 and 0.02 radians.

In case of long links, web stiffeners should be placed at a distance of 1.5 times b from each end of the link where a plastic hinge would form (clause 6.8.2(12)b).

Intermediate links should be provided with intermediate web stiffeners meeting the requirements of both short and long links (clause 6.8.2(12)c).

Similarly to what is required for CBFs (see section 3.6.1), also for EBFs EC8-1 aims at enforcing the formation of a global mechanism by controlling the variation of the design overstrength of the dissipative elements (i.e. the links) at each storey of the structure. The design overstrength factors Ω_i differ for short, intermediate and long links, because they depend on the link failure mode, and they should be evaluated as follows:

$$\Omega_i = \frac{1.5 V_{p,link,i}}{V_{Ed,i}} \quad \text{for short links} \qquad (3.38)$$

$$\Omega_i = \frac{1.5 M_{p,link,i}}{M_{Ed,i}} \quad \text{for both intermediate and long links} \qquad (3.39)$$

where $V_{Ed,i}$ and $M_{Ed,i}$ are the design values of the shear force and the bending moment in link "i" in the seismic design situation, while $V_{p,link,i}$ and $M_{p,link,i}$ are the shear and bending plastic design resistances of "i^{th}" link as given by equations (3.25) and (3.24), respectively.

Hence, the design overstrength Ω of the seismic links (in line with the general criterion expressed by equation (3.6)) is the minimum among all links belonging to the considered alignment of frames, as required by clause 6.8.3(1) and given by:

$$\Omega = \min\left(\frac{1.5 V_{p,link,i}}{V_{Ed,i}}\right) \text{among all short links} \qquad (3.40)$$

3.7 Design Criteria and Detailing Rules for Eccentrically Braced Frames

$$\Omega = \min\left(\frac{1.5 M_{p,link,i}}{M_{Ed,i}}\right) \text{ among all intermediate and long links} \tag{3.41}$$

Once the ratio Ω is calculated, clause 6.8.2(7) mandates that the values of all other Ω_i should be in the range Ω to 1.25Ω. Consequently, this criterion requires to fit the distribution of the seismic links strengths $V_{p,link,i}$ (or $M_{p,link,i}$) to the distribution of the computed action effect $V_{Ed,i}$ (or $M_{Ed,i}$), inevitably leading to changes of the cross section of the diagonals at each storey.

3.7.2 Code requirements for members not containing seismic links

Clause 6.8.3(1) recommends to verify in compression all members not containing seismic links, such as the columns and the diagonal members, if horizontal links are used in the beams, and also the beam members, if vertical links are used, considering the most unfavourable combination of axial force and bending moments. In accordance with the general hierarchy criterion expressed by equation (3.4) (see section 3.4.5), the action effects on connections, beams and columns, calculated by using an elastic model under the design seismic action, have to be amplified using the minimum overstrength factor Ω, given by equation (3.40) for short links and equation (3.41) for intermediate and long links. Subsequently, the design check of a member of the frame is based on:

$$N_{pl,Rd}\left(M_{Ed}, V_{Ed}\right) \geq N_{Ed,G} + 1.1 \cdot \gamma_{ov} \cdot \Omega \cdot N_{Ed,E} \tag{3.42}$$

where:

$N_{pl,Rd}(M_{Ed}, V_{Ed})$ is the design axial resistance of the column or the diagonal member calculated in accordance with EN 1993-1-1, taking into account the interaction with the design value of the bending moment, M_{Ed}, and the shear force, V_{Ed}, in the seismic design situation. It is important to highlight that for braces and columns in compression equation (3.42) intends to refer to the buckling strength, because it is crucial to avoid the buckling of the braces after yielding of the link;

3. EN 1998-1: Design Provisions for Steel Structures

$N_{Ed,G}$ is the compression force in the column or diagonal member due to the non-seismic actions included in the combination of actions for the seismic design situation;

$N_{Ed,E}$ is the compression force in the column or diagonal member due to the design seismic action;

γ_{ov} is the material overstrength factor (see section 3.2.1).

In any case the axial forces ($N_{Ed,E}$, $N_{Ed,G}$) induced by seismic and non-seismic actions are directly provided by the elastic numerical model.

3.7.3 Code requirements for connections of the seismic links

Clause 6.8.4(1) recommends that the connections of the links or of the element containing the links should be designed to resist the action effects E_d computed as follows:

$$E_d \geq E_{d,G} + 1.1 \cdot \gamma_{ov} \cdot \Omega_i \cdot E_{d,E} \quad (3.43)$$

where:

$E_{d,G}$ is the action effect in the connection due to the non-seismic actions included in the combination of actions for the seismic design situation;

$E_{d,E}$ is the action effect in the connection due to the design seismic action;

γ_{ov} is the material overstrength factor (see section 3.2.1 of this Handbook);

Ω_i is the overstrength factor computed as given by equation (3.38) for short links or equation (3.39) for intermediate and long links.

It is noted that in this case the designer must use the plastic strength of the link to design the connection, never the minimum overstrength ratio as done for the members. This verification is in line with the criterion expressed by equation (3.3) (see section 3.4.2), where the plastic resistance of the connected dissipative member R_{fy} (based on the design yield stress of the material) is given by $\Omega_i \cdot E_{d,E}$.[1]

---

[1] N.B. for short links the main transferred effect is the shear force, thus $E_{d,E} = V_{Ed,i}$ so that $\Omega_i \cdot E_{d,E} = (1.5 V_{p,link,i} / V_{Ed,i}) \cdot V_{Ed,i} = 1.5 V_{p,link,i}$

Chapter 4

DESIGN RECOMMENDATIONS FOR DUCTILE DETAILS

4.1 INTRODUCTION

Ductile detailing is crucial to guarantee the expected design performance of steel structures. Unfortunately, it is still quite common to be confronted with opinions of European designers that believe that steel buildings are not susceptible to earthquake-induced structural damage, thus paying limited attention to the appropriate design of the ductile components of buildings.

EC8-1 provides capacity design criteria to enforce overall and local ductile and dissipative behaviour of steel structures. However, there are a number of issues related to conceptual design and constructional detailing that are missing or not sufficiently addressed by the code. This design manual tries to compensate these shortcomings.

The objective of this chapter is to provide a concise overview about the best practice to implement the requirements and design rules for ductile details in light of the state-of-the-art of international standards, research and professional experience. In particular, emphasis is given to detailing for DCH structures and practical information is provided and discussed regarding the following aspects: (i) the design of floors as diaphragms; (ii) the design of connections for MRF, CBFs and EBF; and (iii) the recommendations for the stability of bracing and base fixities.

It is important to highlight that the issues discussed in this chapter are not exhaustive because of the huge variety of possible design choices and allowable details that make it impossible to address them systematically. Hence, the purpose of this chapter is to provide designers and students with consolidated and widely accepted criteria that have been already successfully adopted in existing structures. Therefore, designers should be aware that some detailing rules may also change with the continuous improvement of knowledge.

4.2 SEISMIC DESIGN AND DETAILING OF COMPOSITE STEEL-CONCRETE SLABS

The floors of modern steel buildings are generally made of composite steel-concrete slabs. Indeed, this type of composite deck is very effective to improve the performance of the floor under both gravity and seismic actions. In addition, the man hours necessary to erect the floors are generally smaller than those usually necessary for "all-steel" (i.e. non-composite) structural decks that require in-plan bracing systems (namely additional connections to be fastened on site) to ensure the diaphragm behaviour.

As discussed in chapter 2, most of the mass of the building structure is affected to the floors at each storey, where significant inertial forces develop in the plane of the floor when an earthquake occurs. If conceived properly as horizontal diaphragms, the floor systems have a key role in seismic resisting structures, because they collect the inertia forces and distribute those lateral loads to the vertical frames that provide the overall structural stability.

In addition to inertial forces, diaphragms must be able to transfer the internal forces developing between different vertical seismic resisting systems (Sabelli et al, 2011). To clarify this aspect, it is helpful to refer to the case of dual frames made of MRF and BF. Both sub-systems are characterized by different displacement profiles under lateral loads, if acting independently. On the contrary, when they are interconnected by horizontal diaphragms, they are subjected to the same storey displacements and the diaphragms must develop internal forces in order to impose compatibility of storey displacements. Figure 4.1a shows in a schematic way (illustrating the diaphragm effect of the floors as equivalent connecting ties) how the floor systems become subject to additional internal forces that may be larger than those estimated considering only the primary diaphragm actions due to inertial forces. This effect does not only occur in case of dual systems but also within a floor because of the in-plane redistribution that the diaphragm action provides in order to achieve a uniform distribution of forces to the vertical components (columns), as it is illustrated in Figure 4.1b.

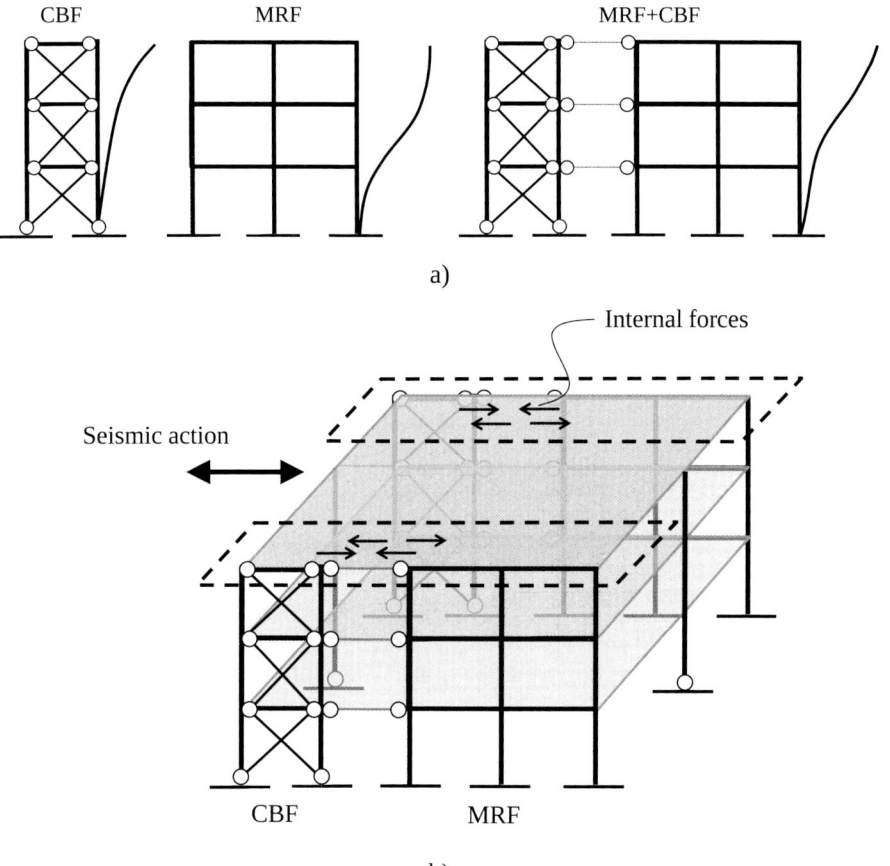

Figure 4.1 – Compatibility transfer forces into the diaphragms

With this regard, EC8-1 does not explicitly account for this effect. Indeed, as shown in section 2.11.1.4, section 4.4.2.5 of EC8-1 recommends verifying the floor systems against seismic action effects magnified by an overstrength factor γ_d (assumed equal to 1.3 or 1.1 for non-ductile or ductile diaphragms, respectively), where the seismic effects are directly obtained from the analysis. However, the application of this rule is not straightforward.

An internationally recognized method to estimate the diaphragm design forces is provided by ASCE 7-10, which recommends to verify the resistance of the floor and the roof diaphragms against the following effects:

4. Design Recommendations for Ductile Details

$$0.2S_{DS}I_e w_{px} \leq F_{px} = \frac{\sum_{i=x}^{n} F_i}{\sum_{i=x}^{n} w_i} w_{px} \leq 0.4S_{DS}I_e w_{px} \qquad (4.1)$$

where F_{px} is the diaphragm design force at x-th level; F_i is the design force applied to i-th level; w_i is the weight tributary to i-th level; w_{px} is the weight tributary to the diaphragm at level x-th; I_e is the importance factor; S_{DS} is the design spectral response acceleration parameter, which is equivalent to the EC8 elastic spectral acceleration between T_B and T_C (see equation. 2.20)

In addition, ASCE 7-10 requires that the collector elements (see Figure 4.2) should be designed to resist the seismic forces originating in other portions of the structure to the element providing the resistance to those forces. These forces should be magnified by the relevant overstrength factor Ω of the primary seismic resisting system.

Once the design forces acting on each diaphragm are obtained it is necessary to evaluate the internal forces within the components of the floor system. Diaphragms in steel structures are generally spanning between the vertical seismic resisting frames, which can be considered as lateral supports that restrain the floor against horizontal actions. Therefore, single-span diaphragms can be effectively studied according to an analogy with a deep beam (Sabelli et al, 2011; Cowie et al, 2013). Consequently, as shown in Figure 4.2, it is possible to evaluate the internal forces in the diaphragm that may be split as the tension and compression acting in the longitudinal chords and the web resisting shear.

The compression chord is basically made of the concrete slab (also the compression part of the steel beams may be considered if shear connectors are used to obtain composite beams). If there are discontinuities (e.g, voids, openings, setbacks) an appropriate load path around those discontinuities should be determined, in accordance with the following principles:

1. The internal loads follow the stiffest load path;
2. If the compression load changes direction it is compulsory to consider that the relevant component of forces may introduce tension forces. In line with this consideration, it is necessary to design the slab reinforcement in order to resist those tension forces (e.g. as required to flow around an opening in the slab).

4.2 Seismic Design and Detailing of Composite Steel-concrete Slabs

These principles apply also to the tension chord, but in this case the tension forces may be absorbed by both the steel frame and/or the steel reinforcement in the slab.

Regarding the shear forces acting at the supports (namely in the alignments of the seismic resisting systems), those effects can be considered to be either uniformly distributed along the depth of the diaphragm, or concentrated near the vertical elements of the lateral load-resisting system. In case of a composite metal deck slab, the first load path depends on the presence of ductile shear connectors (e.g. Nelson studs) applied continuously along all girders of the diaphragm depth. In that case, the distributed shear forces result in a linear accumulation of longitudinal shear forces into the collectors. On the contrary, if connectors are missing on collectors, all shear forces are only concentrated in the spans of the seismic resisting frames, where shear connectors should always be present.

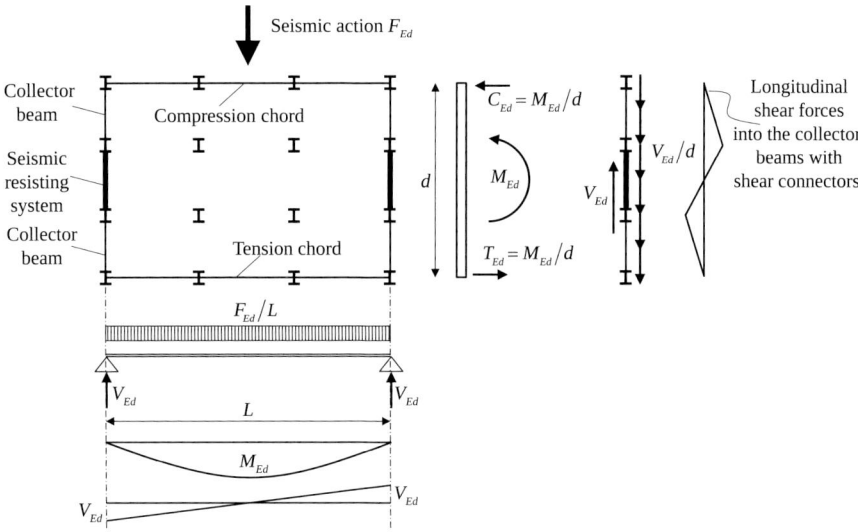

Figure 4.2 – Force distribution into the floor systems: deep beam analogy

In light of the above considerations, the longitudinal shear forces due to the transfer mechanism between the floor diaphragm and its supporting steel beams require steel reinforcement and mesh that should be added to those calculated for gravity load effects of the seismic design conditions.

However, the longitudinal shear transfer mechanism also influences the design of the steel beams and their connections. Indeed, as highlighted in

4. Design Recommendations for Ductile Details

Figure 4.3, the effect of the vertical offset between the plane of the diaphragm and the centreline of the supporting beam imposes additional bending moments on the beams (Burmeister and William, 2008). In particular, if uniform horizontal shear is induced by the seismic loads, uniformly distributed bending moments shall be applied on the steel beams, resulting in zero increment in terms of flexural engagement but with substantial increase of transverse shear forces in the beam and consistently an increase of shear forces applied to the connections. The free body diagram illustrated in Figure 4.3 clearly shows that the collector beams are subjected to a linearly variable axial load that is not considered in the gravity design approach usually adopted for the members that do not belong to the primary seismic resisting system. These considerations enforce the designer to verify those beams considering the combined effects of the axial forces due to the lateral loads and the flexural forces due only to the gravity loads. In addition, the increase of shear forces combined with axial loads must be considered in the design of connections at both ends of the member.

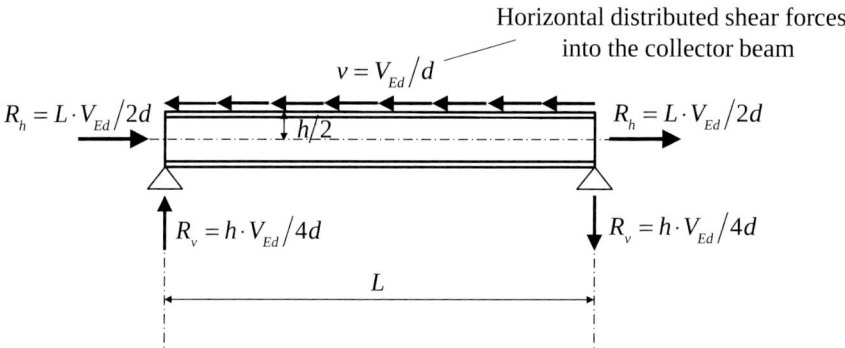

Figure 4.3 – Secondary effects induced by horizontal distributed shear forces in collector beams (Burmeister and William, 2008)

The detailing of the floor systems in order to adequately absorb the connection/distribution of forces between the vertical framing system and the floor slabs very much depends on the specific type of floor system and it is outside the scope of this design manual. Specific detailing solutions are required for conventional concrete floors or composite floors, slim or deep, and wet, dry or prefabricated floor systems. Example 4.1 illustrates the quantification of the diaphragm actions and the distribution of forces within the diaphragms. It further exemplifies the detailing for a typical composite metal deck slab.

4.2 SEISMIC DESIGN AND DETAILING OF COMPOSITE STEEL-CONCRETE SLABS

Example 4.1: Consider the floor diaphragm illustrated in Figure 4.4 that corresponds to the worked example shown in chapter 8. Calculate the diaphragm actions and the distribution of forces within the diaphragm. Detail the reinforcement for the composite metal deck slab.

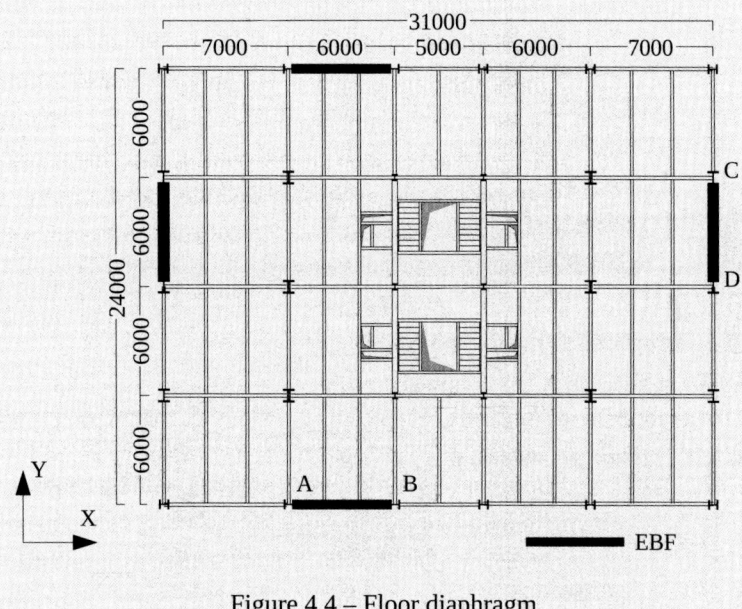

Figure 4.4 – Floor diaphragm

a) Calculation of the diaphragm actions and the distribution of forces within the diaphragm

The maximum seismic action F_{Ed} is at roof level and it is equal to 407.26 kN in the X direction and 406.85 kN in the Y direction. According to clause 4.4.2.5(2) of EC8-1 these seismic effects should be multiplied by an overstrength factor $\gamma_d = 1.3$, since non-ductile diaphragms are assumed. Hence the equivalent distributed diaphragm forces are obtained as follows:

$$\gamma_d \frac{F_{Ed,X}}{L_X} = 1.3 \cdot \frac{407.26 \text{ kN}}{31 \text{ m}} = 17.08 \text{ kN/m}$$

$$\gamma_d \frac{F_{Ed,Y}}{L_Y} = 1.3 \cdot \frac{406.85 \text{ kN}}{24 \text{ m}} = 22.04 \text{ kN/m}$$

4. DESIGN RECOMMENDATIONS FOR DUCTILE DETAILS

The corresponding forces acting in the slab are calculated as follows:

$$M_{Ed,X} = \gamma_d \cdot \frac{F_{Ed,X}}{L_X} \cdot \frac{L_X^2}{8} = 2051.56 \text{ kNm}$$

in X direction

$$V_{Ed,X} = \gamma_d \cdot \frac{F_{Ed,X}}{L_X} \cdot \frac{L_X}{2} = 264.45 \text{ kN}$$

$$M_{Ed,Y} = \gamma_d \cdot \frac{F_{Ed,Y}}{L_Y} \cdot \frac{L_Y^2}{8} = 1586.72 \text{ kNm}$$

in Y direction

$$V_{Ed,Y} = \gamma_d \cdot \frac{F_{Ed,Y}}{L_Y} \cdot \frac{L_Y}{2} = 264.45 \text{ kN}$$

b) Detailing of the reinforcement for the composite metal deck slab

It is clear that the longitudinal shear resistance of the composite metal deck slab must be checked to ensure that the diaphragm forces can be effectively transferred from the slab to the seismic resisting system and vice versa. Hence, it is necessary to guarantee the minimum transverse reinforcement area as indicated by section 6.6.6.3 of EN 1994-1-1 (CEN, 2004e). In addition, the longitudinal shear forces acting on collector and chord girders is calculated, it is necessary to verify the shear resistance of the composite slab along any potential shear surface as recommended by section 6.6.6.4 of EN 1994-1-1. Figure 4.5 shows the potential surfaces of shear failure, and A_{sf} is the cross-sectional area of transverse reinforcement, s_f is the longitudinal spacing centre-to-centre of the stud shear connectors, A_t is the cross-sectional area of the top transverse reinforcement, and A_b is the cross-sectional area of the bottom transverse reinforcement.

Longitudinal and transverse steel reinforcements should also be detailed in order to avoid any potential slipping of the rebar. Hence, on the spandrel beams and along the borders of the floor plan it is advisable to add transverse hooked bars together with a lapped longitudinal edge trimmer rebar to the mesh (longitudinal and transverse) reinforcements, as depicted in Figure 4.6. In addition, the lower leg of the bonding hook should be extended beyond the shear stud, crossing the half of the beam flange (i.e. the vertical symmetry axis of the cross section) by at least 50 mm. This detail allows avoiding shear splitting at the base of the stud that impairs the crushing of a shear stud into the concrete slab.

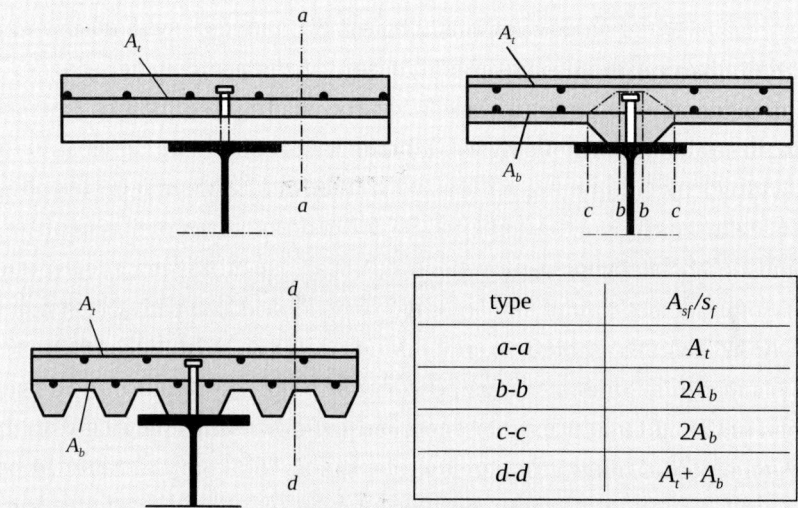

Figure 4.5 – Typical potential surfaces of shear failure for composite slab according to section 6.6.6.4 of EN 1994-1-1

Another important detail is the seating of the steel sheeting, which should be at least 50 mm on the top of the beam flange, if the deck is oriented both orthogonally and longitudinally to the spandrel beams (see Figure 4.6).

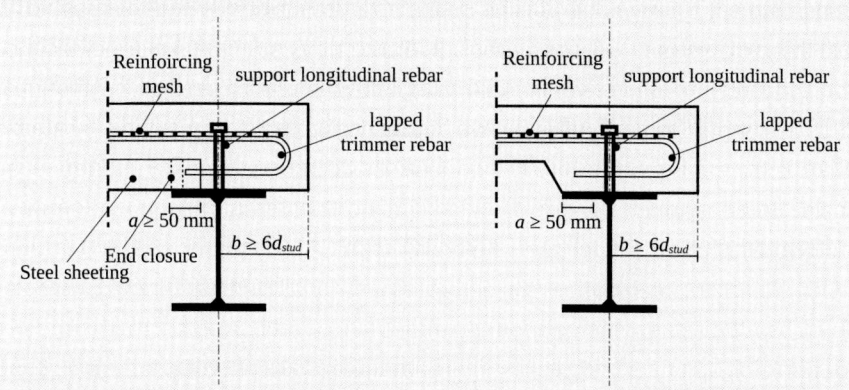

Figure 4.6 – Details of reinforcements and seating of the steel sheeting oriented both orthogonally and longitudinally to the spandrel beams (Clifton and El Sarraf, 2005)

4. Design Recommendations for Ductile Details

4.3 DUCTILE DETAILS FOR MOMENT RESISTING FRAMES

4.3.1 Detailing of beams

As discussed in section 3.5.1, beams of MRFs are the favoured elements where plastic hinges should form under seismic actions.

EC8-1 imposes that the design of the beams avoids any possible bending-shear interaction that could impair the flexural ductility (i.e. the shear demand calculated according to equation (3.9) should be less than half the shear strength of the beam). In addition, it is necessary to verify the stability of the beams when plastic hinges form and rotate in order to guarantee adequate rotation capacity. With this regard, clause 6.6.2(1) states that: *"Beams should be verified as having sufficient resistance against lateral and lateral torsional buckling in accordance with EN 1993, assuming the formation of a plastic hinge at one end of the beam. The beam end that should be considered is the most stressed end in the seismic design situation"*. Considering the span lengths of common MRF structures, which range from 5 m to 10 m, for "all-steel" dissipative beams (namely without composite action) it is necessary to introduce stability bracings within a distance of $h/2$ along the length of the member, where h is the overall depth at the plastic hinge location (clause 6.3.5.2(4)B of EN 1993-1-1) to satisfy this requirement. An example of this detail is given in Figure 4.7, whereby it is noted that the segment containing plastic hinges, of length equal or smaller than a stable length, L_{stable}, must be effectively braced at both ends. The stability of members containing plastic hinges is specified in clauses 6.3.5.1 to 6.3.5.3 of EN 1993-1-1 (2005). Sub-chapter 5.3 of the ECCS Eurocode Design Manual dedicated to Part 1-1 of EC3 (Simões da Silva *et al*, 2016) details and exemplifies the verification of the stable length, L_{stable}, and the calculation of the local force Q_m applied at the plastic hinge location that must be resisted by the bracing system, given by:

$$Q_m = 1.5 \cdot \frac{N_{f,Ed}}{100} \qquad (4.2)$$

where $N_{f,Ed}$ is the axial force in the compressed flange of the stabilized member at the plastic hinge location. For seismic conditions it can be assumed as:

4.3 DUCTILE DETAILS FOR MOMENT RESISTING FRAMES

$$N_{f,Ed} = 1.1 \cdot \gamma_{ov} \cdot \frac{M_{pl,Rd}}{d_b} \quad (4.3)$$

where γ_{ov} is the material overstrength factor (see section 3.2.1), $M_{pl,Rd}$ is the beam plastic bending resistance and d_b is the beam depth.

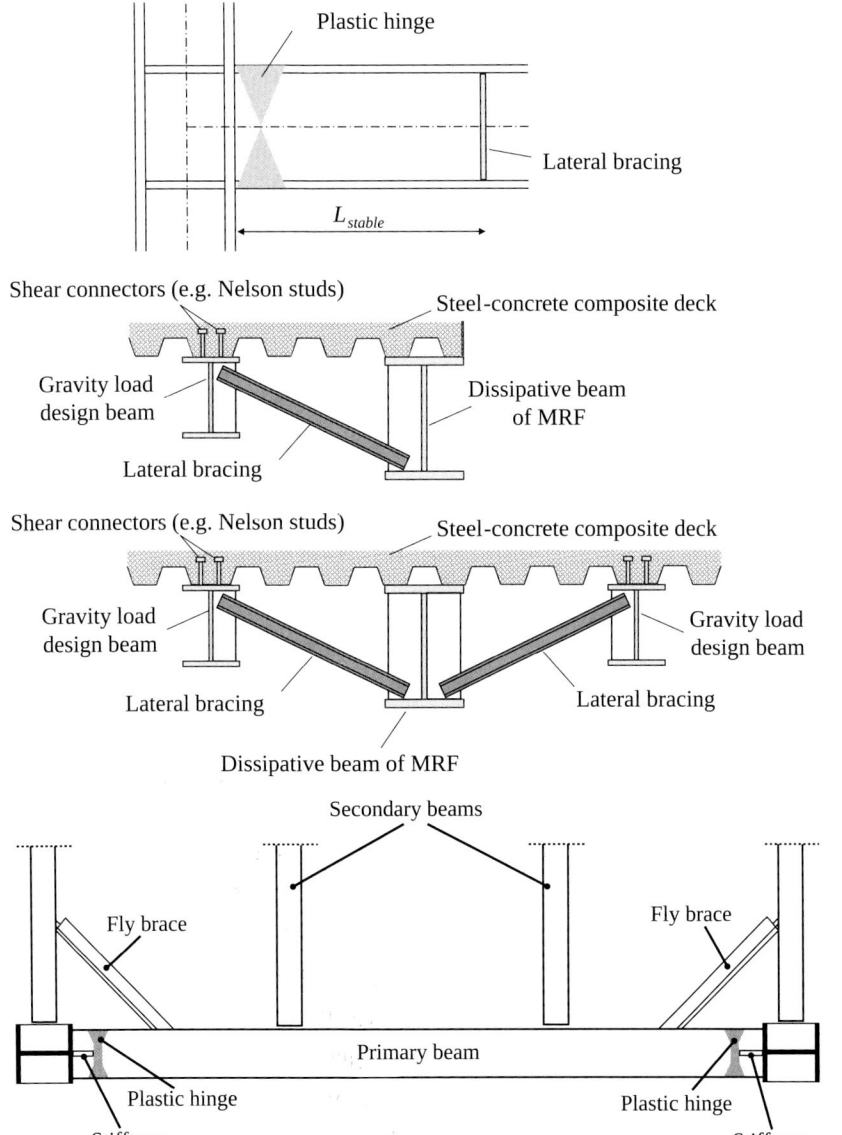

Figure 4.7 – Beam of MRF with fly bracings and composite slab

4. Design Recommendations for Ductile Details

In many cases, the slabs are designed to act compositely with the steel beams, because this option has several benefits, e.g. increasing the stiffness against gravity and seismic actions, reducing the depth of beams, etc. However, composite behaviour may induce poor seismic performance if not properly accounted for. A composite beam is characterized by larger bending strength than the corresponding bare steel profile, even accounting for the asymmetric behaviour that depends on the direction of the bending moment. It may thus fail the hierarchy of resistances (strong column-weak beam) and impair the performance of the "all-steel" beam-to-column joints. In order to avoid composite action and to inhibit the load transfer from the slab to the column, the shear connectors may be applied only on the beams of the gravity load designed bays (see Figure 4.7). Alternatively, it is possible to have both the benefit of composite action and satisfy the strong column-weak beam criterion by applying the shear connectors everywhere along the beam length except at the end parts where plastic hinges should form (see Figure 4.8). Clause 7.7.5(1)P of EC8-1 states that to avoid composite action it is sufficient to disconnect the slab (i.e. no shear connectors) within a circular zone around the MRF columns of diameter $2b_{eff}$, where b_{eff} is the larger of the effective widths of the beams connected to that column. This prescription results from the work done in the European ICONS research project (Plumier and Doneux, 2001).

It should be observed that, even though the beam and the slab are disconnected, this arrangement cannot guarantee (at very large rotation of plastic hinges) that no interaction occurs between the slab and the columns. Therefore, in order to have totally disconnected beam-to-column joints clause 7.7.5(2) of EC8-1 recommends to avoid any contact between the slab and any vertical side of any steel element (e.g. columns, shear connectors, connecting plates, corrugated flange, steel deck nailed to flange of steel section). This requirement can be satisfied by keeping a gap from both sides of both column flanges (see Figure 4.8a), or from other protruding elements associated with the beam-to-column joints (see Figure 4.8b), and the structural slab. This gap can be effectively filled with a compressible material (e.g. polystyrene), and it should be obviously applied around any protruding part of the joints prior to concrete placement. Experimental tests carried out by Sumner and Murray (2002) showed that this arrangement does not modify the "all-steel" beam-to-column hierarchy criterion. This slab detailing is also prescribed by recent US provisions (AISC 358, 2016).

It is noted that the use of composite beam-to-column joints with continuity rebars around the column adds additional complexity to this issue. The use of composite beam-to-column joints is outside the scope of this design manual and will be addressed in a forthcoming design manual dedicated to the seismic design of steel-to-concrete composite structures.

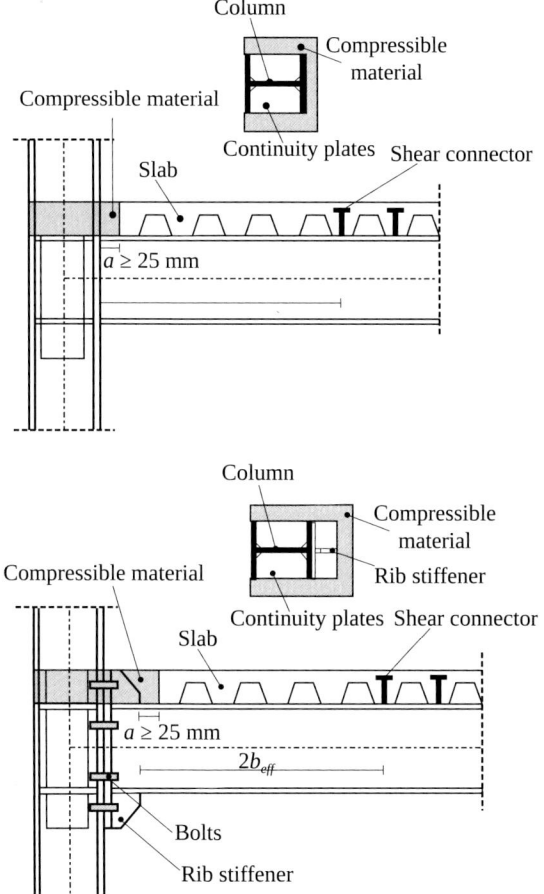

Figure 4.8 – Details of MRF beams to avoid composite action

If the details in Figure 4.8 are adopted, the segment containing the plastic hinge in hogging bending moment is considered restrained against lateral-torsional buckling by the connected slab, provided that the length $2b_{eff}$ is smaller that L_{stable} (see equations (4.2) and (4.3)) and the compression flange of the beam at the end of the length $2b_{eff}$ is in contact with the floor slab.

4. DESIGN RECOMMENDATIONS FOR DUCTILE DETAILS

Testing carried out within the SAC program (FEMA 350, 2000) indicated that composite floor slabs provide adequate restrain up to a storey drift angle of 0.04 rad. Therefore, the FEMA 350 recommendations do not require the placement of lateral bracings at plastic hinge locations adjacent to column connections for beams with composite slab, while bracings are recommended for conditions where drifts larger than 0.04 rad are expected.

In any case, considering that the existing tests were not carried out on European profiles and due to the alternate path of seismic induced bending moment it is advisable to add lateral bracings as shown in Figure 4.7 within the length L_{stable}.

Another aspect to be accounted for is the proper detailing of beam-to-beam splices if welded column-tree joints are adopted. Indeed, the portion of beam stub where the plastic hinge is presumed to form is welded to the column element. Therefore, the length of the welded beam segment should be adequate to guarantee the formation of the plastic hinge out of the splice as shown in Figure 4.9.

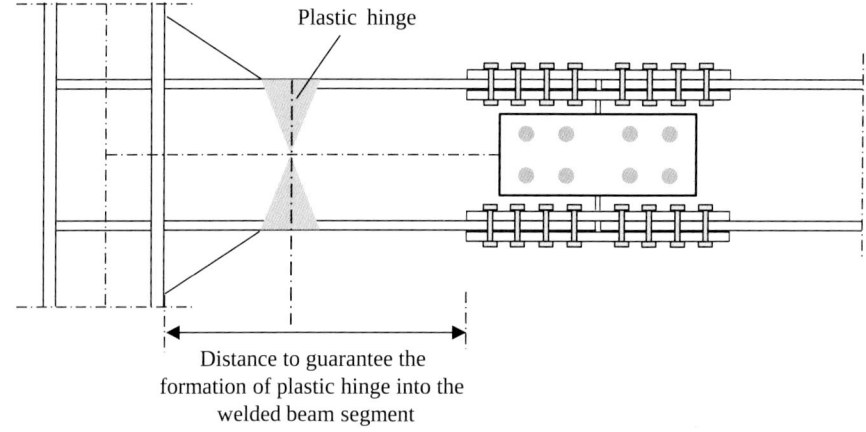

Figure 4.9 – Details of beam-to-beam splices in plastic hinge zone

4.3.2 Detailing of beam-to-column joints

4.3.2.1 Introduction

A large number of different types of joints have been extensively studied in the literature and most of them are commonly used in real structures. Beam-to-column joints of steel MRFs are generally conceived as rigid full-strength, usually of fully-welded or bolted configurations (see Figure 4.10) and design provisions aim at ensuring that the joints have adequate overstrength such that plastic deformations occur mainly in the beams. This requirement is fulfilled for both welded and bolted joints by strengthening the connection (e.g. through haunches, rib stiffeners, cover plates, etc.) and the column web panel (e.g. through supplementary welded web plates).

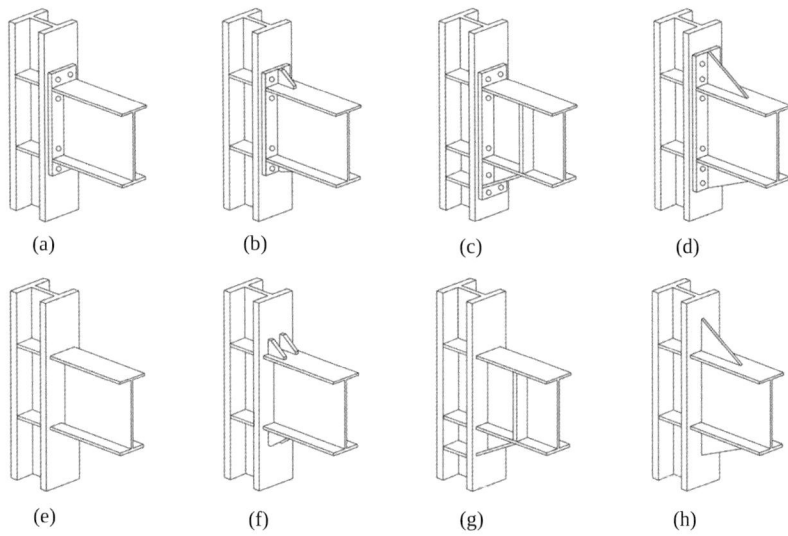

Figure 4.10 – The types of welded and bolted joints most commonly used in European practice

However, it is also possible to adopt partial strength or full strength joints but without sufficient overstrength that behave as dissipative joints. If properly detailed, the typologies of Figure 4.10a,b,e may potentially satisfy the requirements of clause 6.6.4.

As an alternative that enables the use of simpler joints that satisfy the overstrength criterion, it is possible to use the dog-bone or reduced beam section. These three alternatives are detailed and discussed in the following sub-sections.

For all cases, the resistance of the joint can be evaluated by means of the component method, as implemented in EN 1993-1-8. This methodology can be applied to both welded and bolted joints and it is thouroughly described in the ECCS Eurocode design manual devoted to the design of joints (Jaspart and Weynand, 2016).

4. Design Recommendations for Ductile Details

4.3.2.2 Full strength non-dissipative joints

As discussed in section 3.5.3, full strength beam-to-column joints are often designed to enforce the formation of all plastic deformations in the beam. Hence, in accordance with section 6.5.5 of EC8-1 (see equation (3.3) in section 3.4.2), the capacity design criterion to obtain full strength joints is expressed by the following inequality:

$$M_{j,Rd} \geq M_{j,Ed} = 1.1 \cdot \gamma_{ov} \cdot \left(M_{pl,Rd} + V_{Ed} \cdot s_h\right) \tag{4.4}$$

where $M_{j,Rd}$ is the joint bending resistance, $M_{j,Ed}$ is the design bending moment at the column face, $M_{pl,Rd}$ is the plastic flexural strength of the connected beam, V_{Ed} is the shear force corresponding to the formation of a plastic hinge in the connected beam, calculated according to equation (3.9) (see section 3.5.1) and s_h is the distance between the column face and the probable position of the plastic hinge (see Figure 4.11).

It should be noted that the EC8 design overstrength factor $1.1\gamma_{ov}$ may be non-conservative. Indeed, the amount of hardening developed by a plastic hinge may be larger than 1.1. This is the reason why AISC 358-16 recommends using the hardening overstrength factor γ_{sh} instead of a fixed value of 1.1, given as follows:

$$\gamma_{sh} = \frac{f_y + f_u}{2 \cdot f_y} \leq 1.20 \tag{4.5}$$

Based on the characteristic yield and ultimate stress of European mild carbon steel grades, the first term of the inequality (4.5) varies between 1.2 (for S355) and 1.3 (for S235). Hence, considering that S355 is the most used grade for seismic applications, $\gamma_{sh} = 1.2$ may be reasonably assumed in Europe.

The required joint resistance given by equation (4.4) strictly depends on the details of the stiffeners and the assumed position of the plastic hinge. According to AISC 358-16, the beam plastic hinge is simply assumed lumped at the section at the end of any stiffeners protruding from the connection zone (see Figure 4.11a). Hence, under this assumption, the lever length s_h is known once the geometry of the stiffeners is defined.

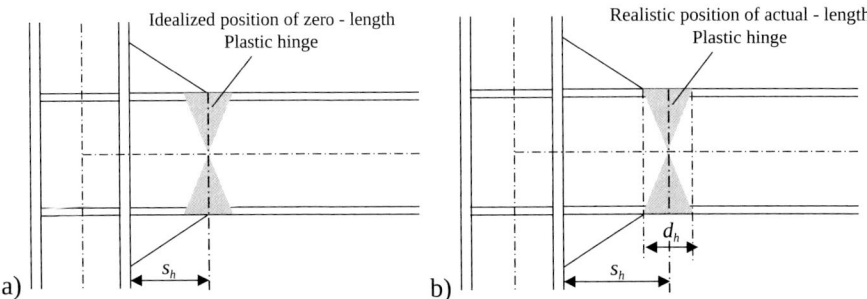

Figure 4.11 – Position of plastic hinge and corresponding design lever arm s_h

In general, the stiffeners adopted to restrain the connection zone may have a slope ranging between 30°- 45°. The lower slope limit is the most favourable detail because of two main reasons: 1) concentrated stresses applied to an unsupported edge of a plate tend to propagate from that point towards the supported edge through an angle of about 30° (i.e. Whitmore distribution model); 2) the smaller is the slope of the stiffener the smaller is the stress concentration in the beam and the force transfer mechanism from the beam to the joint does not differ from classical beam theory (see Figure 4.12a). On the contrary, increasing the slope of the stiffeners, the force transfer mechanism tends to a strut and tie truss behaviour (see Figure 4.12b), leading to stress concentrations in the compressed zone that may impair the ductility of the beam plastic hinge.

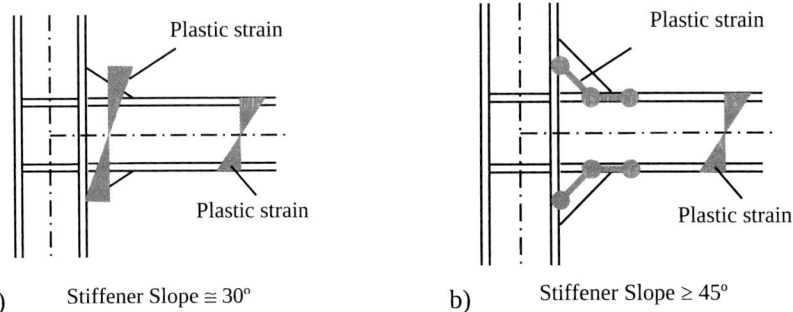

Figure 4.12 – Influence of the stiffener slope on the force transfer mechanism from beam to joint

In real beam-to-column double haunched joints, the plastic hinge forms out of the stiffened zone and has finite dimensions (see Figure 4.11b). Its length d_h may be approximatively calculated as follows (Gioncu and Pectu, 1997a,b):

4. Design Recommendations for Ductile Details

$$d_h = 2\beta c \quad \text{where} \quad \beta = 0.6 \cdot \left(\frac{t_f}{t_w}\right)^{3/4} \cdot \left(\frac{d}{c}\right)^{1/4} \tag{4.6}$$

where t_f is the flange thickness, t_w is the web thickness and d is the beam depth. The parameter c was derived experimentally, and was recently recalibrated for European profiles by D'Aniello et al (2012) as follows:

$$c = 0.5 \cdot \left(b_f - t_w\right) \quad \text{for IPE and HE profiles} \tag{4.7a}$$

$$c = b_f - 2 \cdot \left(t_w + r_i\right) \quad \text{for RHS and SHS profiles} \tag{4.7b}$$

where b_f is the flange width and r_i the inner corner radius of the profiles.

It is clear that considering the real position of the centre axis of the plastic hinge leads to a larger lever arm s_h and, consequently, larger design forces. However, this assumption is more consistent with the actual behaviour.

4.3.2.3 Dog-bone or reduced beam section

A very efficient and quite easy solution to avoid the development of plastic deformations in non-dissipative connections is to reduce the beam section in the portion of the beam where it is presumable to form the plastic hinge. This type of joint is known as dog-bone or reduced beam section (RBS) joint (see Figure 4.13). According to Annex B.5.3.4 of EN 1998-3 (CEN, 2005c), the details for cutting the beam flange of RBS joints are the following:

i) The distance "a" between the column face and the beginning of the RBS is fixed as:

$$a = 0.60 \cdot b_f \tag{4.8}$$

ii) The length "b" over which the flange will be reduced is given as:

$$b = 0.75 \cdot d_b \tag{4.9}$$

where b_f is the flange width and d_b is the beam depth.

4.3 Ductile Details for Moment Resisting Frames

iii) The depth of the flange cut (g) on each side should be not greater than $0.25 \cdot b_f$. The recommended value is:

$$g = 0.20 \cdot b_f \quad (4.10)$$

Once the geometry of the dog bone is obtained, it is possible to compute the plastic modulus ($W_{pl,RBS}$) and the plastic moment ($M_{pl,Rd,RBS}$) of the plastic hinge section at the centre of the RBS as:

$$W_{pl,RBS} = W_{pl,b} - 2 \cdot g \cdot t_f \cdot (d - t_f) \quad (4.11)$$

$$M_{pl,Rd,RBS} = W_{pl,RBS} \cdot f_{yb} \quad (4.12)$$

where $W_{pl,b}$ is the plastic modulus of the beam and f_{yb} is the yield strength of the steel in the beam.

In addition, the distance s_h between the column face and the intended plastic hinge section at the centre of the RBS is easily obtained as follows:

$$s_h = a + \frac{b}{2} \quad (4.13)$$

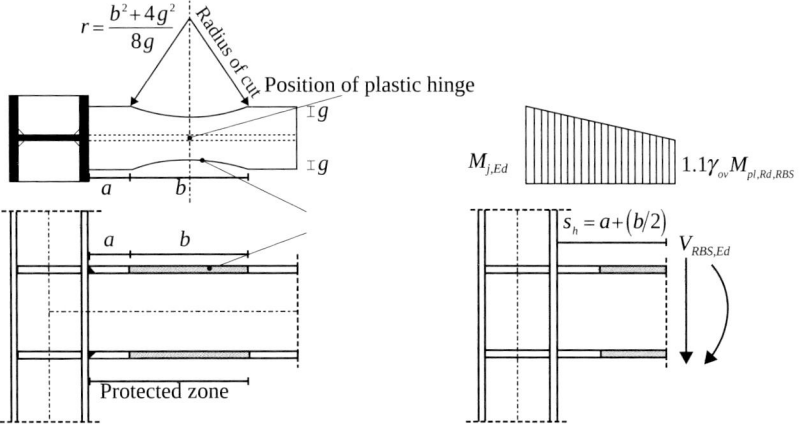

Figure 4.13 – Geometry of dog-bone or reduced beam section (RBS)

Finally, once the design bending action is obtained, the resistance of the joint can be evaluated by means of the component method.

4. Design Recommendations for Ductile Details

It is important to highlight that large out-of-plane deformations can be developed when a plastic hinge forms in a RBS segment. Therefore, it is necessary to design lateral bracing to restrain the beam section close to the dogbone zone against lateral torsional buckling as shown in Figure 4.7. In this case, L_{stable} can be assumed equal to the protected zone (i.e. $a+b$, see Figure 4.13).

It is also important to give a proper surface aspect to the beam reduced section by means of a complete grinding of that surface. Indeed, the section reduction is generally made by flame cutting which creates indentations from which cracks could start while the plastic hinge is activated.

4.3.2.4 Dissipative joints

The criteria for dissipative beam-to-column joints in MRFs were described in section 3.5.3. Their adoption implies that the connections have a rotation capacity consistent with the global deformations and the effect of connection deformation on global drift is assessed by non-linear analysis (clause 6.6.4(2)). Secondly (clause 6.6.4(3)), the connection design should be such that a minimum plastic rotation capacity of 35 mrad for DCH or 25 mrad for DCM with $q > 2$ is available, respectively, with a maximum degradation of stiffness and strength of 20 % under cyclic loading. Also (clause 6.6.4(4)), the contribution of the column web panel shear deformation to the plastic rotation capacity θ_p should not exceed 30 % and the column web panel shear resistance should satisfy (see discussion in section 3.5.3) equation (4.14):

$$\frac{V_{wp,Ed}}{V_{wp,Rd}} \leq 1.0 \qquad (4.14)$$

where $V_{wp,Ed}$ is the design shear force in the web panel due to the plastic resistance of the adjacent dissipative zones (namely beams or connections) and $V_{wp,Rd}$ is the shear resistance of the web panel, given by equation (3.18):

$$V_{wp,Rd} = V_{wc,Rd} + V_{wp,add,Rd} \qquad (4.15)$$

with:

$$V_{wc,Rd} = \frac{0.9 f_{y,wc} \cdot A_{vc}}{\sqrt{3} \cdot \gamma_{M0}} \text{ and } V_{wp,add,Rd} = \min\left(\frac{4M_{pl,fc,Rd}}{d_s}; \frac{2M_{pl,fc,Rd} + 2M_{pl,st,Rd}}{d_s}\right),$$

d_s is the distance between the centrelines of the stiffeners, $M_{pl,fc,Rd}$ is the design plastic moment resistance of a column flange and $M_{pl,st,Rd}$ is the design plastic moment resistance of a stiffener.

Demonstrating by calculation that a dissipative joint satisfies all these requirements is not easy, despite the recent advances in the extension of the component method to cyclic loading conditions (Simões da Silva *et al*, 2016). Hence, a pragmatic solution that is in line with the recommendations of clauses 6.5.5(6) and 6.5.5(7) is the pre-qualification of joint configurations for use as dissipative joints. The European research project EQUALJOINTS (Landolfo, 2016) undertook the task of prequalifying the following beam-to-column joint configurations, summarized in Figure 4.14.

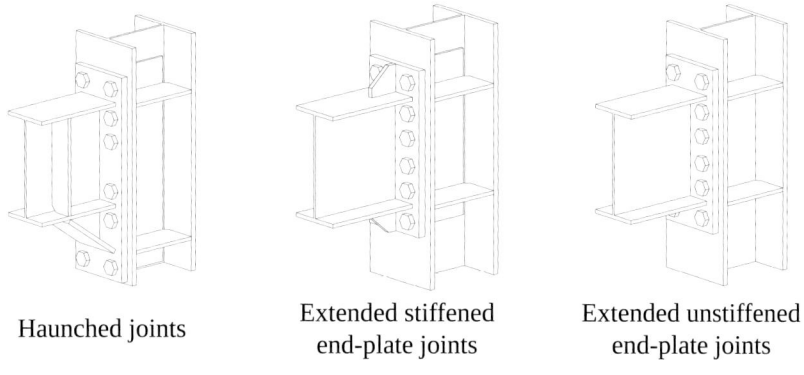

Haunched joints   Extended stiffened end-plate joints   Extended unstiffened end-plate joints

Figure 4.14 – Prequalified beam-to-column joints (Landolfo, 2016)

The joints in Figure 4.14 have been classified on the basis of both the connection and the column web panel strength. For what concern the connection, three performance levels are considered:

- Full strength connection: plastic deformations should occur only in the beam, while the connection should behave elastically; the connection is designed to provide a moment capacity at least 1.5 times larger than that of the connected beam;
- Equal strength connection: plastic deformations may contemporary occur both in the connection zone and in the connected beam, namely the connection strength is equal to the beam strength;
- Partial strength connection: plastic deformations occur in the connection zone before the plastic resistance in the connected beam

is reached; in 0.8 partial strength, the connection should be able to develop a moment capacity equal to the 80 % of the plastic strength of the connected beam (projected on the column face).

For what concerns the column web panel, the relative capacity of the column web panel in shear respect to the capacity of the connection zone and/or connected beam allows to identify three different behaviours as described in the following:

- Strong web panel joints: the column web panel behave elastically, while the first nonlinear event is reached in the connection zone and/or in the connected beam (depends on the connection strength criteria);
- Balanced web panel joints: the first nonlinear event is reached concurrently in the column web panel, and in the weakest component(s);
- Weak web panel joints: the column web panel experiences nonlinear events before plastic hinge(s) occur in the connection zone and/or in the connected beam.

These criteria have been validated on the basis of experimental tests, analytical and finite element simulations for both external and internal joints with beam ranging from IPE 360 to IPE 600 and column varying from HEB 280 and HEB 500. Detailed prequalification guidance is expected in 2018, in line with the ongoing revision of the structural Eurocodes.

4.3.2.5 Detailing of supplementary web plates

By applying the component method, designers will recognize that the thickness of the column web panel plays an important role in providing both strength and stiffness because it affects several components, as the web panel in shear, in compression and in tension and it may also influence the resistance of the T-stub on the column flange side. As a consequence, additional web plates could be necessary in several cases to satisfy the strength requirement. Supplementary web plates can be placed either onto the column web panel or spaced away from the column web. Both arrangements can be used with or without continuity plates. In light of experimental results by Ciutina and Dubina (2008) it is suggested to disregard the limitation imposed by clause 6.2.6.1(6) of EN 1993-1-8 for the maximum thickness of supplementary web plates.

Accordingly, the resisting shear area A_v can be reasonably assumed as the sum of column shear area $A_{v,c}$ and the gross area of the additional web plates $A_{v,p}$. On the other hand, also the limitation on minimum thickness (i.e. the width of the unrestrained part of the supplementary web plate should be less than $40 \cdot \varepsilon \cdot t_s$, where $\varepsilon = \sqrt{235/f_y}$ and t_s is the thickness of the supplementary web plate) given by clause 6.2.6.1(13) of EN 1993-1-8 may be neglected for the case of supplementary plates placed against the column web only if additional plug welds are used to stabilize their out-of-plane deformations. In this case, in order to avoid constructional problems, it is convenient to avoid using additional plates having thickness larger than the root radius of the column cross section.

Another aspect needing some remarks is the depth of additional web plates. Clause 6.2.6.1(10) of EN 1993-1-8 recommends that the depth of the supplementary web plates should be such that these stiffeners extend throughout the effective width of the column web in tension and compression. This requirement is valid for both welded and bolted connections. Anyway, if bolted extended end-plate joints (either stiffened by plates or haunch) are used, it is suggested to assume the depth of the end-plate as the minimum depth for these supplementary plates. This assumption should guarantee that the shear resisting area is sufficiently far from the heat affected zone of the horizontal fillet welds connecting the additional plate to the column web panel.

4.3.2.6 Detailing of welds

Another crucial feature is the proper detailing of the welds. Indeed, all design considerations previously discussed require that the failure of welds should be avoided in any case. As part of the calculation of resistance that should comply with EN 1993-1-8, the joint details should be conceived by adopting the most appropriate type of weld depending on the component to be connected and its relevant plastic engagement. Unfortunately, EN 1993-1-8 does not provide specific details for seismic resistant joints. Hence, the designer is theoretically free to select the type of weld base material and details that are nominally able to withstand the design forces, but this approach does not guarantee the fulfilment of the design performance. On the contrary, US practice based on the qualification procedure given by AISC 358-16 avoids any subjective choice by imposing specific details to guarantee the design objectives. Hence, it is reasonable to extend the main types of weld details given by AISC 358-16 to European joints.

4. Design Recommendations for Ductile Details

Consequently, four types of weld details can be rationally adopted for the types of bolted joint configurations reported in Figure 4.14, namely: fillet welds (FW), plug welds (PW), groove welds (GW) and full penetration welds (FPW).

FPWs may be adopted for end-plate stiffeners and haunch, because of the large stress concentration and strain demand induced by the strut transfer force mechanism. FPWs are also suggested for beam flange to end-plate splices with reinforcing fillet welds (see Figure 4.15a,b). This choice is crucial to ensure the appropriate T-Stub mechanism in the connection zone where the larger demand is expected. On the other hand, both beam web to end-plate welds (i.e. for bolted connections) and beam web to column face (for welded connections) may be specified as FWs (see Figure 4.15c), since less concentration of stress and relevant strain demand is expected. FPWs should also be used for plates stiffening beam flange, end plate and column flange (e.g. rib stiffeners, haunch, etc.), as shown in Figure 4.15d.

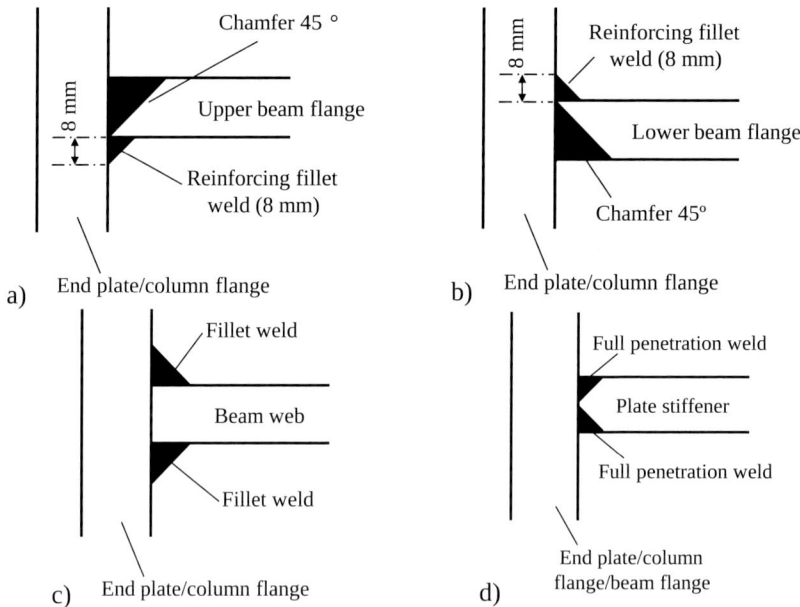

Figure 4.15 – Details of welds on the beam side

For full welded beam-to-column joints, weld access hole are generally used to facilitate welding operations and to minimize the induction of thermal stresses by improving cooling conditions. However, several studies and research activities (Chi et al, 1997; Lee et al, 2000; Ricles et al, 2000; Roeder, 2002) have shown that the shape of the weld access hole can have a significant effect on the behaviour of moment connections. Therefore, AISC 341 and AISC 358 recommend to adopt weld access hole configuration pre-qualified for specific joint typology. This implies that the use of different weld access holes other than those prescribed by AISC specifications should be avoided. AISC 341 also highlights that there are some joint typologies, as end-plate moment connections, for which weld access holes are undesirable.

An example of allowed weld access hole prescribed by AISC 358-16 (2016) for welded unreinforced flange-welded web (WUF-W) joints is depicted in Figure 4.16.

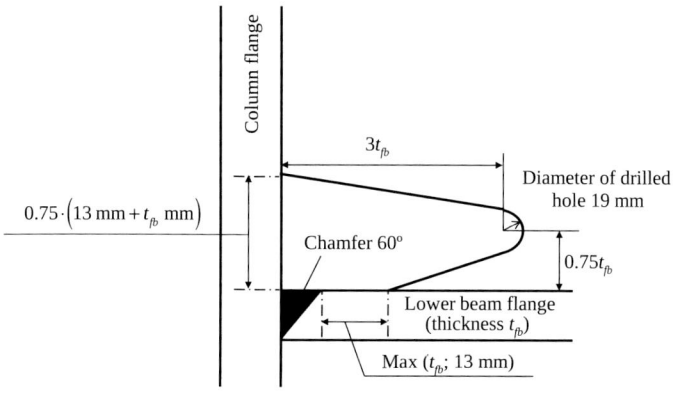

Mandatory details:
1. Hole to be drilled
2. All flame cut surfaces should be grinded

Figure 4.16 – Details of weld access hole in accordance to AISC 358-16

The details of the welds of continuity plates (CP) to the column flange depend on the design criterion adopted for the resistance of the column web panel. If the column web panel is designed neglecting the contribution of the transverse stiffeners, $V_{wp,add,Rd}$, (see equation (3.16) in section 3.5.3) no plastic engagement is expected for CP and FWs can be used (see Figure 4.17a). On the contrary, if the contribution $V_{wp,add,Rd}$ is accounted for, FPWs should be used (see Figure 4.17b) in order to avoid brittle failure due to their large plastic engagement. CPs should be also longitudinally welded to the column web. In this case FWs can be used (see Figure 4.17c).

In the case of supplementary web plates placed against the column web, GWs should be used for connecting the vertical edges of additional plates to the column (see Figure 4.17d) and also PWs to prevent the buckling or separation of the

4. Design Recommendations for Ductile Details

lapped parts (see Figure 4.17e). In this case, a minimum of four PWs is advisable because a single PW would create a boundary condition that substantially differs from the continuously restrained edge. Hence, as recommended by AISC 341-16, PWs should be placed in pairs or lines, dividing the plate into approximately four equally sized rectangles. If the supplementary web plates are spaced away from the column web zone, GWs should be adopted to connecting the vertical edges of the additional web plates to the column flange (see Figure 4.17f).

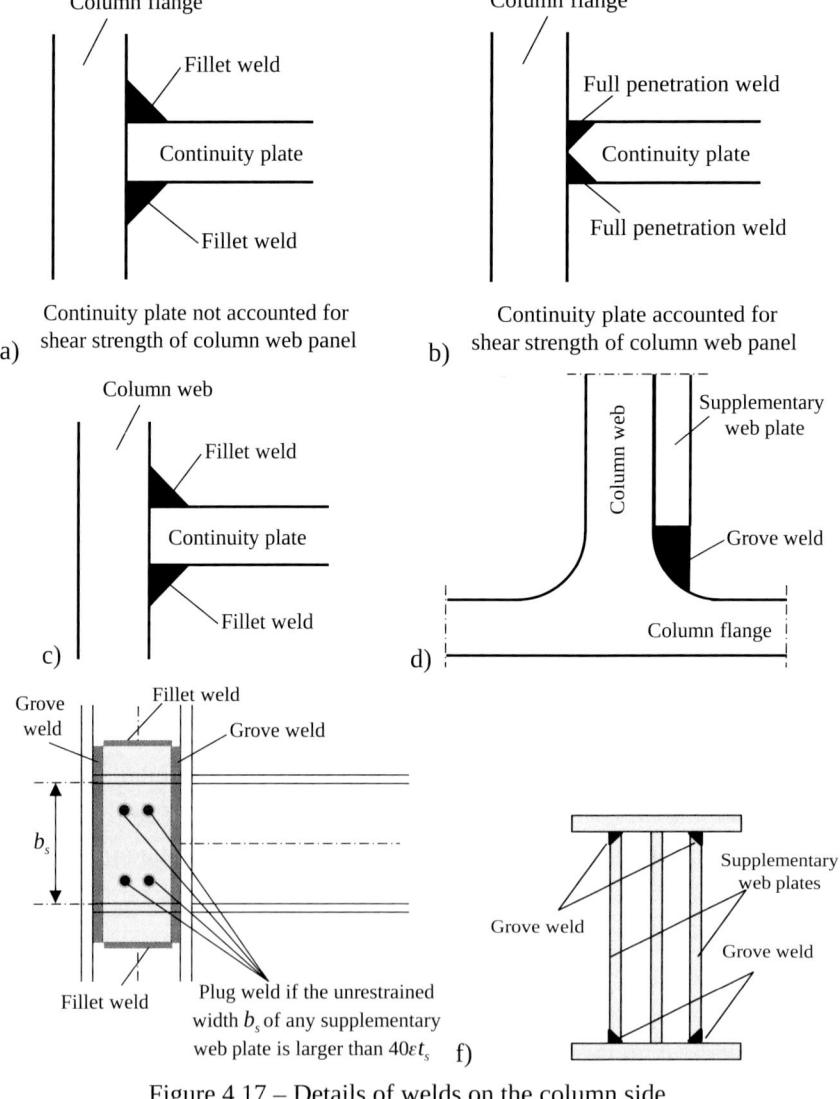

Figure 4.17 – Details of welds on the column side

4.3.2.7 Examples

Example 4.2: Non-dissipative beam-to-column joint (Jaspart *et al*, 2017).

Consider the haunched joint depicted in Figure 4.18. The beam consists of a IPE 450 and the column is a HEB 340, both in steel grade S355, and a haunch at 45° with 6 rows of M30 10.9 bolts. Calculate the resistance of the joint and check that it can be classified as non-dissipative.

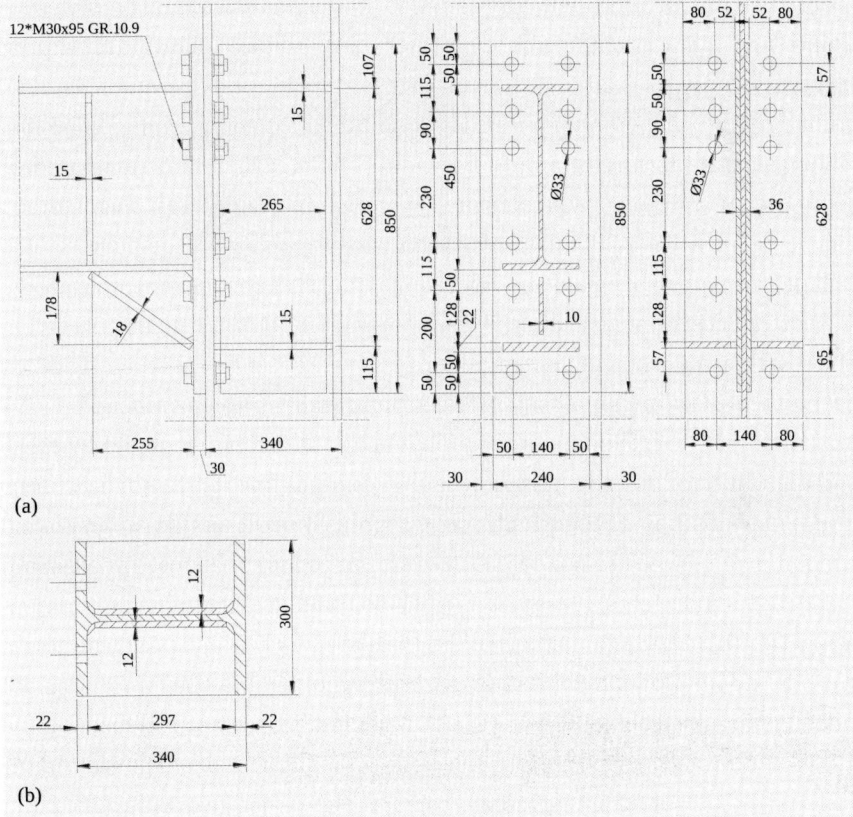

Figure 4.18 – Haunched beam-to-column joint

a) Design criteria

As it intended to check that the joint is non-dissipative, it is required to achieve a full strength joint. This implies that the plastic deformations should only occur in the beam and the plastic hinge forms at the section at the end of the

4. DESIGN RECOMMENDATIONS FOR DUCTILE DETAILS

haunch. The joint consists of a connection, column web panel and the connected member (beam). The connection is designed for the bending moment and shear force at the column face corresponding to the formation of the plastic hinges in the beam (near to the haunch), accounting for material overstrength and strain hardening (Jaspart et al, 2017). For hogging moment, the centre of compression is located at a distance Δ_c above the haunch flange. It is assumed that the centre of compression is shifted up by 40 % of the haunch depth ($\Delta_c = 0.4\ h_b$, see Figure 4.18). For sagging moment, the usual assumption of centre of compression located at the middle of the compression flange is adopted (Figure 4.18). On the other hand, bolt rows located close to the compression centre develop negligible tension forces, due to flexibility of the end plate and limited ductility of the bolt rows at the tension flange. Consequently, only the bolt rows located beyond the mid-depth of the beam are considered active for the purpose of computing the flexural capacity of the connection.

b) Design bending moment at the column face and corresponding shear force

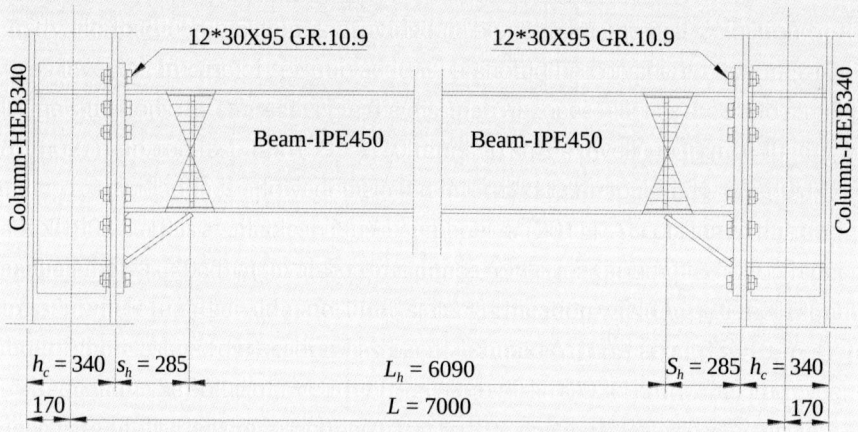

Figure 4.19 – Haunched beam-to-column joint: position of plastic hinge

The design bending moment at the column face, corresponding to yielded and fully strain hardened plastic hinge at the end of the haunch is (Figure 4.19):

$$M_{j,Ed} = M^*_{pl,Rd} + V^*_{Ed} \cdot s_h = 830.79 + 272.84 \cdot 0.285 = 908.55\ \text{kNm}$$

The design shear force in the connection $V_{j,Ed}$ is determined based on the assumption that fully yielded and strain hardened plastic hinges form at the both ends of the beam:

$$V_{j,Ed} = V_{Ed}^* = 272.84 \text{ kN}$$

Also,

$$M_{pl,Rd}^* = \gamma_{sh} \cdot \gamma_{ov} \cdot W_{pl,beam} \cdot f_{y,beam} = 1.20 \cdot 1.25 \cdot 1702000 \cdot 355 =$$
$$= 830.79 \text{ kNm}$$

where:

$$\gamma_{sh} = \frac{f_y + f_u}{2 \cdot f_y} \leq 1.20;$$

$$\gamma_{ov} = 1.25;$$

$$W_{pl,beam} = 1702000 \text{ mm}^3;$$

$$f_{y,beam} = 355 \text{ N/mm}^2;$$

$$V_{Ed}^* = V_{Ed,M} + V_{Ed,G} = 272.84 + 0 = 272.84 \text{ kN};$$

$$V_{Ed,M} = \frac{2 \cdot M_{pl,Rd}^*}{L_h} = \frac{2 \cdot 830.79}{6.09} = 272.84 \text{ kN};$$

$V_{Ed,G} = 0$ (shear force in the column neglected);
$s_h = 285 \text{ mm};$

$$L_h = L - (h_c + 2 \cdot s_h) = 7000 - (340 + 2 \cdot 285) = 6090 \text{ mm};$$

$h_c = 340 \text{ mm};$
$L = 7000 \text{ mm}.$

c) Check of beam end including the haunch

The beam end including the haunch is checked according to EN 1993-1-1 for the expected design bending moment at the column face:

$$\frac{M_{j,Ed}}{M_{bh,Rd}} = \frac{908.55}{947.24} = 0.96 < 1.0$$

4. DESIGN RECOMMENDATIONS FOR DUCTILE DETAILS

where:

$$M_{bh,Rd} = W_{pl,bh} \cdot f_{y,beam} = 2668272 \cdot 355 = 947.24 \text{ kNm};$$

with $W_{pl,bh}$ calculated considering an equivalent double T-section at the face of the end plate connection, neglecting the intermediate beam flange (Figure 4.20).

$$W_{pl,bh} = 2668272 \text{ mm}^3$$

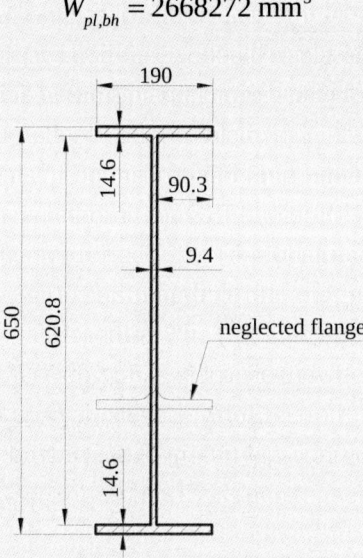

Figure 4.20 – Equivalent double T-section at haunch

d) Bending resistance of end-plate connection

The following components are used to obtain the moment resistance of the connections: (i) CFB - Column flange in bending; (ii) EPB - End-plate in bending; (iii) BWT - Beam web in tension; (iv) CWT - Column web in tension; (v) CWC - Column web in compression.

$M_{j,Rd}$ is determined according to EN 1993-1-8, with the following modifications:

- only the bolt rows located beyond the mid-depth of the beam (with respect to the centre of compression) are considered active for the purpose of in computing the flexural capacity of the connection;

- for hogging bending moment the centre of compression is shifted up by 40 % of the haunch depth ($\Delta_c = 0.4\, h_h$, see Figure 4.20);

The following components are not taken into account: column web panel in shear, beam flange and web (and haunch) in compression.

The resistance of each bolt row is given by:

- First bolt row: $F_{t1,Rd} = 738$ kN – column flange in bending;
- Second bolt row: $F_{t2,Rd} = 737$ kN – column flange in bending;
- Third bolt row: $F_{t3,Rd} = 509$ kN – column flange in bending group.

The design moment resistance is evaluated as follows:

$$M_{j,Rd} = \sum_r h_r F_{tr,Rd} = 738 \cdot 0.6335 + 737 \cdot 0.5185 + 509 \cdot 0.4285$$

$$M_{j,Rd} = 1067.76 \text{ kNm}$$

where:
    $h_1 = 633.5$ mm – lever arm for the first bolt row;
    $h_2 = 518.5$ mm – lever arm for the second bolt row;
    $h_3 = 428.5$ mm – lever arm for the third bolt row.

Finally, the resistance of the connection in bending, under hogging moment, is checked as follows:

$$\frac{M_{j,Ed}}{M_{j,Rd}} = \frac{908.55}{1067.76} = 0.85 < 1.0$$

e) Column web panel

The design shear force in the column web panel is determined based on the bending moments and shear forces acting on the web panel.

$$V_{wp,Ed} = \frac{M_{cwp,Ed}}{d_s} - V_{C,Ed}$$

4. DESIGN RECOMMENDATIONS FOR DUCTILE DETAILS

For this example the shear force in the column ($V_{C,Ed}$) is disregarded.

For a *strong column web panel*, the design shear force should be obtained accounting for the development of fully yielded and strain hardened plastic hinges in the beam:

$$M_{cwp,Ed} = \sum M_{j,Ed} = 908.55 \text{ kNm}$$

leading to:

$$V_{wp,Ed} = \frac{M_{cwp,Ed}}{d_s} = \frac{908.55}{0.628} = 1446.74 \text{ kN}$$

The shear resistance of the column web panel is given by:

$$V_{wp,Rd} = V_{wp,c,Rd} + V_{WAP,Rd} = 2207.84 \text{ kN}$$

where:

$$V_{wp,c,Rd} = \frac{0.9 \cdot A_{vz} \cdot f_{y,wc}}{\gamma_{M0} \cdot \sqrt{3}} = \frac{0.9 \cdot 5609 \cdot 355}{1.0 \cdot \sqrt{3}} = 1034.66 \text{ kN}$$

with:

$A_{vz} = 5609$ mm² – shear area of the column HEB 340, parallel with the web according EN 1993-1-1.

and the shear resistance of two additionally web plates with 12 mm thickness and width 265 mm is given by:

$$V_{wp,c,Rd} = \frac{0.9 \cdot A_{vz} \cdot f_{y,wc}}{\gamma_{M0}} = \frac{0.9 \cdot 6360 \cdot 355}{1.0 \cdot \sqrt{3}} = 1173.19 \text{ kN}$$

$$A_{AWP} = 2 \cdot 12 \cdot 265 = 6360 \text{ mm}^2$$

The resistance of the column web panel is checked with the following relation:

$$\frac{V_{wp,Ed}}{V_{wp,Rd}} = \frac{1446.74}{2207.84} = 0.66 < 1.0$$

4.3 DUCTILE DETAILS FOR MOMENT RESISTING FRAMES

Example 4.3: Dissipative beam-to-column joint (Jaspart *et al*, 2017).

Consider the unstiffened end-plate joint depicted in Figure 4.21. The beam consists of a IPE 450 and the column is a HEB 340, both in steel grade S355, with 4 rows of M30 10.9 bolts. Calculate the resistance of the joint and check that it can be classified as dissipative.

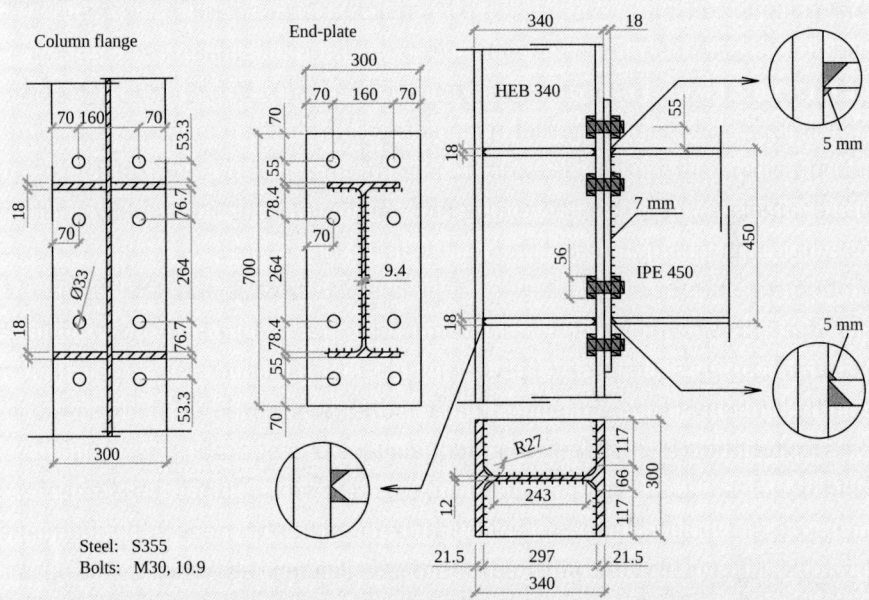

Figure 4.21 – Dissipative double-extended end-plate beam-to-column joint

a) Design criteria

As it intended to check that the joint is dissipative, it is required to achieve a partial strength joint. In this case, it is necessary to check that the resistance of the connection is less than the beam cross section resistance and whether the web panel is weak or not. Hence, the joint will dissipate the energy through one or more of the following elements: (i) web panel (with the value limited by EC8-1); (ii) end-plate or column flange in bending. The ductility of the web panel is in line with the requirements so that it will be required to ensure that the end-plate and the column flange in bending also meet the ductility requirements.

4. Design Recommendations for Ductile Details

b) Beam cross section resistance

The beam cross section resistance, $M_{b,pl,Rd}$, is given by

$$M_{b,pl,Rd} = \frac{W_{b,pl,Rd} \cdot f_{y,b}}{\gamma_{M0}} = \frac{1702 \cdot 10^3 \cdot 355 \cdot 10^{-6}}{1.0} = 604.21 \text{ kNm}$$

where:
- $W_{b,pl,Rd}$ plastic modulus of the beam;
- $f_{y,b}$ yield stress of the beam;
- γ_{M0} partial safety factor (=1.0).

c) Resistance of the column web panel

The design plastic shear resistance, $V_{wp,Rd}$, of the column web panel (EC3-1-8, 6.2.6.1(2)) is given by:

$$V_{wp,Rd} = \frac{0.9 \cdot f_{y,wc} A_{vc}}{\sqrt{3} \cdot \gamma_{M0}} + V_{wp,add,Rd}$$

where:
- A_{vc} shear area of the column = 5609 mm²;
- $V_{wp,add,Rd}$ contribution to the web panel shear resistance due to the plastic hinge occurring in the beam flanges or continuity plates.

The contribution to the web panel shear resistance due to the plastic hinge occurring in the beam flanges or continuity plates is given by (EC3-1-8, 6.2.6.1 (4)):

$$V_{wp,add,Rd} = \frac{4M_{pl,fc,Rd}}{d_s} \text{ but } < \frac{2M_{pl,fc,Rd} + 2M_{pl,st,Rd}}{d_s}$$

where:
- $M_{pl,fc,Rd}$ design plastic moment resistance of a column flange;
- $M_{pl,st,Rd}$ design plastic moment resistance of a stiffener;
- d_s distance between the centre lins of the stiffeners.

$$M_{pl,fc,rd} = W_{pl,fc,Rd} \cdot f_{y,c} = \frac{b_c t_{fc}^2}{4} f_{yc} = \frac{300 \times 21.5^2}{4} \times 10^{-9} \times 355$$

$$= 12.31 \text{ kN}$$

$$M_{pl,st,rd} = W_{pl,st,Rd} \cdot f_{y,c} = \frac{b_{st} t_{st}^2}{4} f_{yc} = \frac{243 \times 18^2}{4} \times 10^{-9} \times 355 = 6.99 \text{ kNm}$$

$$V_{wp,add,Rd} = \min\left(\frac{4 \times 12.31}{435.4 \times 10^{-3}}, \frac{2 \times 12.31 + 2 \times 6.99}{435.4 \times 10^{-3}}\right) = 88.63 \text{ kN}$$

$$V_{wp,Rd} = \frac{0.9 \times 355 \times 56.09 \cdot 10^{-4}}{\sqrt{3} \times 1.0} + 88.63 = 1034.7 + 88.63 = 1123.3 \text{ kN}$$

The capacity in bending (moment at the column face) for simple sided joints is given by (Jaspart *et al*, 2017):

$$M_{p,wp} = \frac{V_{wp,Rd}}{\dfrac{1}{d_s} - \dfrac{L}{H \cdot (L - h_c)}} = \frac{1123.3}{\dfrac{1}{435.4 \cdot 10^{-3}} - \dfrac{6.8}{3.4 \cdot (6.8 - 340 \cdot 10^{-3})}}$$

$$= 565.3 \text{ kNm}$$

where $H = 3.4$ m and $L = 6.8$ m.

d) Resistance of the connection zone

The following components are active: (i) BFC – Beam flange in compression; (ii) CWC – Column web in compression; (iii) BT – Bolts in tension; (iv) EPB– End-plate in bending; (v) CWT – Column web in tension and (vi) CFB – Column flange in bending. Table 4.1 summarizes the components resistance.

Table 4.1 – Resistance of components

Shear	CWS	$V_{wp,Rd}$ [kN]	1123.3
Compression	BFC	$F_{c,fc,Rd}$ [kN]	1387.7
	CWC	$F_{c,wc,Rd}$ [kN]	2870.6
	Compression resistance	$F_{c,Rd}$ [kN]	1123.3
Tension			
Bolts (per bolt)	Bolts in tension	$F_{bolt,Rd}$ [kN]	403.9

4. Design Recommendations for Ductile Details

Table 4.1 – Resistance of components (continuation)

Bolt row 1	EPB	$F_{t,ep,Rd}$ [kN]	386.2
	CWT	$F_{t,wc,Rd}$ [kN]	1084.7
	CFB	$F_{t,fb,Rd}$ [kN]	674.9
	Tension resistance	$F_{t,1,Rd}$ [kN]	386.2
	Lever arm	Z_1 [mm]	497.7
Bolt row 2	EPB	$F_{t,ep,Rd}$ [kN]	572.4
	CWT	$F_{t,wc,Rd}$ [kN]	1051.5
	CFB	$F_{t,fb,Rd}$ [kN]	663.9
	Tension resistance	$F_{t2,Rd}$ [kN]	572.4
	Lever arm	Z_2 [mm]	349.7
Bolt row 3	EPB	$F_{t,ep,Rd}$ [kN]	572.4
	CWT	$F_{t,wc,Rd}$ [kN]	1051.5
	CFB	$F_{t,fb,Rd}$ [kN]	663.9
	Tension resistance	$F_{t,3,Rd}$ [kN]	189.8
	Lever arm	Z_3 [mm]	85.7
Group 2+3	EPB	$F_{t,ep,Rd}$ [kN]	1094.5
	CWT	$F_{t,wc,Rd}$ [kN]	1420.4
	CFB	$F_{t,fb,Rd}$ [kN]	1296.5
	Tension resistance	$F_{t,2,Rd}$ [kN]	547.3
Plastic moment resistance		$M_{pl,Rd}$	399.9

4.3 Ductile Details for Moment Resisting Frames

e) Check of ductility of end-plate in bending

Since the thickness of the end-plate is smaller than the thickness of the column flange, the end-plate is the critical component for ductility. The overstrength of the end-plate is given by:

$$f_{y,ep,ov} = f_{y,ep}\gamma_{ov} = 355 \cdot 1.25 = 443.75 \text{ N/mm}^2$$

The computation of the resistance (mode 1) of the end-plate in bending taking into account the overstrength factor leads to (Landolfo, 2016)

$$F_{over,Rd,1,r1} = F_{T,1,Rd,1,r1} = 482.74 \text{ kN}$$

$$F_{over,Rd,1,r2} = \min\left[F_{T,1Rd,r2}, F_{T,1Rd,g1}\right] = 977.94 \text{ kN}$$

Similarly, the computation of the resistance (mode 1 or mode 2) of the end-plate in bending leads to (Landolfo, 2016)

$$F_{over,Rd,r1} = F_{T,Rd,1,r1,\min} = 482.74 \text{ kN}$$

$$F_{over,Rd,r2} = \min\left[F_{T,Rd,r2,\min}, F_{T,Rd,g1,\min}\right] =$$

$$= \min\left[615.68, 1168.50\right] = 615.68 \text{ kN}$$

Hence the ductility factors β and η for bolt rows 1 and 2 are:

$$\beta_{r1} = F_{over,Rd,1,r1}/F_{bt,Rd} = 482.74/(2 \cdot 406.89) = 0.59$$

$$\beta_{r2} = F_{over,Rd,1,r2}/F_{bt,Rd} = 977.94/(2 \cdot 406.89) = 1.20$$

$$\eta_{r1} = F_{over,Rd,r1}/F_{bt,Rd} = 482.74/(2 \cdot 406.89) = 0.59$$

$$\eta_{r2} = F_{over,Rd,r2}/F_{bt,Rd} = 615.68/(2 \cdot 406.89) = 0.76$$

The ductility of the end-plate is verified, sinceductility degree 2 is achieved ($\beta > 1$ and $\eta \leq 0.95$).

f) Classification of the joint

The ratio of joint resistance versus beam resistance is given by:

4. DESIGN RECOMMENDATIONS FOR DUCTILE DETAILS

$$\frac{M_{j,Rd}}{M_{b,pl,Rd}} = \frac{399.86}{604.21} = 0.662$$

Hence, the joint is classified as partial strength and dissipative.

4.3.3 Detailing of column bases

4.3.3.1 Introduction

Column bases have an important influence on the behaviour of MRFs. The specific rules for MRFs allow that plastic hinges may form at the base of the frame (clause 6.6.1(1)P) because the rotational demand is generally quite high. This means that if the pastic hinge is to form in the column, the column base joint should be designed as full strength non-dissipative, while if the plastic hinge develops in the column base joint, the column should exhibit adequate overstrength to allow the development of cyclic yielding in the joint (clause 6.5.2(5)P) and the column base joint should satisfy the specific requirements for connections in dissipative zones. EC8-1 does not provide specific guidance for dissipative column bases so that the general requirements for dissipative beam-to-column connections of clause 6.6.4(2) are recommended in this case.

In most cases, designers opt to consider the column bases as fully fixed and non dissipative, so that the plastic hinge forms in the column. However, in case of multi-storey building in medium/high seismicity areas it is quite difficult and also quite expensive to guarantee appropriate restraint conditions for column bases.

In reality, most column base joints are flexible and provide some degree of rotational restraint. The alternative of assuming the column base joints as perfectly pinned results in a conservative design, overestimating column flexibility, building period and mostly drift at the first storey, therefore penalising the economy of the design. When pinned bases are assumed, the column base anchorage must be designed with adequate capacity to transfer the shear and the axial forces to the foundation, while allowing the rotations that will occur at the column bases. This may be achieved by a physically pinned column base or by appropriate detailing that results in functionally pinned behaviour.

The following sub-sections discuss and detail both alternative options.

4.3.3.2 Detailing of full strength non-dissipative column bases

When rigid and full-strength column base joints are adopted, the column base components should be designed to be strong enough to enforce the formation of a plastic hinge in the column. The design actions for column base connections are in accordance to equation (3.4) (see section 3.4.3), namely the plastic bending strength of the column, taking into account the interaction with axial force, and magnified by $1.1\gamma_{ov}$ (i.e. 1.375 time larger). Usually, the columns at the first level are often quite large due to drift control and stability criteria. Hence, the design bending moment for column base joints is very severe and it is not easy to detail a typical column base joint that satisfies the overstrength requirements.

An efficient way to satisfy the overstrength criterion recommended by the commentary of AISC 341-16 is to extend the column below the assumed seismic base into a basement (if any) or encased into a grade beam, or embedded in a stocky concrete wall, thus ensuring the column fixity without the need for rigid base plate connections. Figure 4.22 illustrates this solution as, in case of a basement, it is very practical to extend the column below the ground floor (i.e. the seismic base, from which the seismic forces acting on the structure should be calculated) and the rigid flexural restaint is given by the following combined effects:

1) The horizontal displacements of the ground floor are fixed by the reinforced concrete structure of the basement;
2) The columns are continuous from the ground floor (i.e. seismic base) to the basement;
3) The simple connections at the base of the column guarantee equilibrium with the horizontal reaction forces developed by the horizontal displacement restraints at ground floor level. This binary of forces is illustrated in Figure 4.22, highlighting the lever arm that consists on the extended portion of the column that guarantees the flexural restraint.

4. DESIGN RECOMMENDATIONS FOR DUCTILE DETAILS

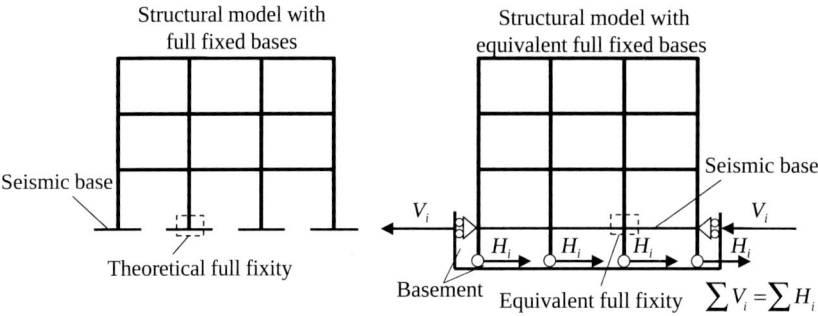

Figure 4.22 – Example of rigid base assemblies for MRFs with extended columns into the basement

A second possibility is to use either reinforced concrete or composite steel-concrete grade beams to improve the column base restraint. It should be noted that grade beams are commonly used as components of building foundations, also in non seismic areas, to transmit the vertical loads into spaced foundations or to tie the footings. Hence, their strength and stiffness can be wisely used to guarantee the adequate flexural restraint to the column bases.

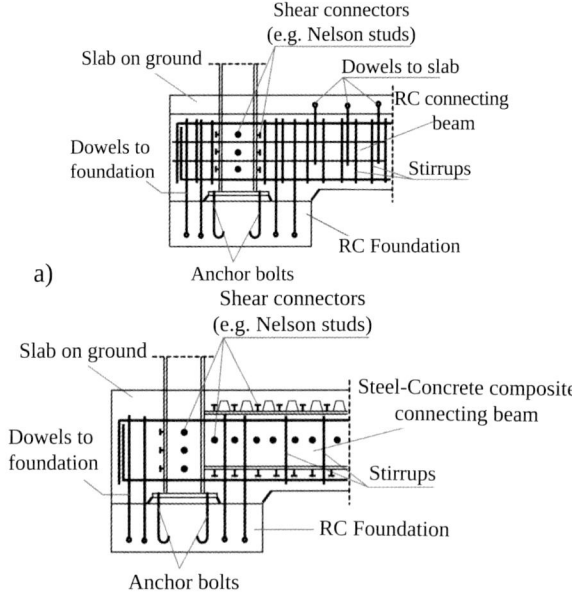

Figure 4.23 – Examples of rigid base assemblies for MRFs with grade beams: a) column base embedded in a concrete grade beam (adapted from Cochran, 2003); b) column rigidly connected to a steel moment resisting grade beam

When reinforced concrete grade beams are used (see Figure 4.23a), the column base is totally embedded in the reinforced concrete grade beam. Steel rebars or Nelson shear connectors must be welded to the column to help the transmission of bending moments and shear forces from the steel column to the concrete grade beam. In addition, transverse and longitudinal rebars should be designed to provide the required strength. Finally, an efficient detailing requires that the grade beam and the footing must be tied together (see Figure 4.23) with a set of dowels.

If a composite solution is adopted (see Figure 4.23b), the grade beam should be made of a steel beam rigidly attached to the column (in this case, full strength fully rigid connections should be designed as for beam-to-column joints). The steel beam is fully encased in concrete and composite action is guaranteed by Nelson studs welded to its flanges and web. Also in this case it is fundamental to tie the grade beam and footing with dowels.

If a column base joint with base plate is adopted, stiffened configurations are usually necessary to ensure full strength non dissipative behaviour. Figure 4.22 illustrates several possible alternatives. Usually, this leads to the use of thick base plates, stiffeners, haunches, or any other types of strengthening systems to increase the bending strength of the connection as required to develop the plastic hinge in the column. Sections 6.2.8 and 6.3.4 of EN 1993-1-8 specify the procedures for the calculation of the design resistance and stiffness of column bases with base plates. However, when the columns are quite large (e.g. HEM or HD profiles), it may not be possible to achieve a full-strength non dissipative joint with the simpler base plate detail. In addition, such design are more likely to fail underground and would cause problem because they are difficult to check after an earthquake.

The embedment of a column stub in the concrete block, illustrated in Figure 4.24b may improve this situation. It is important to note that in all cases the concrete components should present some degree of overstrength with respect to the ductile steel components of the joint. Guidance for the design of the concrete components is available in EN 1992-4 (CEN, 2017) and in the design guides produced in the framework of the European project INFASO+ (Wald *et al*, 2014). Finally, it is further emphasized that piled foundations may be necessary to achieve a rigid foundation in flexible soils.

These considerations imply that the modelling assumptions generally adopted by designers should match the real conditions or they may lead to unconservative results because they underestimate the column flexibility, the fundamental period and the drift at the first storey. Hence, it is advisable to avoid

4. Design Recommendations for Ductile Details

the fixed base assumption unless the bases and the relevant foundation elements (e.g. the footings) can effectively provide adequate restraint conditions.

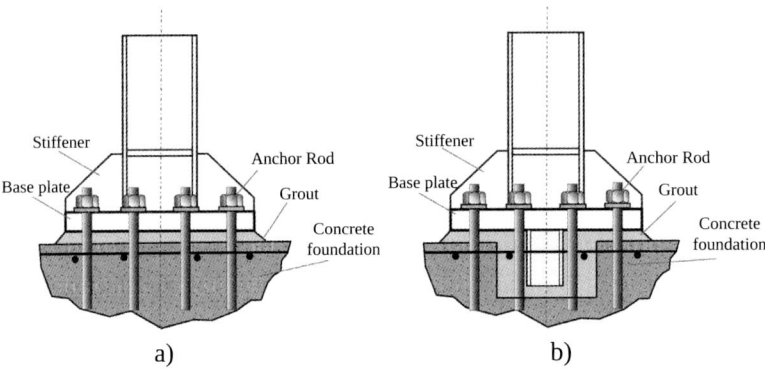

Figure 4.24 – Examples of rigid base assemblies for MRFs with plate directly sitting on top of foundation

4.3.3.3 Detailing of dissipative column bases

In case it is desired to achieve pinned behaviour at the base of the frame, the column base joint may be designed as a true pinned joint. It is noted that in this case there is no difficulty to detail pinned joints even for massive column sections, as this detail is often used in bridge piers whenever zero bending moment is sought at the base of the pier.

Alternatively, a column base with base plate may be designed such that is functionally pinned, with low effective rotational stiffness, as shown in Figure 4.25. Compression transfer is through the grout beneath the base plate. Tension is transferred by the holding down bolts and the embedded plates. Shear may be transferred by the holding down bolts, by friction between the base plate grout and foundations or by alternative means. In this case, the rotation capacity under cyclic conditions must be explicitly demonstrated. Prequalification of appropriate configurations would be a welcome practical solution for this case. It should be emphasized that the cyclic rotation supply should come exclusively from the ductile steel components and concrete should not govern the joint capacity. Ductility is provided by bending of the anchor plate but it is noted that the anchorage resistance is reduced by cracks.

Cast-in place bolt groups are used to connect the column base plate to the concrete foundation. Columns are set to level using steel shims. The bolt pockets and the space between the base plate and the foundation are filled with non-shrinking grout or drypack. Cementitious grout is placed by enclosing the area to be filled beneath the base plate and pouring grout into the enclosed area. Care must be taken to ensure that the bolt pockets are properly filled.

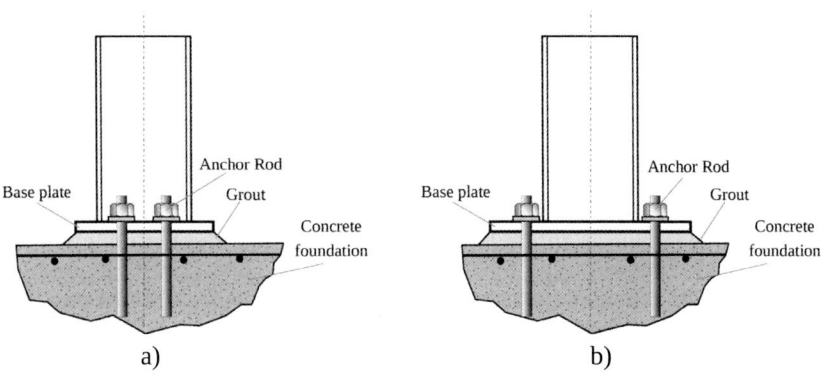

Figure 4.25 – Examples of pinned base assemblies for MRFs

4.4 DUCTILE DETAILS FOR CONCENTRICALLY BRACED FRAMES

4.4.1 Introduction

The seismic performance of bracing members of CBFs is characterized by cyclic buckling under compression and subsequent tension yielding when the diagonals are straightened. The buckling mode induces the formation of plastic hinges at both the centre and the ends of the braces and the relevant plastic engagement depends on the orientation of the brace cross section and its connection to the beams and columns, which can be designed as rotationally rigid or pinned. In both cases, the brace end connections must be able to transfer the axial cyclic tension and compression to the rest of the structure and they must be detailed to confine plastic rotation to the ends of bracing (if designed as rigid) or to guarantee adequate flexural ductility to allow end rotations of the braces (if designed as pinned).

In general, three different types of joints must be detailed to connect the braces: (i) brace-to-beam/column joints; (ii) brace-to-beam midspan joints; and (iii) brace-to-brace joints. In addition, as discussed in section 3.6.1, several potential failure mechanisms may occur (plastic hinge develops into the brace or plastic lines forms in the gussets), particularly for braces working in

4. Design Recommendations for Ductile Details

compression that result in different detailing rules. These issues are discussed in the following sections.

4.4.2 Detailing of brace-to-beam/column joints

EC8-1 addresses the design of brace-to-frame joints only conceptually without providing specific detailing rules and technological prescriptions. Indeed, the designer has only to comply with the overstrength condition given by equation (3.4) (see section 3.4.3), which essentially accounts for the axial force demand transferred by the yielded brace in tension as expressed by the following inequality:

$$N_{j,Rd} \geq 1.1 \gamma_{ov} \cdot N_{pl,Rd} \qquad (4.16)$$

where $N_{j,Rd}$ is the joint axial strength; $N_{pl,Rd}$ is the plastic axial strength of the brace; γ_{ov} is the material overstrength factor (see Section 3.2.1).

Unfortunately, if the braces are designed as fixed at both ends, this criterion disregards the flexural strength that the connections should have to confine inelastic rotation to the bracing member only. On the other hand, if pinned restraints are assumed at both ends of the braces, the hierarchy criterion given by equation (4.16) does not provide any ductility requirements to ensure brace end rotations and to prevent fracture due to flexural strain concentrations.

In contrast, American codes as AISC 341-16 provide comprehensive design requirements and provisions to detail properly bracing-to-frame connections. In particular, on the basis of the experimental study carried out by Astaneh-Asl et al (1984), AISC 341-16 assumes that for brace buckling in the plane of the connection (e.g. if gusset plates are used buckling is intended in their plane), the end connections are conceived as fixed restraints that should be designed to resist to the brace yield strength in tension (see equation (4.16) and the *M-N* interaction forces transferred by the buckled brace. The latter are assumed as the expected compressive strength (i.e. the brace buckling strength $N_{b,Rd}$) and the expected flexural strength of the brace (i.e. $1.1 \gamma_{ov} M_{pl,Rd}$, where $M_{pl,Rd}$ is the plastic bending strength of the brace cross section). This type of bracing end connections are less used in building frames because double gusset-plates (i.e. one for each side of the brace cross section) are necessary to provide adequate flexural strength.

Most commonly, single gusset plates are used to connect bracings to the main frame and out-of-plane buckling is consequently accepted, so that brace end rotations induce weak-axis bending in the gusset plate, as shown in Figure 4.26.

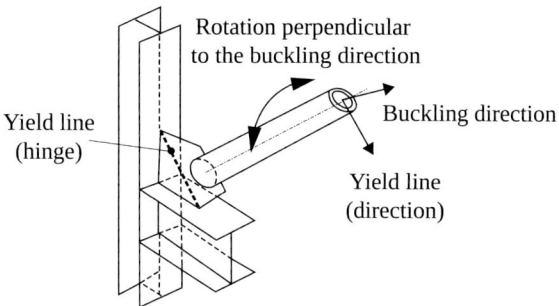

Figure 4.26 – Weak-axis bending in the gusset plate due to out-of-plane buckling of braces (adapted from Cochran, 2003)

In this case, suitable performance can be guaranteed by preventing the buckling of the gusset plates and enforcing restraint-free plastic rotations about a hinge line (i.e. yield line) in the gusset plate, which acts as an equivalent pin connection and allows the brace rotating and yielding about itself, as shown in Figure 4.27.

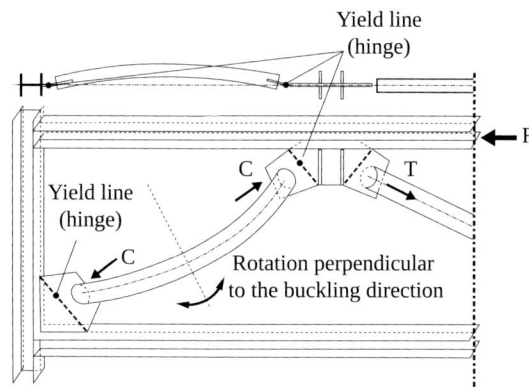

Figure 4.27 – Out-of-plane buckling mode of braces and formation of yield line in the gusset plate (adapted from Cochran, 2003)

The yield lines at each end of the brace must be perpendicular to the brace axis (see Figure 4.28) and, on the basis of tests and recommendations by Astaneh-Asl *et al* (1982, 1985, and 1986), AISC 341-16 recommends to assume for the gusset-plate hinge-zone (namely the zone where the yield line can form) a minimum free length equal to $2t$ (where t is the thickness of the gusset plate) between the end of the brace and the assumed geometric line of the gusset restraints that is drawn from the point on the gusset plate nearest to

4. Design Recommendations for Ductile Details

the brace end that is constrained from out-of-plane rotation (see Figure 4.29 where gusset-plate hinge-zone is indicated as "*a*" segment). Hence, it should be short enough to avoid the occurrence of plate buckling prior to member buckling, but sufficiently long to accommodate plastic end rotations.

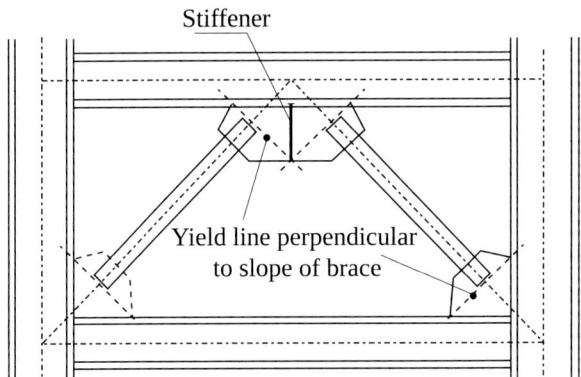

Figure 4.28 – Bracing centrelines and gusset plate yield lines (adapted from Cochran, 2003)

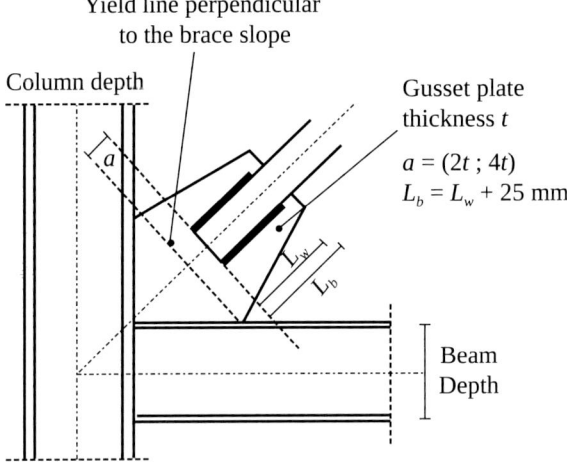

Figure 4.29 – Gusset plate yield line and offset requirements (adapted from Cochran, 2003)

It should be noted that $2t$ is the minimum offset for the yield line. As shown by Cochran (2003), due to erection tolerances larger offsets are recommended (e.g. $3t$). However, according to Astaneh-Asl *et al* (2006), the actual hinge-zone length at either end of the brace should not be larger than $4t$ in order to guarantee that the detail works properly.

The yield line off-set should properly extend across the width of the gusset plate to both free edges of the gusset plate. Typically, the yield line intersects the beam or the column flange with one end of the gusset plate (as shown in Figure 4.29). Conversely, the yield line should not be extended into the area of the gusset plate welding to the column flange or beam flange, as shown in Figure 4.30, because such detail will result in restraint to the gusset plate buckling out-of-plane that is potentially prone to tearing along these edges due to high plastic strain concentrations when the buckled brace starts rotating about this hinge line.

Similar problems may arise also when interaction occurs between gusset plate and concrete slab. Indeed, the presence of concrete around the gusset plate represents a flexural restraint against out-of-plane rotations that avoids the formation of the yield line at the intersection with the beam flange. Therefore, the gusset plate should be isolated from the slab if the offset of the yield line is located within the thickness of concrete slab. To decouple this potential interaction, coherently to what is shown in the previous section for beam-to-column joints of MRFs, the gusset plate can be isolated from the slab by means of an interposed zone filled with a compressible material (e.g. polystyrene, fire caulking, etc.) as shown in Figure 4.31 for the case of a bracing member with either circular or square hollow cross section, and in Figure 4.32 for a bracing member with wide flange cross section.

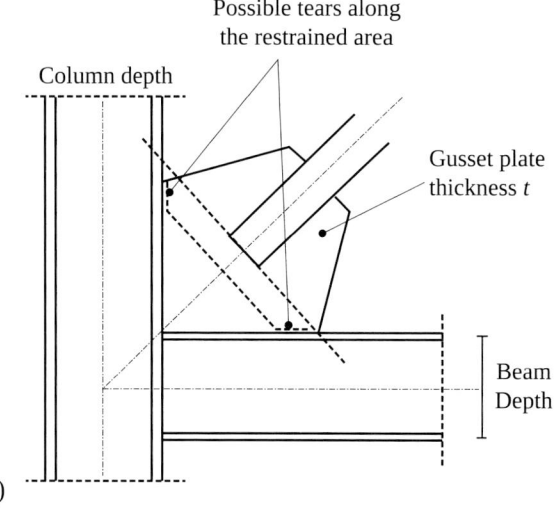

Figure 4.30 – Not recommended gusset plate yield line in both welded (a) and bolted (b) gussets: inadequate off-set and potential tears along gusset plate restraints (adapted from Cochran, 2003)

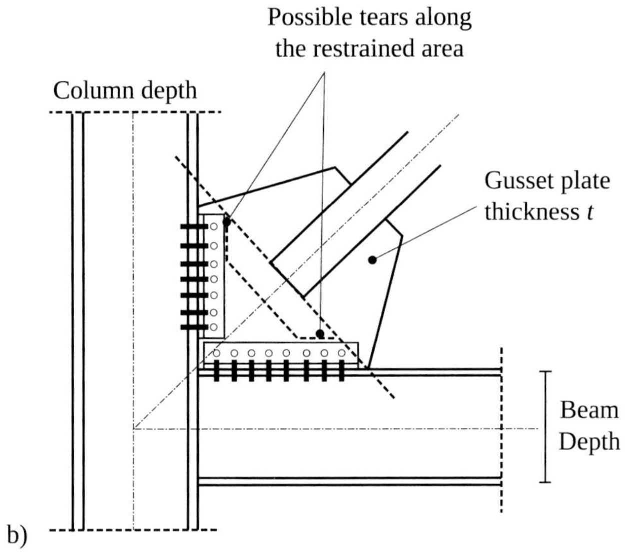

4. Design Recommendations for Ductile Details

Figure 4.30 – Not recommended gusset plate yield line in both welded (a) and bolted (b) gussets: inadequate off-set and potential tears along gusset plate restraints (adapted from Cochran, 2003) (continuation)

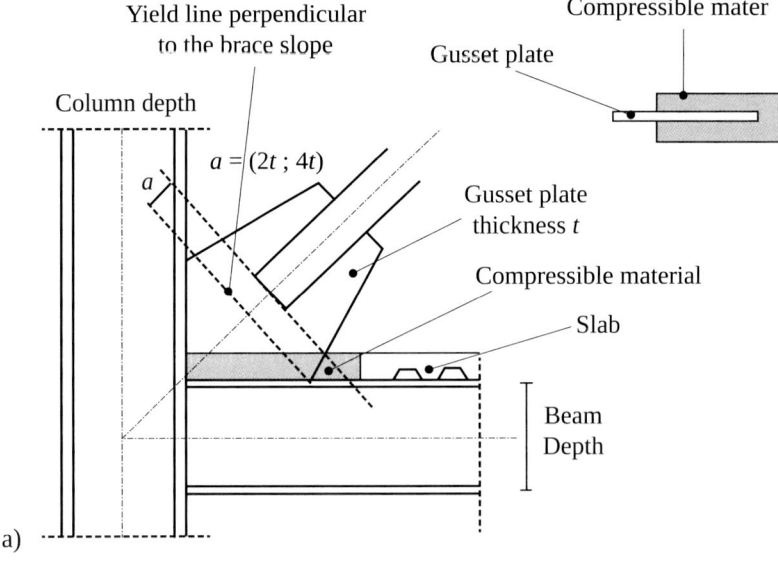

Figure 4.31 – Welded (a) and bolted (b) gusset plate with yield line isolated from concrete slab: brace with circular or square hollow cross section

4.4 DUCTILE DETAILS FOR CONCENTRICALLY BRACED FRAMES

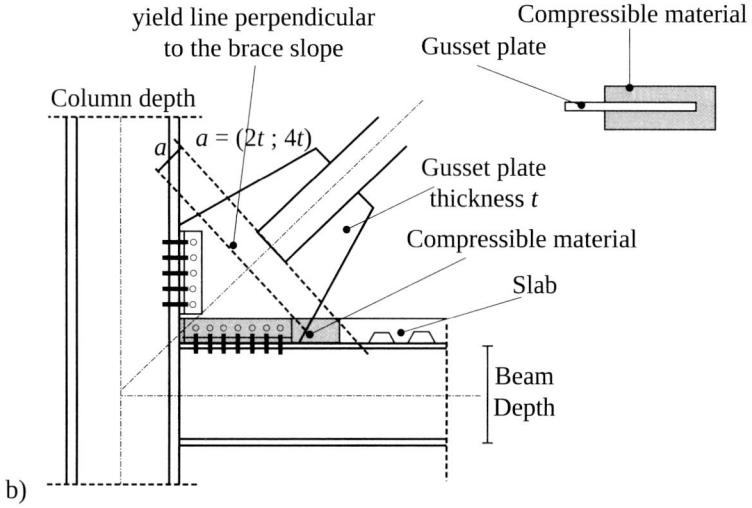

Figure 4.31 – Welded (a) and bolted (b) gusset plate with yield line isolated from concrete slab: brace with circular or square hollow cross section (continuation)

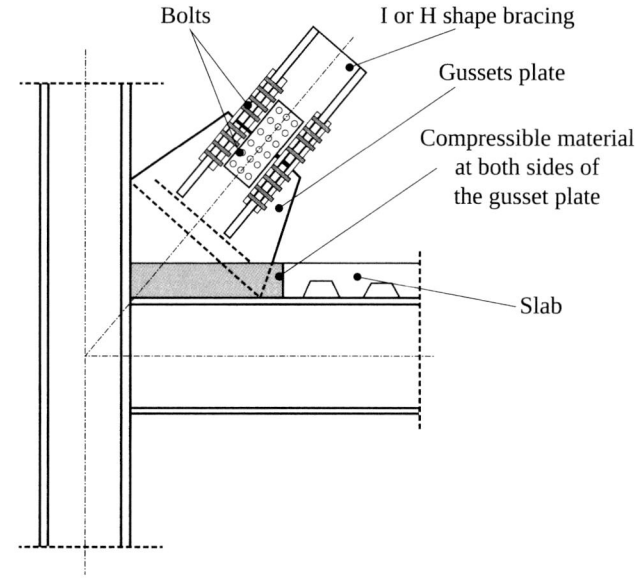

Figure 4.32 – Welded (a) and bolted (b) gusset plate with yield line isolated from concrete slab: brace with wide flange cross section (adapted from Astaneh-Asl et al, 2006)

4. DESIGN RECOMMENDATIONS FOR DUCTILE DETAILS

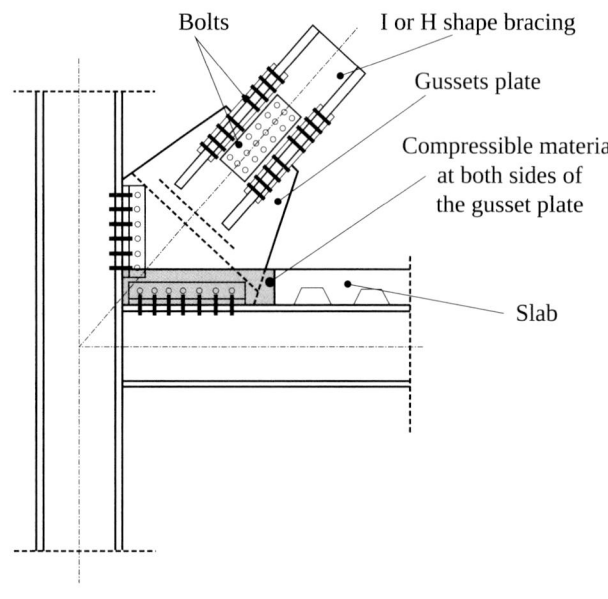

b)

Figure 4.32 – Welded (a) and bolted (b) gusset plate with yield line isolated from concrete slab: brace with wide flange cross section (adapted from Astaneh-Asl et al, 2006) (continuation)

When the free edge of the gusset plate is quite large (generally more than $4t$ beyond the theoretical yield line), it is necessary to design an edge stiffener to stabilize the gusset plate and to restrain the yield line. However, the stiffener should terminate outside the gusset-plate hinge. This offset from the yield line is very important, because it avoids the presence of welds in the zone of the gusset plate where a plastic hinge should form when the brace buckles, thus preventing any possible fracture initiation in or near the gusset plate hinge-zone. An example of this detail is shown in Figure 4.33, where the gusset plate is restrained by the presence of the slab (i.e. no compressible material is considered) and the yield line is anchored to the topping of the slab.

It is worth noting that the design of the gusset plate can be very sensitive to changes in building bay sizes, because any slight variation in bay length or interstorey height corresponds to a change of the slope of the brace and, consequently, the dimensions of the gusset plate in order to satisfy the geometric requirement of yield line offset between $2t$ and $4t$ from the end of the brace.

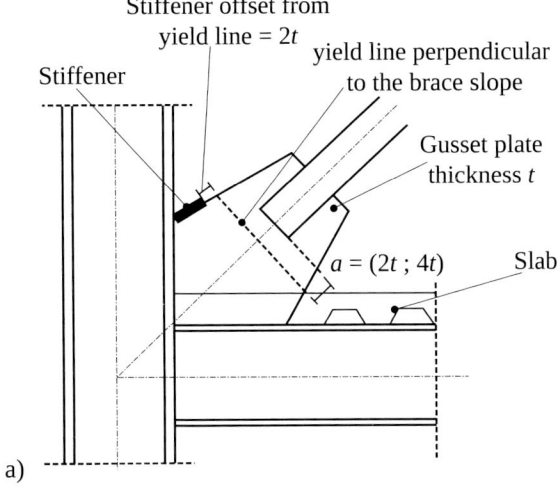

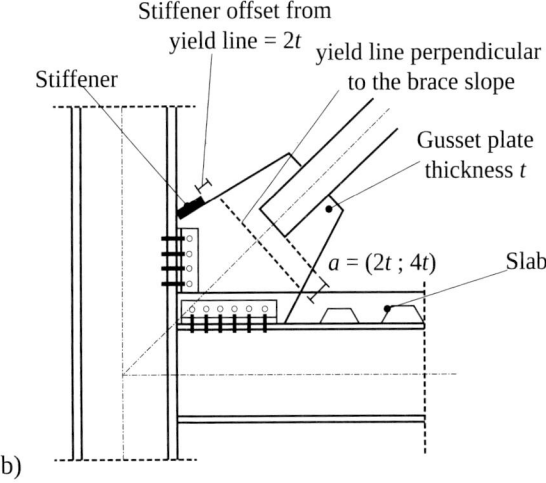

Figure 4.33 – Edge stiffener for welded (a) and bolted (b) gusset plate (adapted from Cochran, 2003)

With this regard, US design and constructional practice developed the so called "critical angle" concept. The critical angle is assumed as the minimum between brace-to-beam and brace-to-column angles, which corresponds to the side of the brace where yield line restraint occurs on, namely where the yield line intersects either the column or beam flange at one end, and the other end of the yield line intersects the free edge of the gusset plate, as shown in

4. DESIGN RECOMMENDATIONS FOR DUCTILE DETAILS

Figure 4.34. Only in the rare case that both column and beam have the same depth, and the brace slope is 45°, each end of the yield line would intersect both column and beam flange simultaneously.

The critical angle is a very important geometric datum from the constructional point of view, because US designers should provide only this information in the structural drawings and the steelwork company or the detailer will complete/update the gusset plate geometry on the basis of the actual building dimensions even with small changes or variation of erection tolerances. It is desirable to introduce this concept in the European practice, due to the potential advantages in terms of reduction of design time. In Figure 4.34a, the critical angle is on the beam side since the first restraint of the yield line occurs at the beam flange and the opposite end occurs at the free edge of the gusset plate. On the contrary, Figure 4.34b shows an example of critical angle on column side. In order to determine the critical angle the slope of both sides of gusset plate starting from brace end should be symmetrical with respect to the brace axis. Under this assumption, the gusset plate dimension can be easily determined to account for the yield line off-set too. In addition, Cochran (2003) and Astaneh-Asl et al (2006) provide some useful prescriptions to detail properly the gusset plate, which are summarized as follows:

1. A minimum offset equal to 25 mm from each side of brace to the free edge of gusset plate should be considered, thus improving the gusset plate strength against block shear check;
2. A slope γ of 30 degrees at the edge of the gusset plate with respect to the brace axis is advisable (see Figure 4.34);
3. The brace lap length L_b of the gusset plate when welding is used should be detailed 25 mm longer than the specified weld length L_w as shown in Figure 4.29. This allows for beginning and termination of the weld slightly away from the end of the brace member and end of the gusset plate;
4. To assume a gusset plate thickness between 15 mm to 40 mm to determine the $3t$ offset distance to avoid huge dimensions of connections.

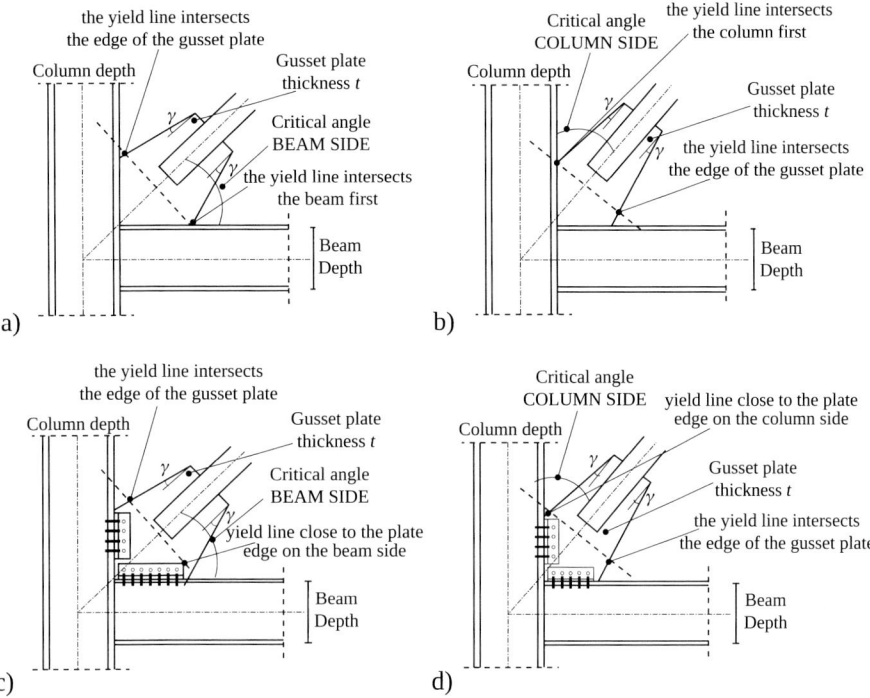

Figure 4.34 – Critical angle concept: a) on beam side; and b) on column side for welded gusset; c) on beam side and d) on column side for bolted gusset (adapted from Cochran, 2003)

Besides the geometric proportioning of the gusset plate that implicitly aims at guaranteeing adequate plastic rotation capacity, it is also necessary to verify both the axial strength of the gusset plate against tensile forces (i.e. calculated according to equation 4.16) and the buckling strength against compression forces (i.e. the buckling strength $N_{b,Rd}$ of the brace) transferred by the connected brace. The effective width W_d of the gusset plate at the hinge zone defines the resisting zone of the gusset plate. The width of the gusset plate to resist the applied axial force can be established by means of the method developed by Whitmore (1952), which assumes for bolted connections that the effective width is comprised by two 30° lines drawn from the first bolts on the gusset that intersect the centreline axis of the last bolt, as shown in Figure 4.35. The portion of gusset plate outside the "Whitmore's width" should not be considered as able to resist design loads.

4. Design Recommendations for Ductile Details

Astaneh-Asl et al (1982) extended the Whitmore's method to the case of welded connections as shown in Figure 4.35. They recommended drawing the 30° lines from the starting point of the weld to intersect a line passing through the end points of the weld itself.

On the basis of these assumptions, the Whitmore's width W_d can be calculated as follows:

$$W_d = b + L_w \cdot 2\sqrt{3} \quad \text{for welded connections} \tag{4.17}$$

$$W = b + L_{bc} \cdot 2\sqrt{3} \quad \text{for welded connections} \tag{4.18}$$

where L_w is the length of the weld connecting the bracing member to the gusset plate; L_{bc} is the length of the bolted connection of the bracing member to the gusset plate; b is the distance between the weld lines or bolt lines.

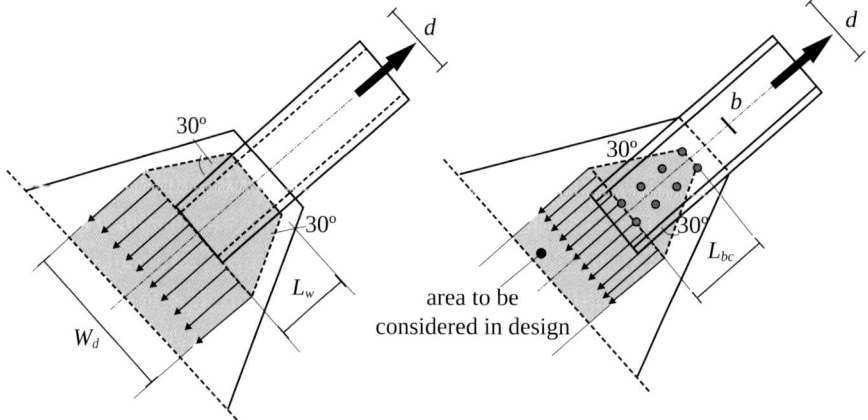

Figure 4.35 – The effective width W_d calculated with Whitmore's method

As a general remark about gusset plates designed using a linear offset rule, to accommodate brace end rotation, the shape and the size of the connections are often quite large, resulting in uneconomical and sometimes unfeasible connections. In order to improve both constructability (i.e. more compact shape of gusset plates) and performance of the connections, Lehman et al (2008) carried out a comprehensive experimental study to develop an alternative detail for gusset plate splices. In light of their experimental results, Lehman et al (2008) proposed an elliptical clearance requirement to replace the linear yield

4.4 Ductile Details for Concentrically Braced Frames

line concept with 2*t* clearance distance. Figure 4.36 shows this type of detail, where the plastic hinge zone is shaped as an elliptical band with a clear 8*t* width. The experimental evidence confirmed that, if the gusset plate is well sized to conform to this elliptical clearance, the connections provide large system ductility and deformation capacity, limiting fracture of the welds or brace.

The actual dimensions of the elliptical band can be easily determined graphically from the gusset plate dimensions. Alternatively, a closed form analytical solution can be directly determined using the notation reported in Figure 4.36, where α is the slope of the brace, c is maximum distance from the centroidal axis to the extreme fiber of the brace, and l^* is the length from the imaginary corner of the rectangular gusset plate to the end of the square cut of the brace.

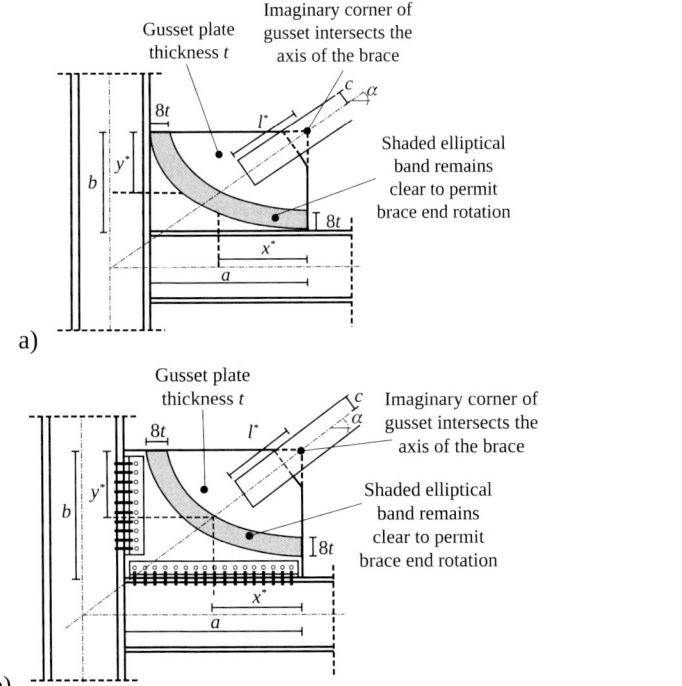

Figure 4.36 – Elliptical clearance with 8*t* band width (Lehman *et al*, 2008) for welded (a) and bolted (b) gussets

The dimensions *a* and *b* should be selected so that the imaginary corner of the gusset intersects the centroidal axis of the brace as shown in Figure 4.36 for both welded and bolted gussets. The radii of the ellipse are fixed as follows:

4. Design Recommendations for Ductile Details

$$a' = a - 8t \text{ and } b' = b - 8t \qquad (4.19)$$

$$\rho = \frac{a'}{b'} \qquad (4.20)$$

The coordinates x^* and y^* shown Figure 4.35 can be derived as the exact centreline dimensions to assure the 8t elliptical clearance, namely in case the brace has no width, as follows:

$$y^* = a'\sin\left(\arctan\left(\rho\tan(\alpha)\right)\right) \text{ and } x^* = a'\sqrt{1-\left(\frac{y^*}{b'}\right)^2} \qquad (4.21)$$

However, the brace has a width equal to 2c (see Figure 4.36) and the design purpose is to assure that the extreme corners of the brace are also out of the elliptical plastic zone. Lehman et al (2008) also derived the equation of the elliptical clearance that guarantees this requirements as follows:

$$l^* = \sqrt{x'^2 + y'^2} + Corr \qquad (4.22)$$

where:

$$Corr = c\sin(\beta)\cos(\alpha) \qquad (4.23)$$

$$\beta = \arctan\left(\frac{-2}{\rho}\sqrt{\frac{a'^2}{x'^2}}\right) \qquad (4.24)$$

Regarding the weld size requirements for gusset plates several experimental and analytical studies are available (Johnson, 2005; Yoo, 2006), showing that, if fillet welds are used, their depth should be equal to or greater than the thickness of the gusset plate.

4.4.3 Detailing of brace-to-beam midspan connections

Under cyclic inelastic response of the connected bracing member, it is expected that the brace-to-beam mid-span connection for chevron CBFs should behave similarly to the brace-to-beam/column connection. Consequently, their geometrical properties and strength requirements are very similar (Astaneh-Asl et al, 2006). This implies that the yield line concept discussed

4.4 Ductile Details for Concentrically Braced Frames

in the previous section (both the $2t$ linear and the $8t$ elliptical offset) should be adopted for this type of connections.

Special attention should be paid to the details of the bottom edge of the gusset plate. Indeed, most designers prefer using a straight free edge in order to simplify the constructional phase. However Astaneh-Asl et al (2006) showed that this portion of the gusset plate must be tapered (see Figure 4.37) to reduce the length of the free edge, because long unstiffened gusset plate free edges are prone to buckling which can occur prior to the brace attaining its compression capacity.

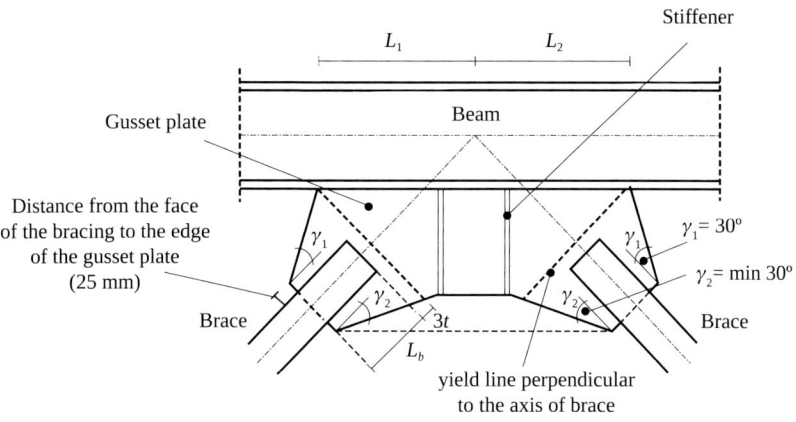

Figure 4.37 – Details of brace-to-beam mid-span connection (adapted from Cochran, 2003)

Another important aspect needing care is that the yield lines should not overlap. Hence the gusset plate should be trimmed back to avoid the overlapping of yield lines. Brace-to-beam mid-span gusset plates are generally quite large and slender. Therefore, transverse stiffeners are necessary to prevent out-of-plane buckling and they should also be positioned at least $2t$ offset back from the plastic zone to prevent welding near this hinging area (i.e. as the stiffeners used for the free edge of brace-to-beam/column gusset plates shown in Figure 4.33). As described by Cochran (2003), the intersection of the two tapered edges of the gusset plate should be radiused to prevent any notches which might lead to fracture initiation due to hinge rotation of the legs of the gusset plate.

Brace-intercepted beams should also be restrained to avoid lateral-torsional buckling. Figure 4.38 shows a suitable detail with fly bracings.

4. Design Recommendations for Ductile Details

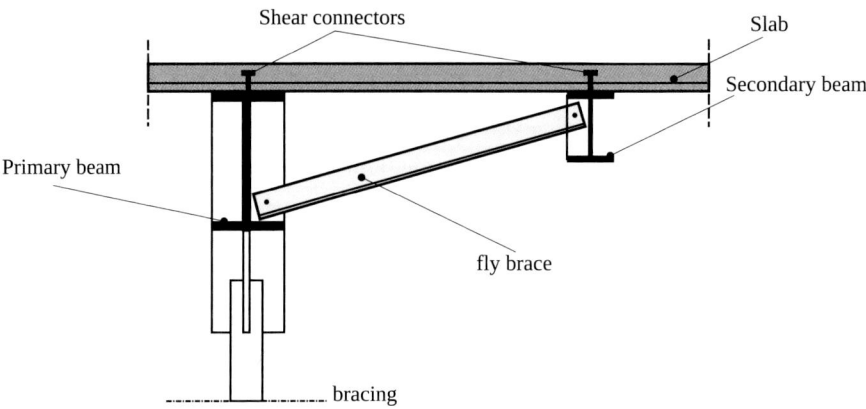

Figure 4.38 – Details of lateral torsional bracing to restrain the brace-intercepted beam

4.4.4 Detailing of brace-to-brace connections

X-CBFs are very common in seismic areas owing to their high strength and stiffness. Indeed, thanks to the mid-length mutual restraint, cross-braced members are characterized by lower effective length factor and slenderness ratio when compared to chevron and single-diagonal braces. This feature makes the system very attractive from the economical point of view because smaller profiles are necessary, and consequently less material consumption.

The efficiency of this bracing system strongly depends on the local details such as brace end connections and brace-to-brace connections. Indeed, the slenderness is highly affected by the brace boundary conditions, which may vary from pinned to fixed. Brace end connections were already discussed in section 4.4.2, while more detailed discussion is needed for brace-to-brace connections that could be the most influential feature affecting the effective buckling length.

Considering the most common details in practice, intersected bracings can be conceived as either continuous or discontinuous. In the first case, the braces are directly welded to each other and continuity plates are used at the mid-length connections, thus behaving as an internal rigid restraint (see Figure 4.39). The degree of flexural restraint at the ends of two brace segments at horseback of the mid-connection depends on both the flexural and the torsional stiffness of the transverse brace.

4.4 DUCTILE DETAILS FOR CONCENTRICALLY BRACED FRAMES

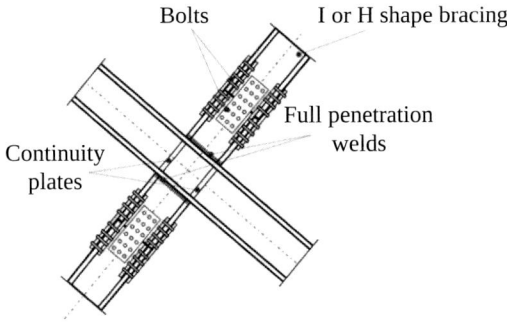

Figure 4.39 – Example of continuous brace-to-brace connection

Discontinuous X-braces typically have two segments of diagonal connected to one continuous brace. Various types of details can be used to transfer the load between the two disconnected brace portions at the mid-length connection (see Figure 4.40). In order to be consistent with this type of detail, the structural model should consider the brace portions as pinned at both segments because the gusset plate has negligible out-of-plane flexural stiffness compared to the brace stiffness.

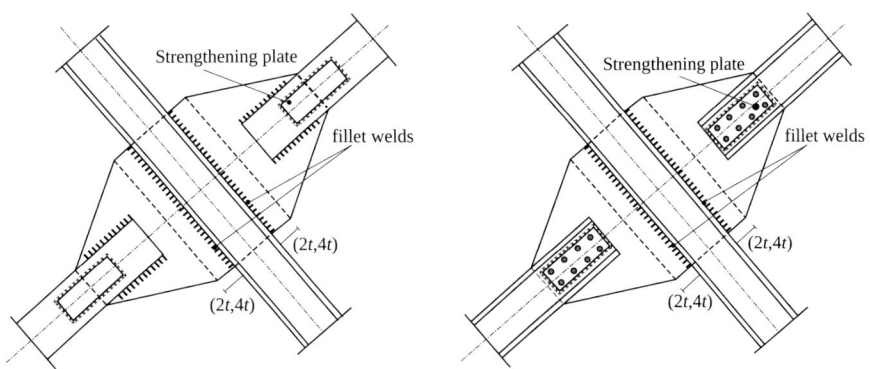

Figure 4.40 – Example of discontinuous X-CBFs using different types of mid-length splices: a) welded and b) hybrid welded/bolted connection

Another possible detail is shown in Figure 4.41 and it consists of two discontinuous braces (i.e. four brace segments), where both ends of each brace portions can be detailed in order to develop the yield line mechanism,

4. Design Recommendations for Ductile Details

as described in section 4.4.2. This type of detail requires a heavy central core (as shown by Ebadi and Sabouri-Ghomi, 2012) in order to restrain effectively the braces and to impose the buckling mode in the free portion of the brace between the yield lines of the gusset plates at both ends of each brace segment. This arrangement guarantees out-of-plane buckling with a buckling length clearly identified (i.e. roughly effective length equal to half length of the entire diagonal). In addition, the ductility and the fracture life of the bracing system is large because the gusset plates are mainly engaged in bending with little torsional interaction.

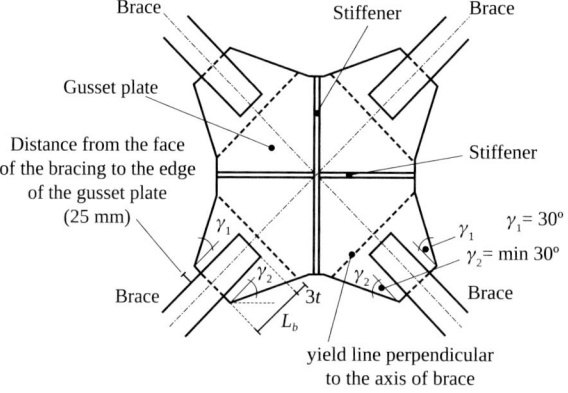

Figure 4.41 – Example of brace-to-brace connection using two discontinuous bracing members and a stiffened central core

    The details reported in Figure 4.39 to Figure 4.41 induce different elastic and inelastic performance of X-braced systems, raising some problems for the designer. Indeed, the upper-bound and lower-bound limits on the non-dimensional slenderness of the diagonals required by EC8-1 imply the need to evaluate accurately the effective buckling length, which depends on the brace boundary conditions, on the flexural and torsional stiffness of the brace and on the level of force acting in the tensioned diagonal.
    Several analytical studies (Picard and Beaulieu, 1989a,b; Stoman, 1989; Sabelli and Hohbach, 1999) proposed effective length factors and developed effective length spectra of cross-bracing members through an elastic buckling analysis theory, assuming identical diagonals with one of them always in tension, and both bracing members either simply supported or rigidly attached at their ends (i.e. the working nodes belonging to the frame). These studies showed

4.4 Ductile Details for Concentrically Braced Frames

that the tension brace can effectively restrain both in-plane and out-of-plane displacements of the diagonal in compression at the brace intersecting point.

In general, the in-plane flexural stiffness of brace end connections is larger than out-of-plane, as well as its ductility. Hence, in most of cases (especially if tubular sections are used for the braces), the most typical brace buckling mode is out-of-plane.

As shown in Figure 4.42 two different types of out-of-plane buckling modes are possible for continuous compression braces in X-bracing: i) sway buckling mode, and ii) S-shape buckling mode.

In contrast, due to the presence of mid-brace splices, discontinuous X-CBFs exhibit four types of out-of-plane buckling modes (see Figure 4.43), namely: i) sway buckling mode, ii) S-shape buckling mode, iii) four-hinge buckling mode, and iv) two-hinge buckling mode.

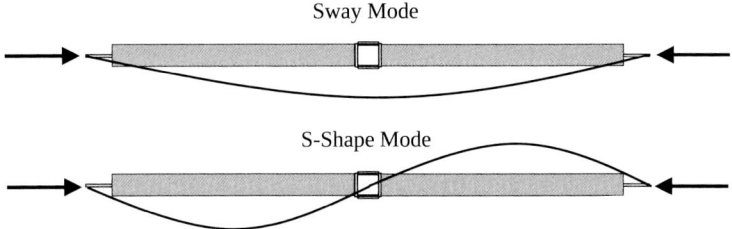

Figure 4.42 – Possible out-of-plane buckling modes for continuous X-Braces

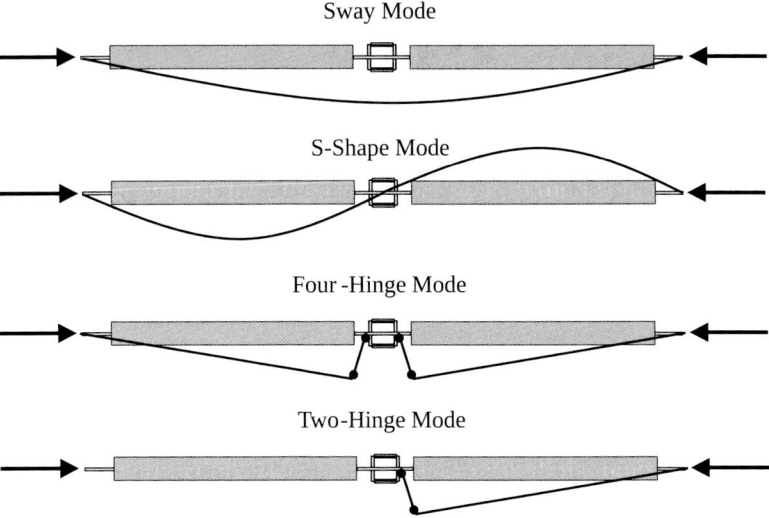

Figure 4.43 – Possible out-of-plane buckling modes for discontinuous X-Braces

4. Design Recommendations for Ductile Details

In light of the remarks discussed in section 4.4.2, for X-CBFs it is rational to control the plane of buckling by favouring out-of-plane modes so that the ductility of end-gussets can be exploited. If continuous bracing is used, Sabelli and Hohbach (1999) clearly showed that it is possible to minimize the uncertainty regarding the plane where buckling occurs. Indeed, it is sufficient to verify that the upper-bound buckling strength in the plane for which ductile detailing is provided is sufficiently smaller than the lower-bound buckling strength in the perpendicular plane, where the connections have poor post-elastic performance. For this purpose, Table 4.2 reports the effective length factor k for continuous bracing, which depends on both the flexural and torsional stiffness of the connected braces, as illustrated by Sabelli and Hohbach (1999).

Table 4.2 – Effective Length Factors K of continuous X-bracings

Strength	In-Plane Buckling		Out-of-Plane Buckling	
	Pinned-end	Fixed-end	Pinned-end	Fixed-end
Upper bound	0.800	0.544	0.354	0.250
Lower bound	0.843	0.590	1	0.699

In case of flexurally stiff diagonals with geometrical slenderness lower than 200, the sway buckling mode of the compression brace is prevented by the tensioned brace, which enforces a S-shape mode that corresponds to an effective buckling length factor K equal to 0.5.

However, this approach does not account for inelastic buckling under cyclic conditions and it is not directly applicable to discontinuous X-bracings systems. At the present time, although several experimental studies have been carried out, more research efforts are necessary to develop design recommendations for efficient and economical brace-to-brace connections. Anyway, some considerations may be drawn on the basis of the existing scientific literature.

Using 3D finite element models, Davaran and Hoveidae (2009) showed that brace-to-brace discontinuity can lead to K values greater than 0.5 and only increasing the flexural stiffness of the mid-connection can lead to K close to 0.5.

Experimental tests carried out by Wakabayashi *et al* (1977) on single storey moment resisting frames with X bracing made with H-shape members showed that the brace effective buckling length was almost equal to the length of the segment of diagonal between the end of the gusset plate and the intersection point among the braces. Nakashima and Wakabayashi (1992) reported the results of some tests performed on X-CBFs that led to effective length factor K ranging between 0.40 and 0.65. Tests carried out by Tremblay et al (2003) on continuous X braced frames, with diagonal members made of square hollow profiles directly welded to each other and continuity cover plates as stiffeners, showed that the slenderness ratio determined assuming that the tension brace provides a full support at mid-length of the compression brace is adequate. However, the efficiency of the support deteriorates due to cyclic damage and to the permanent extension of the brace, which lead to zero restrain at the zero storey drift position.

Palmer (2012) showed that the buckling capacity of continuous X-diagonals is accurately estimated by using one half the brace length. In addition, he showed that out-of-plane buckling of torsionally stiff braces can impose severe torsional demands on end-gusset plates, with reduced ductility offered by the connections. In addition, the plastic deformation capacity of the X-braced system is often smaller than that achievable with a single diagonal at the same working length (i.e. the distance between the two opposite nodes in the diagonal of the frame). This feature was observed experimentally and it is more important for discontinuous X-CBFs (Palmer et al, 2013), because the inelastic deformation tends to be concentrated in one-half the brace length (e.g. two hinge buckling mode shown in Figure 4.43), owing to both constructional and material imperfections that do not guarantee perfect symmetry of both portions of the brace at horseback of the mid-connection. Hence, the more damaged half brace deteriorates while the other half cannot fully develop its capacity.

In light of this brief review of existing studies, it seems reasonable to assume $K = 0.5$ for both in-plane and out-of-plane buckling of X-braces.

4.4.5 Detailing of brace-to-column base connections

The design criteria and the geometric requirements for gusset plates of brace connections at the base of the columns are basically the same as given in section 4.4.2. The only aspect to account for to establish the gusset-plate dimensions is to determine the point of intersection of the re-entrant corner of the gusset plate, which can intersect the base plate (see Figure 4.44a) or the column (see Figure 4.44b).

4. Design Recommendations for Ductile Details

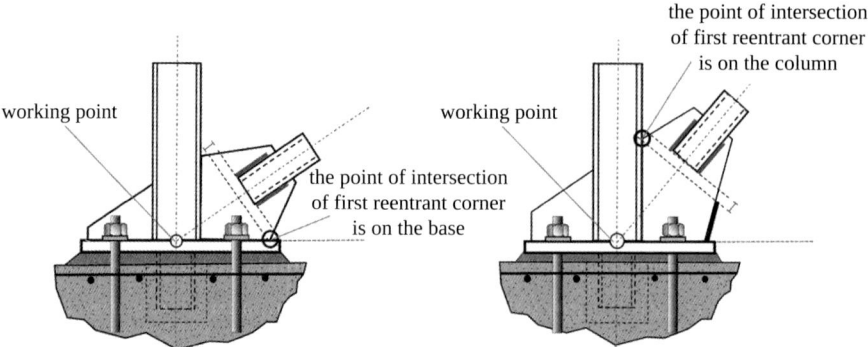

Figure 4.44 – Examples of brace-to-column base connections (adapted from Astaneh-Asl *et al*, 2006)

Brace-to-column base connections are designed to resist significant forces that are the plastic strength directly transferred by the tension brace and the vertical force transferred by the column which can be alternatively tension or compression. In the first case it is crucial to avoid any uplift, which might occur in shallow foundations. If there is this possibility, in case of direct foundation the footings should be ballasted using either spread footing, grade beams, or slab-on-grade, etc.). Anyway, the most effective solution is to adopt deep foundations with piles that can effectively restrain vertical movements.

The horizontal shear force at each column base of the braced span is generally larger than those in MRFs. Hence, it is convenient to adopt socket column stub to transfer the shear force to foundation, while anchor rods are designed to resist mainly vertical forces. Indeed, under high uplift design loads, bending may occur in the base plate between the anchor rods and flanges/web of the column. This mechanism should be avoided and plate stiffeners should be welded to each side of the column in order to restrain the base plate.

4.4.6 Optimal slope, constructional tolerances and local details for braces

It is clear that the geometrical layout of bracing members influences the seismic response of CBFs. Indeed, the smaller the brace-to-beam slope the larger is the lateral stiffness of the structure. Conversely, increasing the verticality of braces, the lateral stiffness decreases and larger cross sections are necessary for the bracings. Anyway, the slope of the braces is important also for technological and constructional reasons. If the relative brace-to-beam angle is smaller than

30° and larger than 60° the gusset plates become quite large (i.e. filling the most of the free space into the vertical plane of the braced span) and are not economical. Therefore, as described by Astaneh-Asl *et al* (2006), it is advisable to change the arrangement of the bracings, by shifting the diagonal into another span such that the brace slope is within the range of 30° to 60°.

In the theoretical model of CBFs the brace, the column and beam centrelines intersect together in the same node. However, this assumption may often impose quite large gusset plate connections that could be impractical and expensive in real structures, thus enforcing designers to detail the end connections of the bracings with some eccentricities when compared to the ideal concentric model (see Figure 4.45). If this detail is adopted, in order to avoid any potential change of predominant inelastic deformation away from the bracing, the designer should account for all secondary moments developing in the connection due to the eccentricity between the axial force transferred by the brace and the frame centrelines. However, this design choice raises some weaknesses and fallacies in the prediction of the actual structural behaviour. If the brace-to-frame eccentricity is too large the structural scheme shifts from CBF to EBF. EC8-1 does not provide any requirements about the allowable constructional tolerances and maximum connection eccentricity for CBFs. Also AISC 341-16 does not give restrictions on the amount of eccentricity allowed in the brace-to-frame connections. As discussed by Cochran (2003), some practical and reasonable requirements about this issue were provided by the Uniform Building Code (1997), which limited the maximum connection eccentricity to the smaller of half of the beam depth or half of the column depth intersected by the brace.

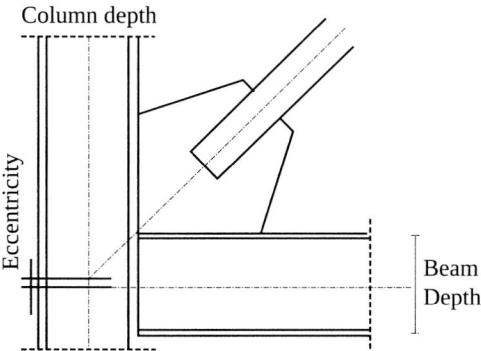

Figure 4.45 – Examples of eccentricity between brace and frame centrelines (adapted from Cochran, 2003)

4. Design Recommendations for Ductile Details

Another key aspect deeply affecting the hysteretic performance of bracing members is the detailing of the net area of their cross section intersecting the gusset plates. Owing to fabrication and erection requirements, tubular bracings may be slotted at the intersection with the gusset plate and bracings made of wide flange profiles may have holes for fasteners. Since braces should operate in the plastic range, significant strain concentration occurs in the net area. Therefore, in order to prevent fracture at the slotted end sections, it is necessary to verify their ultimate strength when the braces yield in tension.

According to clause 6.2.3(2)b of EN 1993-1-1, the design ultimate resistance of the net cross section is given by:

$$N_{u,Rd} = \frac{0.9 \cdot A_{net} \cdot f_u}{\gamma_{M2}} \qquad (4.25)$$

where A_{net} is area of the net cross section, f_u is the ultimate tensile stress of the material and γ_{M2} is the partial safety factor. Since the braces are expected to yield and develop some hardening the verification check for the net area under seismic conditions can be expressed as follows:

$$\gamma_{ov,u} \cdot N_{u,Rd} \geq 1.1 \cdot \gamma_{ov} \cdot N_{pl,Rd} \qquad (4.26)$$

where $\gamma_{ov,u}$ is the material overstrength accounting for the randomness of ultimate stress and $\gamma_{ov,u}$ is the material overstrength accounting for the randomness of yield stress. The correlation between $\gamma_{ov,u}$ and γ_{ov} is still far from being univocally defined. However, it can be roughly assumed that for the material constituting the same member it can be reasonable to consider $\gamma_{ov,u} = \gamma_{ov}$. Under this assumption, the inequality (4.26) can be rearranged as follows:

$$\frac{0.9 \cdot A_{net} \cdot \gamma_{ov,u} \cdot f_u}{\gamma_{M2}} \geq \frac{1.1 \cdot A_{gross} \cdot \gamma_{ov} \cdot f_y}{\gamma_{M0}} \rightarrow \frac{A_{net}}{A_{gross}} \geq 1.2 \frac{\gamma_{M2}}{\gamma_{M0}} \cdot \frac{f_y}{f_u} \qquad (4.27)$$

In addition, if $\gamma_{M2} = 1.25$ and $\gamma_{M0} = 1$, it is possible to express the resistance criterion of the net area as follows:

$$\frac{A_{net}}{A_{gross}} \geq 1.5 \cdot \frac{f_y}{f_u} \qquad (4.28)$$

If the inequality (4.28) is not verified, it is necessary to add some strengthening plates to increase the resisting area of the net area of braces, as shown in Figure 4.40. In case of tubular braces the reinforcing plates should be shop welded on both brace sides and long enough to cover the slotted holes, as shown in Figure 4.46. In case of wide flange profiles, additional drilled plates should be shop welded to the drilled portions of cross section in order to reform the original strength of the member, as shown in Figure 4.47.

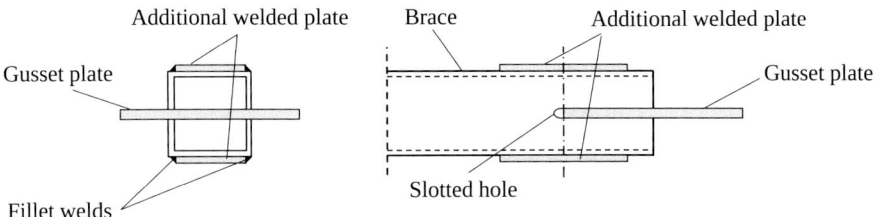

Figure 4.46 – Examples of additional plates to increase the strength of the net area of tubular braces

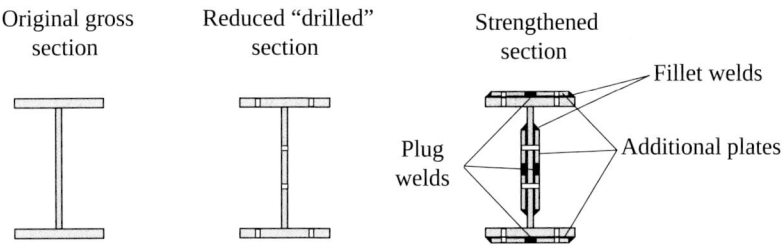

Figure 4.47 – Examples of additional plates to increase the strength of the net area of tubular braces

4.5 DUCTILE DETAILS FOR ECCENTRICALLY BRACED FRAMES

4.5.1 Detailing of links

As discussed in section 3.7.1, link web stiffeners should be designed to prevent inelastic web buckling under large rotation demand. Clause 6.8.2(12a) provides requirements for the spacing of stiffeners in short links that should be determined on the basis of equation (3.36) and equation (3.37) (see chapter 3) depending on the level of expected ductility (see Figure 4.48).

4. DESIGN RECOMMENDATIONS FOR DUCTILE DETAILS

Different details for stiffeners are required for long links. According to clause 6.8.2(12b), web stiffeners should be only placed at a distance of 1.5 times b from each end of the link where it is expected to have the formation of flexural plastic hinge (see Figure 4.49).

The details for stiffeners of intermediate length links should meet the requirements of both short and long links, as indicated by clause 6.8.2(12c).

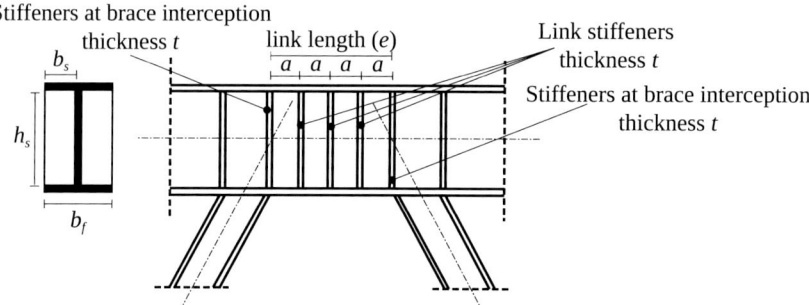

Figure 4.48 – Details for web stiffeners in short links

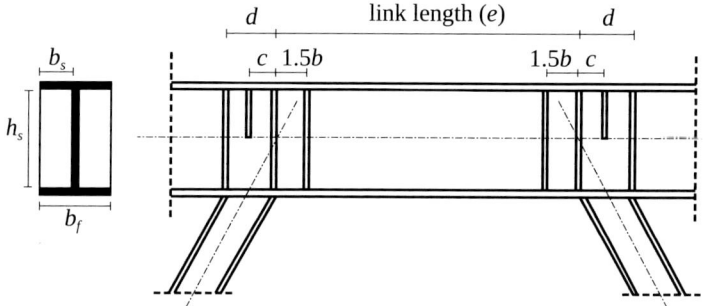

Figure 4.49 – Details for web stiffeners in long links

Due to the high plastic strain concentration, besides the spacing, it is crucial to detail properly the type and the strength of the welds between the stiffeners and the link in order to guarantee a ductile link response. EC8-1 states that full depth intermediate web stiffeners should be considered. For links that are less than 600 mm in depth, stiffeners are required on only one side of the web of the link. The thickness of one-sided stiffeners should not be less than the link web thickness or 10 mm. For links that are 600 mm in depth or greater, similar intermediate stiffeners should be provided on both sides of the web. Moreover, the link stiffeners should be connected to the web of the link by

4.5 Ductile Details for Eccentrically Braced Frames

means of fillet welds having a design strength larger than $\gamma_{ov} f_y A_{st}$, where A_{st} is the area of the stiffener. The design strength of fillet welds connecting the stiffener to the flanges should be larger than $\gamma_{ov} A_{st} f_y /4$ (clause 6.8.2(13)).

In order to strengthen the part of the beam outside the link at the intersection to the diagonal brace ends, full-depth web stiffeners should be provided on both sides of the web (as shown in Figure 4.48 and 4.49). These stiffeners should have a combined width of not less than $(b_f - 2t_w)$ and a thickness t not less than $0.75 t_w$ or 10 mm, whichever is larger, where b_f is the beam flange and t_w is the beam web thickness.

Another important aspect that characterizes the link behaviour is its actual working length related to its theoretical value. The real link length is the distance between the two intersection points between the axis of braces and the central axis of link profile. It is clear that the the slope of bracing is the geometric parameter most affecting the link length. Hence, if "e" is the theoretical length assumed for design calculation, it is necessary to detail the bracing as shown in Figure 4.50a, where the actual link length is equal or slightly smaller than the theoretical "e". Figure 4.50b shows an unsuitable detail. Otherwise, if it is not possible to modify the slope of the braces, it is necessary to consider the actual link length that would result larger than "e" in the design calculations.

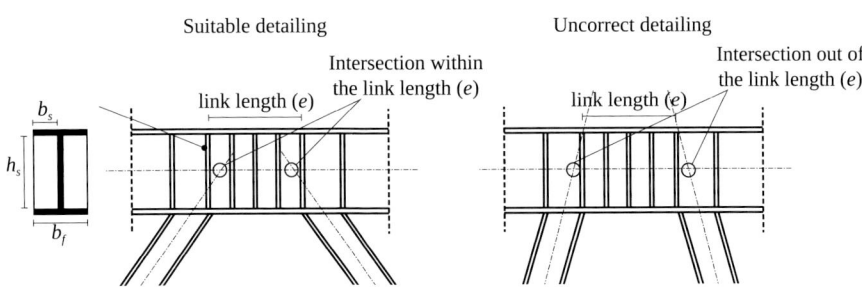

Figure 4.50 – Suitable details for link length

4.5.2 Detailing of link lateral torsional restraints

In order to guarantee stable hysteretic performance, the beam parts outside I-shaped links should be effectively restrained to avoid both shear buckling of the web panels outside of the link and overall lateral-torsional buckling, which might occur prior the exploitation of the available link ductility as shown by Hjelmstad and Lee (1989) and Engelhardt and Popov (1992).

4. DESIGN RECOMMENDATIONS FOR DUCTILE DETAILS

The first buckling mode should be verified by controlling that the shear buckling resistance of the web panels outside of the link conform to EN 1993-1-5 (CEN, 2006c). Regarding the overall buckling mode, clause 6.8.2(14) recommends to design lateral supports at both the top and bottom link flanges at the ends of the link, which should resist axial forces equal to at least 6 % of the expected nominal axial strength of the link flange computed as $f_y \cdot b \cdot t_f$, where f_y is the yield stress of the flange of link cross section, b is the width of the flange and t_f is its thickness. Figure 4.51 shows an example of lateral bracing compliant to EC8-1 recommendations conceived to stabilize the beam segment outside the link in case of an "all-steel" solution, where the beam containing the link is fully disconnected from the slab. In this case, it is necessary to consider double parallel beams, one belonging to eccentric bracings and devoted to resist seismic actions and another supporting the slab and devoted to resist gravity loads only. It is clear that the latter can be effectively designed as composite with the slab, as indicated by the shear connectors in Figure 4.51.

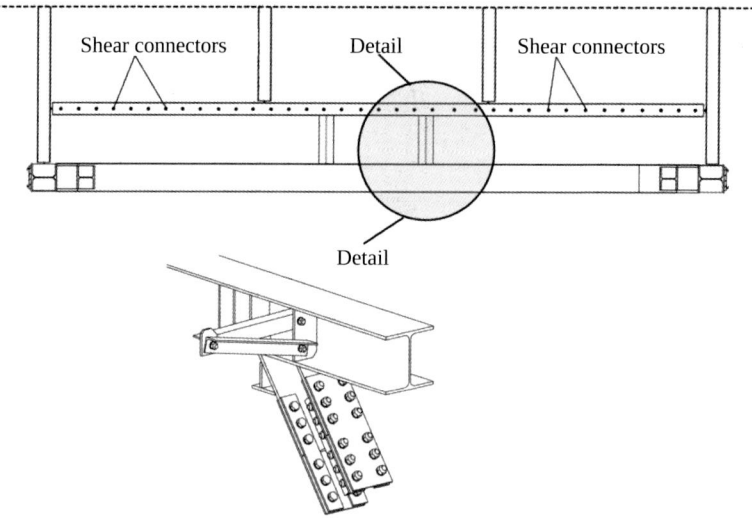

Figure 4.51 – Lateral bracings to restrain the link against lateral torsional buckling: full disconnected slab

In case the beam containing the link works compositely with the slab, it is necessary to disconnect the slab from the link zone, namely no shear connectors should be applied on the upper flange of the link. It is worth noting that even though the concrete slab is connected to the beam segments outside the link, this type of restraint may not be sufficient to provide lateral bracing. Indeed, as shown by Ricles and Popov (1989), the buckling of the composite beam containing the link occurs after shear yielding of the link when a negative moment is applied at one end (i.e. causing compressive stresses in the bottom flange). Therefore, lateral bracings should be designed as recommended by clause 6.8.2(14). An example of this type of detail is shown in Figure 4.52, where the ends of the links are restrained by the slab at the upper flanges, while the lower flanges are restrained by L fly-braces. It is noted that the shear connectors are away from the zone of the beam where the main diagonal braces are attached.

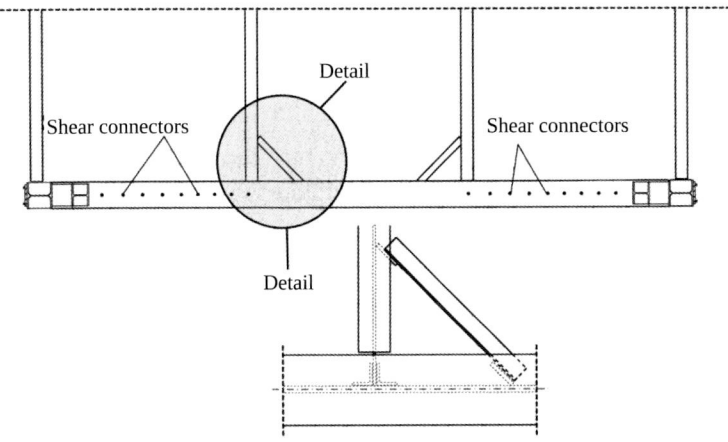

Figure 4.52 – Lateral bracings to restrain the link against lateral torsional buckling: partially connected slab

The case of a slab connected to the beam containing the link deserves additional remarks. Indeed, this type of detail is expected to experience damage in the slab at the interface with the link after a strong earthquake. The slab-to-link interaction may induce significant effects that may impair the seismic performance of eccentric braced frames. Experimental tests carried out by Ciutina *et al* (2013) and by Sabau *et al* (2014) showed that the plastic behaviour of the links may be unsymmetrical and noticeable overstrength may develop due to the presence of the reinforced concrete slab, despite being locally disconnected in the plastic zone, due to the friction between the steel profile and slab. In addition, the experimental tests carried out by Ciutina *et al* (2013) highlighted that simply disconnecting the steel beam from the concrete slab over the dissipative zone is

4. DESIGN RECOMMENDATIONS FOR DUCTILE DETAILS

not sufficient to ensure a pure steel-like behaviour of the link. Their tests showed that the ultimate shear overstrength of short links with disconnected slab was about 20 % larger than that of bare steel links (i.e. $V_{u,link}/V_{p,link} = 1.8 > 1.5$), which was practically very close to that of a full-composite specimen. Therefore, the design overstrength factors Ω for capacity design (see equations (3.36) and (3.37)) should be properly computed adding the plastic strength contribution of the slab to the plastic (either shear or bending) strength of the link.

4.5.3 Detailing of diagonal brace-to-link connections

The design criteria for braces and their relevant connections in EBFs are quite different from those in CBFs. As discussed in section 4.4, the end connections of braces in CBFs must be designed to accommodate the brace buckling and to resist the plastic tensile strength with some hardening, because in that case the braces are the dissipative zones. These requirements are not strictly necessary for end connections of braces in EBFs, as shown in section 3.7.3, because the braces of EBFs must be designed to resist in the elastic range under the design seismic actions (see section 3.7.2). Hence, the brace end connection must be designed to resist the same forces as the braces, as can be seen by comparing equation (3.42) and equation (3.43).

The brace end connections may be designed as full rigid or pinned. In the first case, the braces are generally welded with full penetration welds to the beam containing the link, as shown in Figure 4.53.

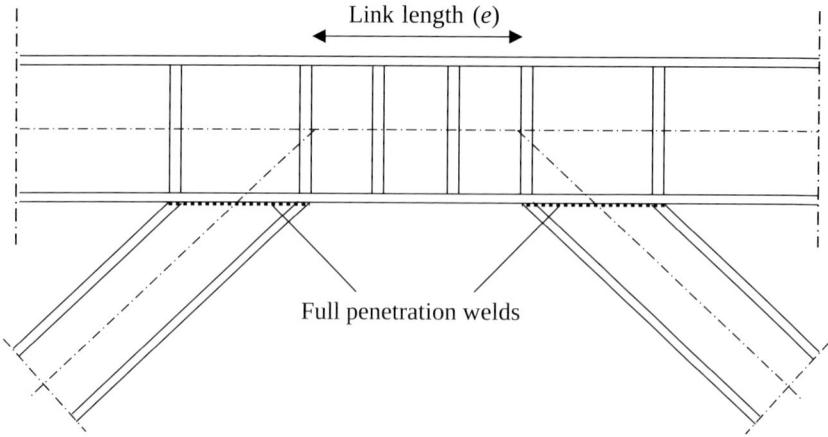

Figure 4.53 – Full rigid brace-to-link connection

4.5 DUCTILE DETAILS FOR ECCENTRICALLY BRACED FRAMES

In case of pinned end, gusset plates are commonly used provided that the gussets do not buckle. Engelhardt and Popov (1989) showed that it is necessary to stiffen the free edges of the gussets and also that the brace end should be extended as much as possible further towards the beam, as shown in Figure 4.54.

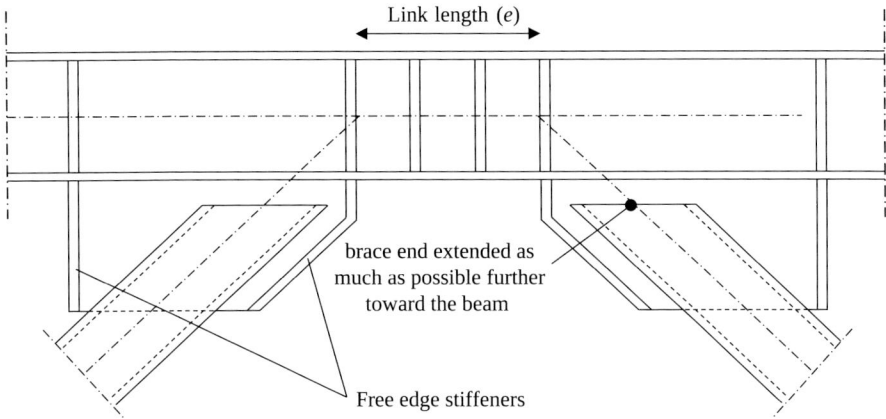

Figure 4.54 – Pinned brace-to-link connection

4.5.4 Detailing of link-to-column connections

In D-EBFs, the link-to-column connection is a key detail that is potentially prone to brittle behaviour occurring before the link exploits its maximum ductility. Experimental tests carried out by Okazaki *et al* (2006) showed that beam-to-column connections usually adopted for ductile moment resisting frames are not efficient for link-to-column connection in EBFs, because they are vulnerable to brittle fracture in the welds. More recent experimental tests carried out by Hong *et al* (2015) showed that it is possible to have excellent performance of link-to-column connections if double supplementary web plates are used to strengthen the end portion of the link next to the column. The supplementary double plates increase both flexural and shear strengths, as well as the rigidity in the reinforced region such that the concentration of secondary flexural stresses in the welds is smoothed out. In addition, Hong *et al* (2015) recommend leaving a small gap above and below the supplementary web plates in order to avoid welds on the link flanges, with welds only along the two vertical edges of the double plates. An example of this type of link-to-column detail is shown

4. Design Recommendations for Ductile Details

in Figure 4.55, where the design forces acting on the connection and the supplementary double web plates are also clarified.

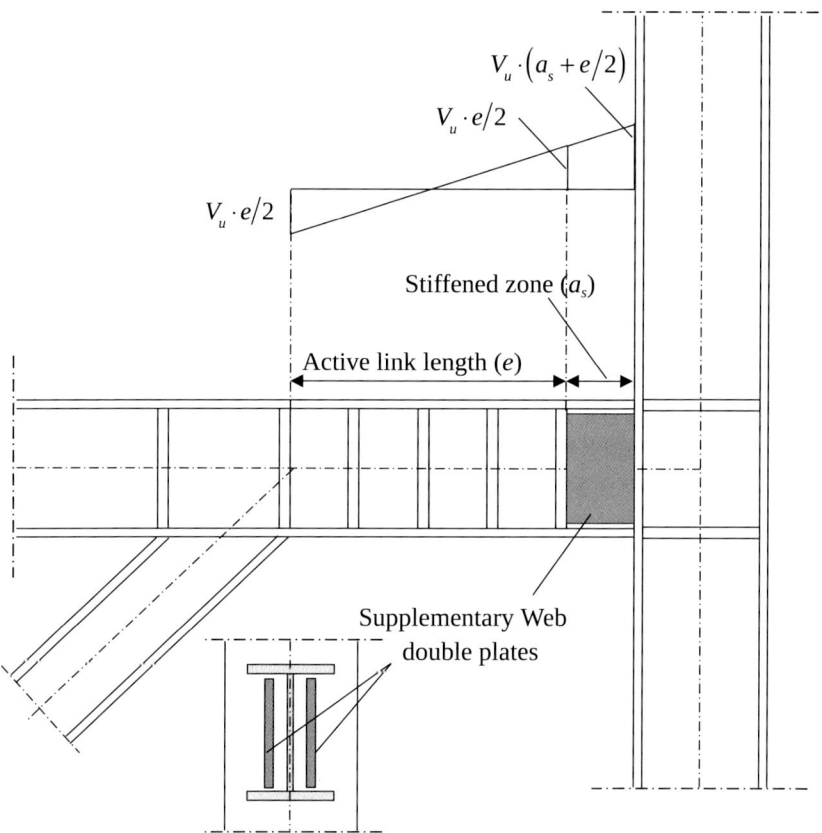

Figure 4.55 – Recommended detail for link-to-column connections

Chapter 5

DESIGN ASSISTED BY TESTING

5.1 INTRODUCTION

Design assisted by testing is a powerful tool for evaluating the performance characteristics of materials, components or kits. In Europe, the principles and application rules are given in EN 1990 (CEN, 2002a), with additional specific requirements in the different parts of the Eurocodes. Design assisted by testing was initially developed for Eurocode 3 (Annex Z of ENV 1993-1-1 (CEN,1998)) but it is now standardized in Annex D of EN 1990 and it is applicable to all kinds of materials and ways of construction (Sedlacek and Muller, 2006). According to EN 1990, all design rules are based on test evaluations using an appropriate test evaluation method.

The type of test depends on the relevant properties to be measured and the conditions of loading or load effect models (Dubina, 2008). The evaluation of the seismic response of structures may enter in this category, as the structure and implicitly the materials can suffer partial or complete damage (breaks, cracks, large deformations, local or global instability). As a result, seismic codes, e.g. EC8 and AISC 341-10 (2010), require that whenever provisions do not adequately cover the calculation of resistance or other parameters, appropriate experimental studies should be performed. Examples of tests that can be included here are the evaluation of the cyclic behaviour of members and connections. For such situations, special provisions have been developed over the last thirty years. An example is the testing procedure for assessing the behaviour of structural steel sub-assemblages under cyclic loads proposed by the European Convention for Constructional Steelwork (ECCS), through its Technical Committee 13 - Structural Safety and Loadings

5. Design Assisted by Testing

(ECCS, 1986). This procedure was mainly intended to enable the assessment of the seismic behaviour of steel elements based on cyclic quasi-static tests using a specified loading history. In the last decade, new systems and devices for seismic resistant structures have been developed. A special category is represented by the anti-seismic devices that are used to modify the seismic response of the structure, for example by increasing the fundamental period of the structure, by modifying the shape of the fundamental mode, by increasing the damping, by limiting the forces transmitted to the structure and/or by introducing temporary connections that improve the overall seismic response of the structure. For such special anti-seismic devices, apart from the requirements and specific rules given in the design code, an attestation of conformity is also required. EN 15129 (CEN, 2009) may be used for the determination of performance characteristics by means of specific experimental tests. Similarly to Europe, testing protocols for determining the seismic performance characteristics of structural and nonstructural components have been developed in the United States (FEMA, 2007; AISC 341-10, 2010). The FEMA 461 guidelines were developed shortly after the 1994 Northridge earthquake, and were limited to the evaluation, repair, modification, and design of welded steel moment frame structures in seismic regions. The AISC 341-10 document specifies loading protocols for qualification testing of connections and other components, based upon the SAC guidelines and subsequent modifications made in 2002 and 2005 editions.

This chapter describes firstly the general guidance provided for design assisted by testing by EN 1990. It is followed by a presentation of specific rules for the testing of seismic components and devices. Finally, an illustrative example of the qualification of a buckling restrained brace is described that details some steps of the procedure for design assisted by testing.

5.2 DESIGN ASSISTED BY TESTING ACCORDING TO EN 1990

5.2.1 Introduction

The general indications regarding design assisted by testing are given in section 5.2 and Annex D of EN 1990.

Clause 5.2(1) of EN 1990 states that design may be based on a combination of tests and calculations. In particular, testing may be carried out in the following circumstances:

- if adequate calculation models are not available;
- if a large number of similar components are to be used;
- to confirm by control checks assumptions made in the design.

Design assisted by testing results shall achieve the level of reliability required for the relevant design situation. The statistical uncertainty due to a limited number of test results shall be taken into account (clause 5.2(2)P). Also, partial factors (including those for model uncertainties) comparable to those used in the appropriate Eurocodes should be used (clause 5.2(3)).

Annex D of EN 1990 details the procedures for design assisted by testing. It should not replace acceptance rules given in harmonised European product specifications, other product specifications or execution standards (clause D1(2)). In the framework of the CE marking requirements laid down in the Construction Products Regulation (EU, 2013), EN 1090-1 (CEN, 2011) and EN 15129 (CEN, 2009) are particularly relevant for design assisted by testing in seismic applications.

EN 1990 provides the procedures for the determination of the characteristic and design material properties, based on the statistical elaboration of test results, as well as the standard procedures evaluating the design resistances of components. If design assisted by testing is used either to replace design by calculation or in combination with the calculation, the testing shall be performed by a testing body approved by the competent authority, or by a body officially notified, or officially accredited, or by a body with equal status and with a quality system. The testing may be realized in a laboratory, under the control of a competent authority. According to EN 1990, testing may be used in design of structures in the following circumstances:

- if the properties of materials are unknown;
- if adequate analytical procedures for designing the component by calculation alone are not available;
- if realistic data for design cannot otherwise be obtained;
- if it is desired to check the performance of an existing structure or structural component;
- if it is desired to build a number of similar structures or components on the basis of a prototype;
- if confirmation of consistency of production is required;

- if it is desired to determine the effects of interaction with other structural components;
- if it is desired to prove the validity and adequacy of an analytical procedure;
- if it is desired to produce resistance tables based on tests, or on a combination of testing and analysis;
- if it is desired to take into account practical factors that might alter the performance of a structure, but are not addressed by the relevant analysis method for design by calculation.

The Annex D procedure is based on the semi-probabilistic approach described in EN 1990. Hence, a brief general overview of EN 1990 principles are presented in the following section, closely following Tankova *et al* (2014).

5.2.2 General overview of EN 1990

All parts of the Eurocodes are based on the partial safety factor method. EN 1990 states the basis of the method. The partial safety factor method recognizes relevant design situations. The safety factors are used on load and on the resistance sides and the design is considered adequate whenever the appropriate limit states are verified:

$$E_d \leq R_d \tag{5.1}$$

where E_d is the design value of the actions and R_d is the design value of the resistance.

The safety factors are either: i) established based on a statistical evaluation of experimental data or ii) on a calibration to experience derived from a long building tradition or iii) both. The partial factors should be calibrated such that the reliability level is as close as possible to the target reliability. The calibration of safety factors can be performed based on full probabilistic methods, or on First Order Reliability Methods. The full probabilistic approach is often not possible to use due to the lack of sufficient statistical data.

The level of safety in EN 1990 is chosen according to Consequence Classes (CC) defined in its Annex B. The Consequence Classes establish the reliability differentiation of the code by considering the consequence of failure or malfunction of the structure. The Consequence Classes (CC) correspond to

5.2 DESIGN ASSISTED BY TESTING ACCORDING TO EN 1990

Reliability Classes (RC), which define the target reliability level through the reliability index β. This index defines the probability of failure, given by:

$$P_f = \Phi(-\beta) \qquad (5.2)$$

where Φ is the cumulative distribution function (CDF) for the standard normal distribution. The reliability index covers the scatter on both resistance and action sides. It can be expressed in terms of the number of standard deviations as shown on Figure 5.1, where σ_E and σ_R are the standard deviations of the actions and the resistance, respectively, (S) is the failure boundary (given as $R - E = 0$) and P is the design point.

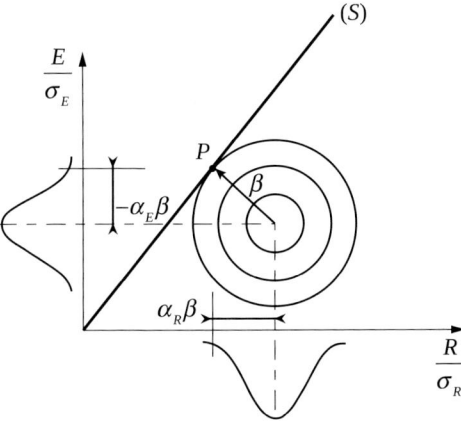

Figure 5.1 – Reliability index β (CEN, 2002a)

According to Gulvanessian *et al* (2002) *"the target reliability index or the target failure probability is the minimum requirement for human safety from the individual or societal point of view when the expected number of fatalities is taken into account. It starts from an accepted lethal accident rate of 10^{-6} per year, corresponding to a reliability index $\beta_1 = 4.7$"*. As the reference period (the design life) depends on the type of structure and its intended use, this leads to $\beta = 3.8$ for most structures (design life of 50 years and Consequence Class 2).

The probability of failure as expressed in equation (5.2) includes the loading and the resistance parts. However, EN 1990 allows separating the scatter due to loading and resistance in terms of coefficients α_E and α_R, respectively (see Figure 5.1), where:

5. Design Assisted by Testing

$$\sqrt{\alpha_R^2 + \alpha_E^2} \approx 1.0 \qquad (5.3)$$

The partial factors related to the resistance are determined based on the following expression:

$$P(r \leq r_d) = \Phi(-\alpha_r \beta) \qquad (5.4)$$

where r stands for resistance and r_d is the design resistance. The factor α_R may be assumed to have a fixed value of 0.8 in case the standard deviation of the load effect and the resistance do not deviate very much ($0.16 < \sigma_E/\sigma_R < 7.6$) (CEN, 2002a). This simplification is crucial for a standardized determination of the partial safety factors for the resistance side without the need to simultaneously consider the loading side (Tankova et al, 2014).

The following sections present general guidelines for the testing phase, followed by a presentation of the procedure for the derivation of design values.

5.2.3 Testing

5.2.3.1 Types of tests

According to Gulvanessian et al (2002), Annex D distinguishes several types of tests (clause D3(1) of EN 1990) depending on their purpose that may be classified into the two following categories: (i) results used directly in design; (ii) control or acceptance tests.

The first category of tests includes the following types:

a) tests to establish directly the ultimate resistance or serviceability properties of structures or structural members for given loading conditions. Such tests can be performed, for example, for fatigue loads or impact loads;
b) tests to obtain specific material properties using specified testing procedures; for instance, ground testing in situ or in the laboratory, or the testing of new materials;
c) tests to reduce uncertainties in parameters in load or load effect models; for instance, by wind tunnel testing, or in tests to identify actions from waves or currents;

5.2 DESIGN ASSISTED BY TESTING ACCORDING TO EN 1990

d) tests to reduce uncertainties in parameters used in resistance models; for instance, by testing structural members or assemblies of structural members (e.g. roof or floor structures).

These test types should lead to design values by applying accepted statistical techniques to the test results (clause D3(2) of EN 1990).

The second category of tests includes the following types:

e) control tests to check the identity or quality of delivered products or the consistency of production characteristics; for instance, testing of cables for bridges, or concrete cube testing;
f) tests carried out during execution in order to obtain information needed for part of the execution; for instance, testing of pile resistance, testing of cable forces during execution;
g) control tests to check the behaviour of an actual structure or of structural members after completion, e.g. to find the elastic deflection, vibrational frequencies or damping.

5.2.3.2 Planning of tests

Prior to the carrying out of tests, a test plan should be agreed with the testing organisation. This plan should contain the objectives of the test and all specifications necessary for the selection or production of the test specimens, the execution of the tests and the test evaluation (clause D4(1) of EN 1990). The test plan should cover:

- objectives and scope;
- prediction of test results;
- specification of test specimens and sampling;
- loading specifications;
- testing arrangement;
- measurements;
- evaluation and reporting of the tests.

In particular, because of its relevance for seismic applications, a list should be made of the procedures for recording results, including time histories

5. DESIGN ASSISTED BY TESTING

of displacements, velocities, accelerations, strains, forces and pressures, required frequency, accuracy of measurements, and appropriate measuring devices.

5.2.4 Derivation of design values

5.2.4.1 Introduction

Annex D of EN 1990 gives a semi-probabilistic procedure for the safety assessment of design methods. The procedures for the statistical determination of resistance models and procedures for deriving design values from tests of type (i) are detailed in the following sub-sections.

In order to use the Annex D procedure, the derivation from tests of the design values for a material property, a model parameter or a resistance should be carried out in one of the following ways (clause D5(1) of EN 1990):

  a) by assessing a characteristic value, which is then divided by a partial factor and possibly multiplied if necessary by an explicit conversion factor;
  b) by direct determination of the design value, implicitly or explicitly accounting for the conversion of results and the total reliability required.

According to Gulvanessian *et al* (2002), EN 1990 recommends method a) with a partial factor taken from the appropriate Eurocode, "*provided there is sufficient similarity between the tests and the usual field of application of the partial factor as used in numerical verifications*" (clause D5(3) of EN 1990).

5.2.4.2 General principles for statistical evaluation

In the evaluation of test results, the behaviour of test specimens and failure modes should be compared with theoretical predictions (a "resistance function"). When significant deviations from a prediction occur, an explanation should be sought: this might involve additional testing, perhaps under different conditions, or modification of the theoretical model (clause D6(1) of EN 1990).

The evaluation of test results should be based on statistical methods, with the use of available (statistical) information about the type of distribution to be used and its associated parameters (clause D6(2) of EN 1990). The methods given in Annex D of EN 1990 may be used only when the following conditions are satisfied:

- the statistical data (including prior information) are taken from identified populations which are sufficiently homogeneous; and
- a sufficient number of observations is available.

In fact (Gulvanessian *et al*, 2002), " *any assessment of probabilities involves basically two types of uncertainty, which are to a certain extent related:*

- *statistical uncertainty due to the limited sample size;*
- *uncertainty due to vague prior information on the nature of statistical distribution.*

These uncertainties may give rise to significant error."

Finally, the result of a test evaluation should be considered valid only for the specifications and load characteristics considered in the tests. If the results are to be extrapolated to cover other design parameters and loading, additional information from previous tests or from theoretical bases should be used (clause D6(3) of EN 1990).

Annex D of EN 1990 presents two procedures for the statistical determination of a single property (section D7 of EN 1990) and the statistical determination of resistance models (section D8 of EN 1990). Both are described in the following sub-sections, closely following Tankova *et al* (2014).

5.2.4.3 Statistical determination of a single property

Section D7 of EN 1990 addresses the derivation of design values from test types: a) tests to establish directly the ultimate resistance or serviceability properties of structures or structural members for given loading conditions; or b) tests to obtain specific material properties using specified testing procedures (clause D7.1(1) of EN 1990).

The single property X may represent: (a) a resistance of a product or (b) a property contributing to the resistance of a product (clause D7.1(2) of EN 1990). In the latter case, the design value of the resistance should also include the effect of other properties, the model uncertainty and other effects such as scaling, volume, etc. (clause D7.1(4) of EN 1990).

5. Design Assisted by Testing

Characteristic or design values of the resistance of a product may be directly obtained from the statistical procedures described in sub-sections D7.2 or D7.3 of EN 1990 that assume that all variables follow either a Normal or a log-normal distribution, that there is no prior knowledge about the value of the mean and that there may be or not prior knowledge about the coefiicient of variation (clause D7.1(5) of EN 1990).

5.2.4.4 Statistical determination of resistance models

Section D8 of EN 1990 addresses the derivation of design values from test types: d) tests to reduce uncertainties in parameters used in resistance models (clause D8.1(1) of EN 1990). In this case, a "resistance function" should be considered. This may be any design rule in the code, and it is henceforth denoted as:

$$r_t = g_{rt}(\underline{X}) \tag{5.5}$$

The theoretical estimate r_t is compared with the experimental one r_e. The procedure considers both types of possible errors, due to model and random uncertainties, which are both briefly summarized below.

Starting with the error related to the design model (Tankova et al, 2014), design models or "resistance functions" are usually theoretical expressions which include as many relevant physical parameters (i.e. "basic variables") as possible and reasonable. As the design model is introduced in terms of r_t, (equation (5.5)), it should be further verified via numerical or experimental tests - r_e. Figure 5.2, also called a scattergram, shows the relationship between the theoretical and the experimental estimates.

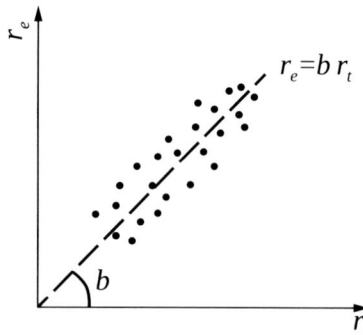

Figure 5.2 – Scatter due to "epistemic uncertainty" (i.e. error of the theoretical design model r_t vs. the experimental evidence r_e)

5.2 Design Assisted by Testing according to EN 1990

where the equation of the regression line passing through the origin is given by:

$$r_e = br_t \tag{5.6}$$

$$b = \frac{\sum_{i=1}^{n} r_{t,i} r_{e,i}}{\sum_{i=1}^{n} r_{t,i}^2} \tag{5.7}$$

The scatter in Figure 5.2 represents the so-called epistemic uncertainty, which is related to the differences that arise between the adopted design model and the reality. This variation is caused by the simplifications of every design model when compared to reality.

The differences are considered in terms of the error δ_i:

$$\delta_i = \frac{r_{e,i}}{br_{t,i}} = \frac{br_{t,i} + \varepsilon_i}{br_{t,i}} \xrightarrow{\varepsilon_i \to 0} \delta_i \approx 1 \tag{5.8}$$

Assuming that the resistance distribution follows a lognormal distribution, the logarithm of the error δ_i is given by the new variable Δ:

$$\Delta_i = \ln(\delta_i) \tag{5.9}$$

The mean value of Δ is found from:

$$\overline{\Delta} = \frac{1}{n} \sum_{i=1}^{n} \Delta_i \tag{5.10}$$

The estimate of the error variance is:

$$s_\Delta^2 = \frac{1}{n-1} \sum_{i=1}^{n} (\Delta_i - \overline{\Delta})^2 \tag{5.11}$$

Finally, the estimator for the coefficient of variation of the error term δ_i is given by:

$$V_\delta = \sqrt{\exp(s_\Delta^2) - 1} \tag{5.12}$$

The error related to the basic input variables, aleatory uncertainty, accounting for the natural randomness, is associated with the basic input variables X_i – yield strength, ultimate strain, geometrical properties, etc.

5. Design Assisted by Testing

Considering the correction factor for the model variance, the resistance function (5.5) may be rewritten as:

$$r = bg_{rt}(X_1, X_2 ... X_n) \tag{5.13}$$

Furthermore, approximations about the moments of functions of random variables, can be found by expansion in a Taylor series about the mean values. Hence the following expressions can be derived, using the notations adopted in EN 1990; If the series is truncated after the linear terms, i.e.,

$$r \cong bg_{rt}(\underline{X_m}) + \sum_{i=1}^{n} b(X_i - X_{m,i}) \frac{\partial g_{rt}}{\partial X_i} \tag{5.14}$$

the first order mean and variance are given by:

$$E(r) \cong bg_{rt}(\underline{X_m}) \tag{5.15}$$

$$Var(r) \cong \sum_{i=1}^{n} \sigma_i^2 \left(\frac{\partial g_{rt}}{\partial X_i} \right)^2 \tag{5.16}$$

Expression (5.16) is based on the assumption that the basic variables X_i and X_j are statistically independent and is obtained from the truncated series of equation (5.14). Whenever statistical dependency is considered, equation (5.16) becomes:

$$Var(r) \cong \sum_{i=1}^{n} \sigma_i^2 \left(\frac{\partial g_{rt}}{\partial X_i} \right)^2 + \sum_{i,j=1 \, i \neq j}^{n} \sum \rho_{i,j} \sigma_{X_i} \sigma_{X_j} \frac{\partial g_{rt}}{\partial X_i} \frac{\partial g_{rt}}{\partial X_j} \tag{5.17}$$

where $\rho_{i,j}$ denotes the correlation coefficient. It can be obtained based on the following expression using experimental data:

$$\rho_{X,Y} = \frac{1}{n-1} \frac{\sum_{i=1}^{n} x_i y_i - n\bar{x}\bar{y}}{s_X s_Y} \tag{5.18}$$

where $\bar{x}$, $\bar{y}$, s_X and s_Y are the sample means and sample standard deviations respectively.

The sensitivity of the resistance function to the variability of the basic input parameters is considered through the coefficient of variation V_{rt}. In case the resistance function is not very complex such as a simple product function, V_{rt} may be obtained as:

$$V_{rt}^2 = \sum_{j=1}^{k} V_{x}^2 \qquad (5.19)$$

However, if the resistance function is expressed by a more complex function, then V_{rt} should be based on equation (5.16), leading to:

$$V_{rt}^2 = \frac{1}{g(X_m)^2} \sum_{j=1}^{k} \left(\frac{\partial g(X_j)}{\partial X_j} \sigma_j \right)^2 \qquad (5.20)$$

or equation (5.17) for statistically dependent variables:

$$V_{rt}^2 = \frac{1}{g(X_m)^2} \left[\sum_{j=1}^{k} \left(\frac{\partial g(X_j)}{\partial X_j} \sigma_j \right)^2 + \sum_{\substack{i,j=1 \\ i \neq j}}^{n} \sum P_{i,j} \sigma_{X_i} \sigma_{X_j} \frac{\partial g}{\partial X_i} \frac{\partial g}{\partial X_j} \right] \qquad (5.21)$$

The partial safety factor – γ_M is obtained by combining both uncertainties. Equation (5.22) combines the effect of scatter due to design model and scatter due to the basic random variables:

$$(V_r^2 + 1) = (V_\delta^2 + 1)(V_{rt}^2 + 1) \qquad (5.22)$$

The second order terms may be ignored if the coefficients of variation are small, leading to:

$$V_r^2 \cong V_\delta^2 + V_{rt}^2 \qquad (5.23)$$

The standard deviation of the lognormal variables is given by:

$$Q_\delta = \sqrt{\ln(V_\delta^2 + 1)} \qquad (5.24)$$

$$Q_{rt} = \sqrt{\ln(V_{rt}^2 + 1)} \qquad (5.25)$$

$$Q = \sqrt{\ln(V_r^2 + 1)} \qquad (5.26)$$

5. Design Assisted by Testing

From a probabilistic stand point, the design value of the resistance should satisfy the following relation, in case of a large number of tests ($n > 100$):

$$\gamma_M = \frac{r_k}{bg_{rt}(X_m)e^{-\alpha_R \beta Q - 0.5Q^2}} \qquad (5.27)$$

In this way, the design value of the resistance considers both uncertainties – the one due to the scatter of the basic input variables and the one related with simplifications introduced in the design model. It also corresponds to the selected reliability level β according to Consequence Classes and the corresponding Reliability Classes.

The characteristic value of the resistance is the resistance function evaluated at nominal values of the basic input variables as follows:

$$r_k = r_{t,nom} = g_{rt}(X_{nom}) \qquad (5.28)$$

In case of a limited number tests (say $n < 30$) allowance should be made in the distribution of Δ for statistical uncertainties according to EN 1990. The distribution should be considered as a central Student's t-distribution leading to equation (5.29a).

The design value of the resistance function based on the mean values of the input parameters is calculated, depending on the sample size used:

$$r_d = \begin{cases} bg_{rt}(X_m)\exp\left(-k_{d,\infty}\dfrac{Q_{rt}^2}{Q} - k_{d,n}\dfrac{Q_\delta^2}{Q} - 0.5Q^2\right) & n \leq 30 \quad (5.29a) \\ bg_{rt}(X_m)\exp\left(-k_{d,\infty}Q - 0.5Q^2\right) & n \to \infty \quad (5.29b) \end{cases}$$

Coefficients $k_{d,\infty}$ and $k_{d,n}$ are design fractile factors, which can be obtained from Table D2 in Annex D of EN 1990.

The partial safety factor is then found:

$$\gamma_M^* = \frac{r_{t,nom}}{r_d} \qquad (5.30)$$

The procedure proposed in Annex D of EN 1990 is summarized in Table 5.1

5.2 Design Assisted by Testing according to EN 1990

Table 5.1 – Annex D procedure

Step 1	Resistance function	$r_t = g_{rt}(X_1, X_2 ... X_n)$
Step 2	Compare experimental and theoretical values	(scatter plot of r_e vs r_t with slope b)
Step 3	Estimate mean vale of correction factor b	$b = \sum_{i=1}^{n} r_{e,i} r_{t,i} \Big/ \sum_{i=1}^{n} r_{t,i}^2$
Step 4	Estimate c.o.v of the error:	$\delta_i = r_{e,i} / b r_{t,i}$ $\Delta_i = \ln(\delta_i)$ $s_\Delta^2 = \dfrac{1}{n-1} \sum_{i=1}^{n} (\Delta_i - \overline{\Delta})^2$ $V_\delta = \sqrt{\exp(s_\Delta^2) - 1}$
Step 5	Analyze compatibility – if the resistance function is acceptable	
Step 6	Determine c.o.v $V_{rt,i}$ of the basic variables	$V_{rt}^2 = \sum_{j=1}^{k} V_{x}^2$ $V_{rt}^2 = \dfrac{1}{g(\underline{X_m})^2} \sum_{j=1}^{k} \left(\dfrac{\partial g(X_j)}{\partial X_j} \sigma_j \right)^2$
Step 7	Determine design value of the resistance	$Q_\delta = \sqrt{\ln(V_\delta^2 + 1)}$ $Q_{rt} = \sqrt{\ln(V_{rt}^2 + 1)}$ $Q = \sqrt{\ln(V_\delta^2 + V_{rt}^2 + 1)}$ $r_d = \begin{bmatrix} b g_{rt}(\underline{X_m}) \exp\left(-k_{d,\infty} \dfrac{Q_{rt}^2}{Q} - k_{d,n} \dfrac{Q_\delta^2}{Q} - 0.5 Q^2 \right) & n \leq 30 \\ b g_{rt}(\underline{X_m}) \exp\left(-k_{d,\infty} Q - 0.5 Q^2 \right) & n \to \infty \end{bmatrix}$

5. Design Assisted by Testing

5.3 TESTING OF SEISMIC COMPONENTS AND DEVICES

5.3.1 Introduction

In case of kits, componentes or devices used for seismic applications, the dynamic nature of the seismic action means that the loading protocol in the testing process plays a crucial role. Hence, in addition to the guidance of Annex D of EN 1990, specific rules need to be followed in the design assisted by testing of seismic componentes or devices.

Among the various strategies to ensure adequate seismic resistance of a structure (see sub-chapter 2.4), improved behaviour may be achieved by modifying the seismic response of the structure by increasing its fundamental period, by modifying the shape of the fundamental mode, by increasing the damping, by limiting the forces transmitted to the structure and/or introducing temporary connections using several types of devices. In this case, the European Standard EN 15129 (CEN, 2009) provides the functional requirements, general design rules, material characteristics, manufacturing and testing requirements, evaluation of conformity, installation and maintenance requirements of anti-seismic devices for use in structures erected in seismic areas in accordance with EC8-1.

In order to simulate the real dynamic conditions of a seismic event in a controlled laboratory environment with the objective of validating the use of a component or device using design assisted by testing methodologies, three main types of experimental testing may be carried out:
- quasi-static monotonic and cyclic testing;
- pseudo-dynamic testing;
- dynamic testing.

These three types of experimental testing are presented and discussed in the following sections.

5.3.2 Quasi-static monotonic and cyclic testing

5.3.2.1 Definition and scope

Quasi-static testing consists of a slow (quasi-static) application of monotonic or cyclic loading, with force or deformation control, applied to a sub-structure, component or device whose behaviour needs to be understood and characterized.

The control parameter may be a force, usually measured with load cells placed at appropriate locations or displacement or other suitable deformation quantity (e.g. rotation). In many situations, in a laboratory test, force and deformation control are both used for different stages of a test.

Cyclic testing, generally performed at progressively increasing amplitudes, aims at simulating the alternating character of the seismic load. When compared to monotonic testing, cyclic testing has the advantage of inducing alternating inelastic deflections in the specimens, thus reproducing more precisely the real stress and deformation levels within tested elements. The use of standardized methods and loading protocols is recommended, as they allow the comparison between test results coming from different laboratories. Quasi-static cyclic testing is generally used to evaluate individual elements (beams, columns), small assemblies (as beam-to-column connections) and also full scale sub-structures. Figure 5.3 shows an example of the hysteretic response of a beam-to-column connection obtained from a quasi-static cyclic test.

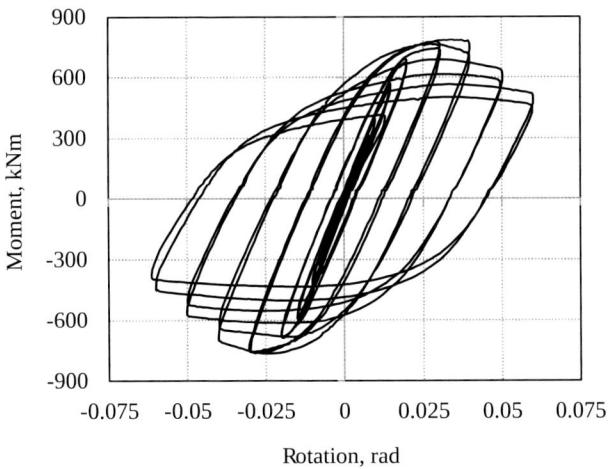

Figure 5.3 – Example of hysteretic response of a beam-to-column connection obtained as result of quasi-static cyclic testing

The low velocity of the input does not allow the application of this type of testing to components whose behaviour depends on velocity. The type of loading history and the number of repetitions of each cycle have an important influence on the cyclic behaviour of the tested components. First, the choice of the loading history is difficult and should be selected such as to replicate as close as possible the real seismic loading. Secondly, the number of cycles is also important, because they may induce low cycle fatigue phenomena, which might be out of the scope of short duration earthquakes. In general, there are several damage states to consider for a specific component and the most common approach is to use a number of specimens that will sustain the complete loading history (see sub-section 5.3.2.6).

5.3.2.2 Specimens

Test specimens should replicate as close as possible the material properties, boundary conditions and loading effects of the real component. In the following, the selection of a specimen for characterization of a beam-to-column joint in a moment resisting frame is used for exemplification. A similar approach can be followed in other cases.

Figure 5.4a shows the deformed shape of a moment resisting frame under seismic loading. The seismic performance of such structures is strongly dependent on the cyclic behaviour of the beam-to-column joints, subjected to large strength demands (in the case of non-dissipative connections) or large plastic deformation demands (in the case of dissipative connections). Consequently, EC8-1 requires that the rotation capacity of the plastic hinge region is guaranteed by experimental testing under cyclic loading. The experimental specimen is typically obtained by isolating a portion of the structure around a beam-to-column joint at the mid-length of the members framing into the joint. An experimental specimen representative of an external (single-sided) beam-to-column joint is shown in Figure 5.4b. A possible replication of the boundary conditions for this specimen is shown in Figure 5.5a. By adopting pinned supports at the column ends and applying a lateral force F at the beam tip, a close match is obtained between the bending moment diagram in the specimen and in the real component, adjusting the beam and column lengths such that the inflexion points correspond to the ends of the beam and columns of the experimental speciment. This scheme neglects the axial force in the column due to gravity loading in the structure, and can be improved by applying a constant axial force at one of the column ends (see Figure 5.5b) using an actuator or pre-tensioned cables. As an alternative that depends on the available laboratory conditions, the specimen may be tested in a rotated position (Figure 5.6a) or in a unrotated position (Figure 5.6c). Figure 5.6 illustrates these two alternative test setups.

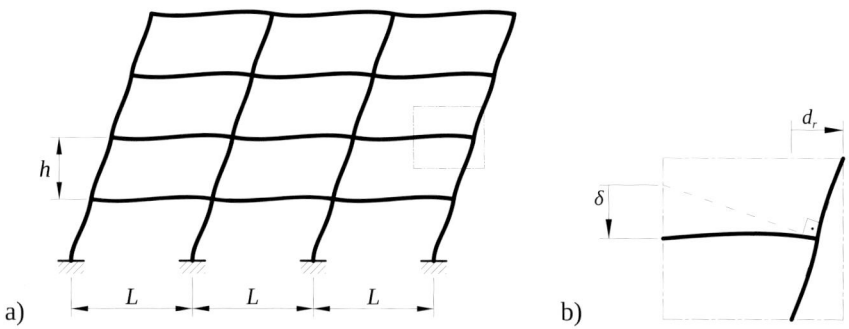

Figure 5.4 – Deformed shape of a moment resisting frame under seismic loading (a) and a close-up view of an exterior beam-column joint (b)

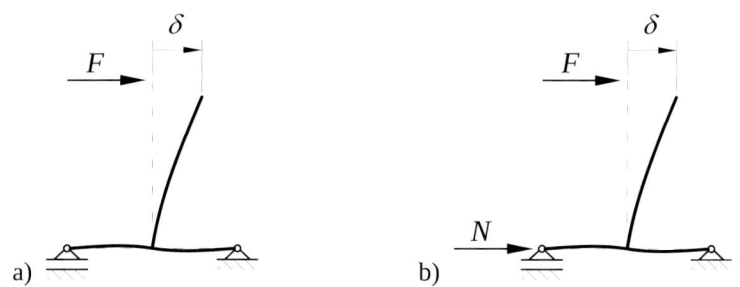

Figure 5.5 – Possible idealizations of boundary conditions of an exterior beam to column joint specimen in moment resisting frames: neglecting axial force in the column (a) or accounting for it (b)

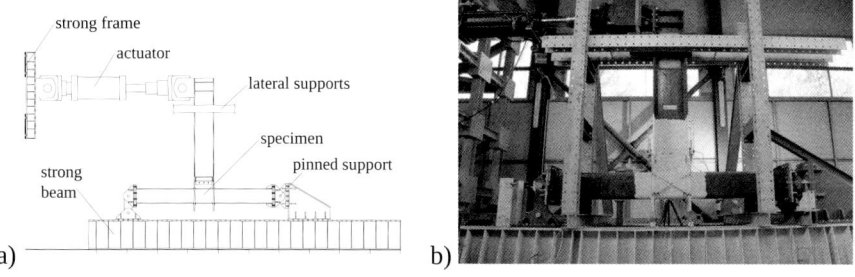

Figure 5.6 – Test setup for a beam-to-column joint specimen: conceptual scheme (a) and its implementation (b) for a rotated position and unrotated position ((c) and (d))

5. Design Assisted by Testing

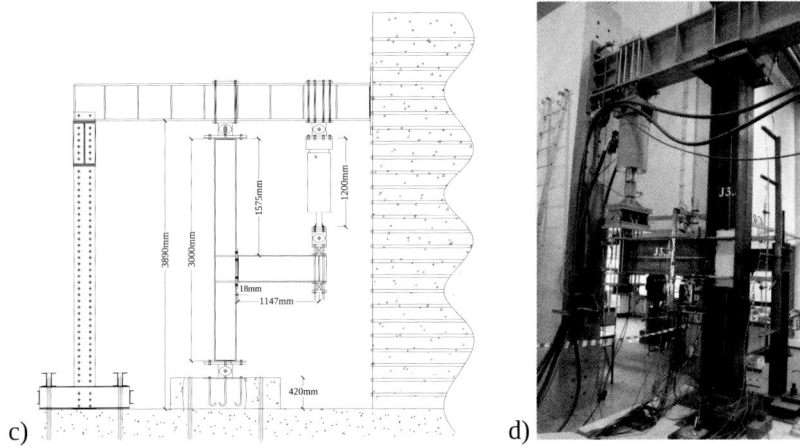

Figure 5.6 – Test setup for a beam-to-column joint specimen: conceptual scheme (a) and its implementation (b) for a rotated position and unrotated position ((c) and (d)) (continuation)

The joint response is characterized by the moment-rotation relationship. The joint rotation is typically defined as (see equation 6.10 in EC8-1):

$$\theta = \frac{\delta}{0.5 \cdot L}$$

where:
- δ is the beam deflection at the mid-span (see Figure 5.4b) and
- L is the beam span.

The beam deflection δ approximates the interstorey drift d_r and is equal to the latter when the beam span is twice the storey height. When this is the case, the joint rotation angle θ is equal to the interstorey drift normalized to the storey height (d_r/h).

Full size specimens are preferred, whenever possible. When specific conditions of the testing do not allow full scale specimens, it is allowed to test down-scaled specimens. The recommendation is to scale the specimen as close possible to the full size specimen in order to reduce the size effects. For example, EN 15129 requires that, if device capacities exceed the feasible range of performance of the existing testing facilities, they can be carried out on reduced scale specimens, whose geometrical scale ratio is not less than 0.5, provided that the pertinent mechanical similitude conditions are fulfilled.

5.3.2.3 Testing equipment

The testing equipment used for quasi-static cyclic testing include reaction frames and/or reaction walls, actuators, the control system, the instrumentation (or sensors) and the data acquisition system.

The reaction frame (or wall) supports the actuators. The reaction frame should have a high strength and stiffness with respect to the test frame and specimens and should be fixed to a strong floor.

The actuator stroke and force capacity should permit the introduction of load until a specific level of damage is reached. When tests are conducted until failure of the specimens, special care should be taken to avoid any risk for the personnel but also to avoid damage to the equipment.

A control system can be either an analog or digital electronic system and is used to control the motions of the actuators.

The instrumentation should allow the measurement of all important parameters that describe the behaviour of the specimen, such as displacements, strains and force. Modern equipment may include also digital image acquisition systems, photographs and video recordings. The exact position and type of each sensor should be specified in order to allow proper interpretation of the results.

A data acquisition is the process of measuring an electrical or physical phenomenon such as voltage, current, temperature, pressure, or sound with a computer. The data acquisition system acts as an interface between the signals from the sensors and a computer that can interpret them.

5.3.2.4 Test plans and procedures

A detailed test plan should be developed prior to the execution of a test program. This test plan should address the following aspects:

- number of specimens to be tested;
- loading history (this should be correlated with the damage states that need to be determined and also with the actuator stroke and force capacity);
- design of a test set-up that permits appropriate load simulation and proper simulation of all important boundary conditions (special care should be taken to prevent specimens from moving out of plane during the test);

5. Design Assisted by Testing

- identification of all important response parameters and the instrumentation plan to measure them.

5.3.2.5 Load control

The quasi-static cyclic tests may be carried out under force control or displacement control. The application of the load under displacement control allows for the evaluation of structural performance in the post-elastic range, as it is possible to track softening behaviour. In order to provide reliable results, the loading rate should respect to following requirements:

- it is sufficiently small such that dynamic effects are negligible and the value of the deformation parameter, at which the onset of the various damage states of interest initiate, is clearly identifiable;
- it is sufficiently large so that the duration of the test is not excessive.

In order to eliminate the dynamic effects throughout a test, the loading application should contemplate regular short stops to allow for the load-displacement curves to settle to the static values (see Figure 5.7)

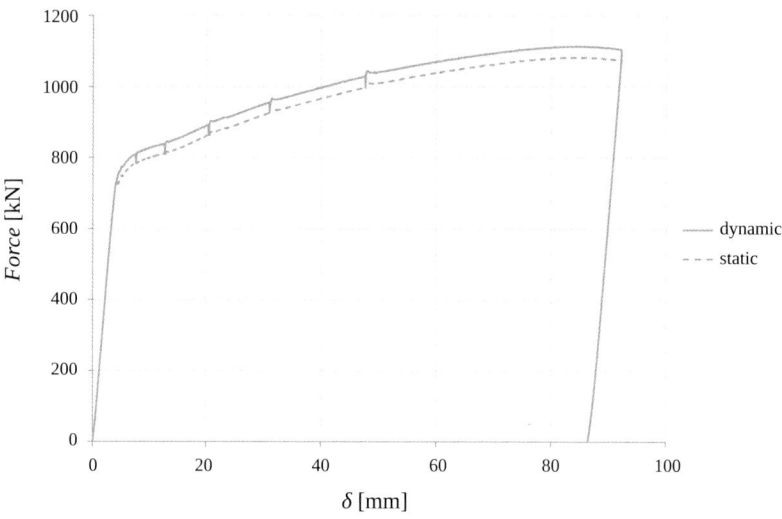

Figure 5.7 – Load-displacement curve illustrating static values occuring at regular stops of loading (Dekker et al, 2015)

5.3.2.6 Loading histories

The cyclic loading protocol requires important considerations because the damage is a cumulative process and is affected by the history of excursions (an excursion is the path from one peak loading value to the next loading peak value). The loading history should be defined such that the number of cycles experienced by the component at the onset of significant damage states is of the same order of magnitude as that experienced by real components in buildings subjected to strong earthquake motion. Particular care should be taken to avoid the introduction of low-cycle fatigue behaviour that is unlikely to be experienced by real components in buildings. Cumulative damage depends mainly on the number and relative amplitudes of the excursions preceding the one at which the damage state is first observed, as well as on the sequence in which the excursions occur and the mean effect (since excursions are not symmetric with respect to the origin). The solution is to develop a loading history, which represents all the cumulative damage effects at all the damage states that are to be quantified in a test (FEMA 461, 2007).

In Europe, most of the past cyclic quasi-static tests of steel components used the loading sequence prescribed in the ECCS Publication 045 (1986). According to this protocol, the seismic effects are replicated by cyclic application of increasing displacements until the relevant damage states are developed. The parameter used for establishing cyclic loading is the yield displacement. Thus, the loading history involves generating four successive cycles for the $\pm 0.25 D_y$, $\pm 0.5 D_y$, $\pm 0.75 D_y$, and $\pm 1.0 D_y$ amplitude ranges, followed further to failure by series of three cycles each of amplitude $\pm 2n \times D_y$, where $n = 1, 2, 3...$, see Figure 5.8a. The reference yield force F_y and the corresponding reference yield displacement D_y are obtained from the recorded force-displacement curve. The reference yield force is defined as the intersection between the tangent modulus S_j at the origin of the force-displacement curve and the tangent that has a slope of $S_j/10$, as indicated on Figure 5.8b. Other conventional definitions of F_y may be used, such as the value corresponding to the 0.2 % offset load at some point in the tested specimen or the maximum load. The end of the test is not defined beforehand. For research purposes the test will probably be continued up to complete failure of the specimens in order to obtain the maximum information. On the other hand a design engineer will probably stop the test as soon as the code requirements are reached.

5. Design Assisted by Testing

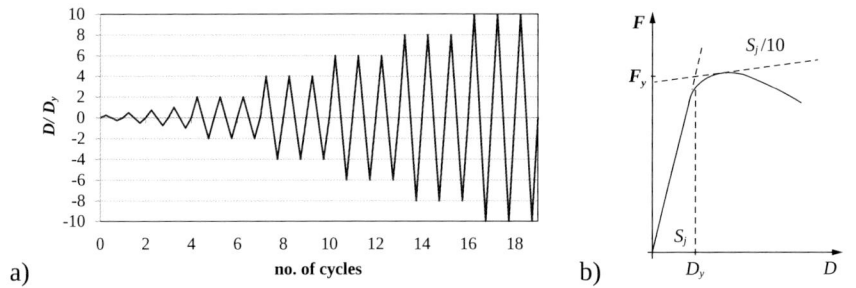

Figure 5.8 – Cyclic loading protocol (a); determination of yielding displacement D_y (b)

With regard to beam-to-column joints in moment resisting frames, EC8-1 states that the rotation capacity of the plastic hinge region should be ensured under cyclic loading without degradation of strength and stiffness greater than 20 %. This requirement is valid independently of the intended location of the dissipative zones. Figure 5.9 illustrates the determination of the ultimate rotation θ_u corresponding to a strength degradation of 20 %.

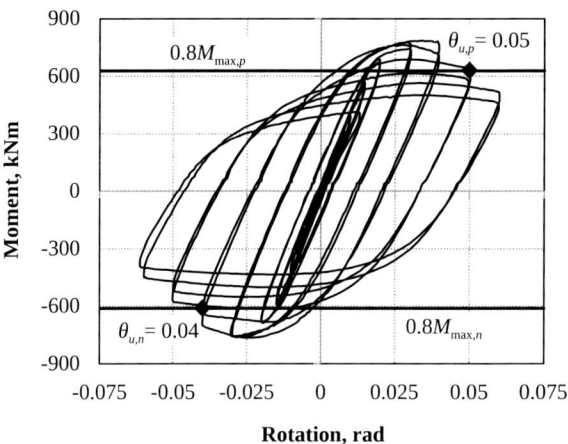

Figure 5.9 – Determination of ultimate rotation according to EC8-1

Another protocol that can be applied to cyclic testing is the test protocol defined in EN 15129. This protocol states that, unless the Structural Engineer prescribes a different program, the test procedure shall include the steps listed below (Figure 5.10):

- evaluation of the force-displacement cycle. Increasing amplitude cycles shall be imposed, at 25 %, 50 % and 100 % of the maximum displacement,

which shall be at least equal to $\pm d_{bd}$, where d_{bd} is the design displacement. Five cycles for each intermediate amplitude and at least ten cycles for the maximum amplitude shall be applied. If the fundamental period of the structural system in which the device has to be used is considerably less than 2 s, a corresponding increase of the number of test cycles at $\pm d_{bd}$ shall be prescribed by the Structural Engineer;
- ramp test for the static evaluation of the failure displacement. Deformations shall be applied at low speed. A displacement not less than d_{bd} multiplied by γ_b and γ_x or a force not less than V_{Ebd} multiplied by γ_b and γ_x, whichever is reached first, shall be imposed. d_{bd} is the design displacement, V_{Ebd} is the force corresponding to d_{bd}, obtained at the 3rd load cycle during a quasi static test, γ_b is the partial safety factor and γ_x is the reliability factor.

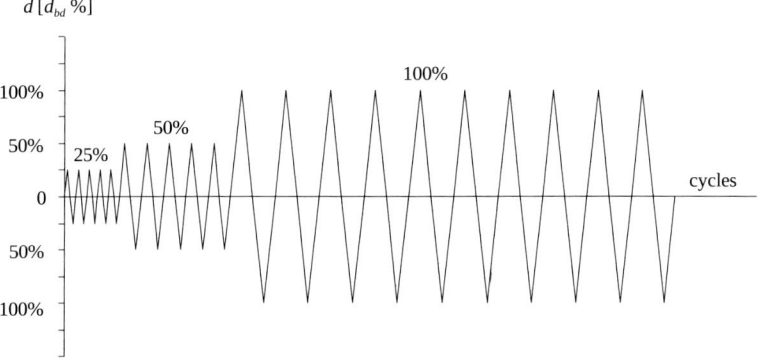

Figure 5.10 – Loading protocol specified in EN 15129

In recognition of different behaviour and demands for structural components, the AISC 341-10 document specifies different loading protocols for beam-to-column connections, link-to-column connections and also buckling restrained braces.

Qualifying cyclic tests of beam-to-column moment connections in special and intermediate moment frames shall be conducted by controlling the *interstory drift angle*, θ, imposed on the *test specimen*, as specified below (Figure 5.11):

- 6 cycles at $\theta = 0.00375$ rad
- 6 cycles at $\theta = 0.005$ rad
- 6 cycles at $\theta = 0.0075$ rad
- 4 cycles at $\theta = 0.01$ rad

5. DESIGN ASSISTED BY TESTING

- 2 cycles at $\theta = 0.015$ rad
- 2 cycles at $\theta = 0.02$ rad
- 2 cycles at $\theta = 0.03$ rad
- 2 cycles at $\theta = 0.04$ rad

Continue loading at increments of $\theta = 0.01$ radian, with two cycles of loading at each step.

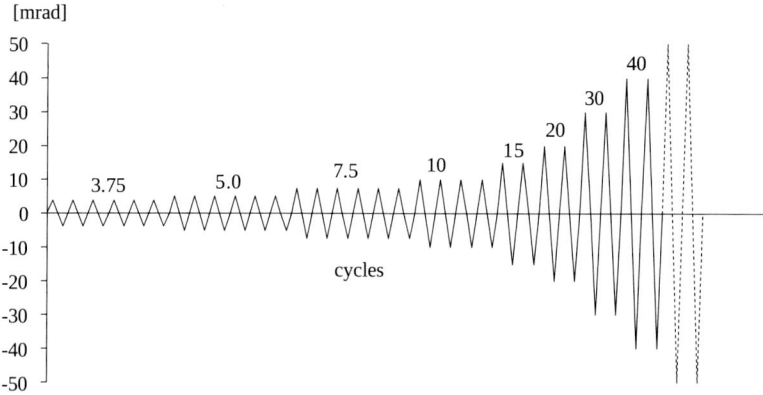

Figure 5.11 – Loading protocol specified in AISC 341-10 for testing beam-to-column connections

Qualifying cyclic tests of link-to-column moment connections in *eccentrically braced frames* shall be conducted by controlling the *total link rotation angle*, γ_{total}, imposed on the test specimen, as follows (Figure 5.12):

- 6 cycles at $\gamma_{total} = 0.00375$ rad
- 6 cycles at $\gamma_{total} = 0.005$ rad
- 6 cycles at $\gamma_{total} = 0.0075$ rad
- 6 cycles at $\gamma_{total} = 0.01$ rad
- 4 cycles at $\gamma_{total} = 0.015$ rad
- 4 cycles at $\gamma_{total} = 0.02$ rad
- 2 cycles at $\gamma_{total} = 0.03$ rad
- 1 cycle at $\gamma_{total} = 0.04$ rad
- 1 cycle at $\gamma_{total} = 0.05$ rad
- 1 cycle at $\gamma_{total} = 0.07$ rad
- 1 cycle at $\gamma_{total} = 0.09$ rad

Continue loading at increments of $\gamma_{total} = 0.02$ radian, with one cycle of loading at each step.

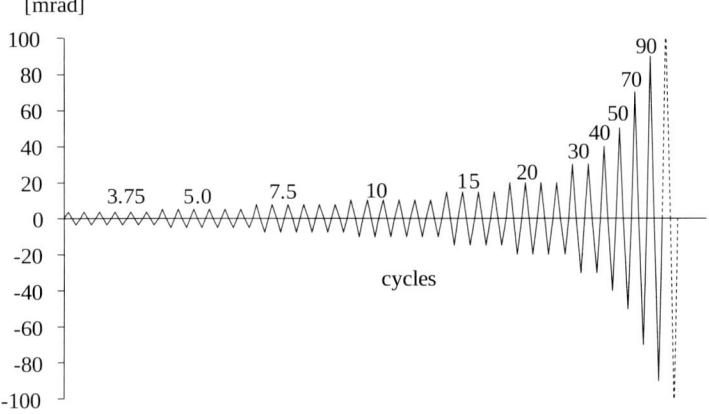

Figure 5.12 – Loading protocol specified in AISC 341-10 for testing link-to-column connections

According to AISC (2005), the cyclic loading sequence for brace test protocol shall be applied to the test specimen to produce the following deformations (Figure 5.13):

- 2 cycles of loading at the deformation corresponding to $\Delta_b = \Delta_{by}$;
- 2 cycles of loading at the deformation corresponding to $\Delta_b = 0.50\Delta_{bm}$;
- 2 cycles of loading at the deformation corresponding to $\Delta_b = 1\Delta_{bm}$;
- 2 cycles of loading at the deformation corresponding to $\Delta_b = 1.5\Delta_{bm}$;
- 2 cycles of loading at the deformation corresponding to $\Delta_b = 2.0\Delta_{bm}$;
- Additional complete cycles of loading at the deformation corresponding to $\Delta_b = 1.5\Delta_{bm}$ as required for the brace test specimen to achieve a cumulative inelastic axial deformation of at least 200 times the yield deformation (not required for the sub-assemblage test specimen).

Δ_{bm} is the value of deformation quantity, Δ_b, corresponding to the design story drift, in mm, Δ_{by} is value of deformation quantity, Δ_b, at first significant yield of test specimen, in mm.

The design story drift shall not be taken as less than 0.01 times the story height for the purposes of calculating Δ_{bm}. Characteristic to this loading

5. Design Assisted by Testing

protocol is the repetition of two cycles at Δ_y, followed by groups of two cycles with increments of $0.5\Delta_{bm}$, until the cumulative inelastic deformation reaches at least 200 times Δ_y.

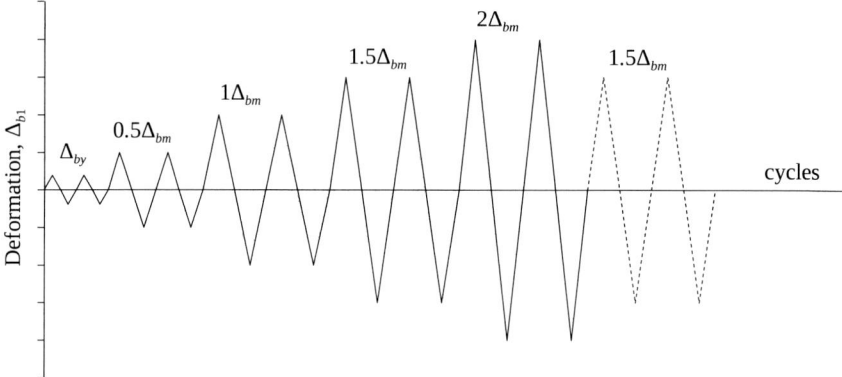

Figure 5.13 – Loading protocol specified in AISC 341-10 for testing buckling restrained braces

5.3.2.7 Reporting

A testing program should be documented in a comprehensive test report. The test report should include information about the testing laboratory, sponsoring agency, test date, date of report, testing protocol (including identification of deviations from this protocol, as appropriate), and the name of the organization reporting the tests (FEMA 461, 2007). The report should also include any witnessing requirements such as a list of independent observers and any required verification/certifications of all test phases. Specific information to be contained in the report also includes:

- a detailed description of the specimens, including geometry, dimensions, materials, and other information needed for understanding the fabrication;
- details of the test equipment, including testing machine, load cell, control system, data acquisition system and sensors;
- loading protocol, including loading history, rate of loading;

- conditions under which the test was carried out, for exemple, ambient conditions;
- measurements methods;
- conditions of the specimens before the beginning of the test and after the test;
- if applicable, the load versus time, displacement versus time and load-displacement relationships (i.e., a complete record in the form of tables, plots, hysteresis curves, and envelope curves) at all measurement locations;
- any abnormal incidents occurring during the test.

5.3.3 Pseudo-dynamic testing

The quasi-static cyclic testing does not describe accurately the deformation history experienced by a structural component in the case of a real earthquake. In the latter case, the component response has a random character and does not increase following a progressive predetermined pattern. An investigating method to assess in a better way the seismic response of structures at a natural scale is pseudo-dynamic testing. The concept was proposed for the first time by Hakuno *et al* (1969), Takanashi (1975) and Takanashi and Nakashima (1987).

The basic concept of pseudo-dynamic testing is that the dynamic response is computed using the experimental result in each time step. During the analysis process, the structural response (displacement) is calculated numerically for each time step. Inertial and damping forces, required during the analysis process for the solution of the equations of motion are modeled analytically. After the calculation of the structural displacements for a specific time step, the numerical calculation module provides this result to the actuator system. In the experimental process, the actuator control system imposes the calculated displacement and then measures and returns the restoring force, $R(t)$, to the computer. With the measured data, the computer can calculate the response for the next time step. With this feedback procedure, the nonlinear inelastic dynamic response can be obtained without shaking table test devices. Figure 5.14 illustrates a flowchart of this process.

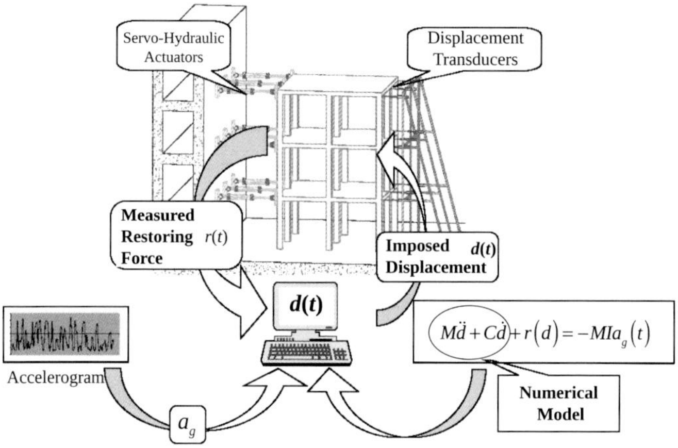

5. Design Assisted by Testing

Figure 5.14 – Scheme of PSD testing principle (Pseudo Dynamic Tests)

For exemplification, recently, a full-scale PSD test programme was carried out on a dual EBF with replaceable links at the European Laboratory for Structural Assessment (ELSA) of the Joint Research Centre (JRC) in Ispra, Italy. The FP7 SERIES project, entitled "Full-scale experimental validation of dual eccentrically braced frame with removable links (DUAREM)" aimed at: (1) validating the re-centring capability of dual structures with removable dissipative members; (2) assessing the overall seismic performance of dual EBFs; (3) obtaining information on the interaction between the steel frame and the reinforced concrete slab in the link region; and (4) validating the link removal technology. Figure 5.15 illustrates the full size 3D three storey composite building tested using pseudo-dynamic methodologies (REF).

Figure 5.15 – Pseudo-dynamic testing on a full-scale 3D steel building Frame (Sabau *et al*, 2014) performed at ELSA, Italy

5.3.4 Dynamic testing

Both quasi-static cyclic testing and pseudo-dynamic testing are not able to reproduce one of the most important structural features under seismic action, i.e. the strain rate effect. One technique which can overcome this limitation is dynamic testing and it requires a shaking table. The shaking table is a platform moved by hydraulic actuators, which is able to induce vibrations similar to those induced by real earthquakes. The platform allows the testing of full scale or scaled down structural models, non-structural components or services. A very large variety of such devices exist, which differ by physical dimensions and the number of degrees of freedom (between 1 and 6) for which displacements or rotations may be imposed.

The shaking table is able to reproduce most accurately the loading conditions on structures during earthquakes. However, there are some drawbacks of this simulation:

- the required cost for construction and maintenance is particularly high;
- owing to the high cost, most shaking tables do not allow for full scale testing;
- the short duration of the tests makes it difficult to perform an accurate observation of the structural response.

Figure 5.16a depicts the two shaking tables of the Department of Structures for Engineering and Architecture at the University of Naples Federico II, while Figure 5.16b,c illustrates a test on a 2-storey light-weight steel building.

a)

Figure 5.16 – Shaking tables at University of Naples Federico II (a) and test on a 2-storey building (b, c)

Figure 5.16 – Shaking tables at University of Naples Federico II (a) and test on a 2-storey building (b, c) (continuation)

5.4 APPLICATION: EXPERIMENTAL QUALIFICATION OF BUCKLING RESTRAINED BRACES

This sub-chapter presents an application of design assisted by testing to a specific anti-seismic device: a buckling restrained braces (BRB). This example corresponds to a real case carried out at the "Politehnica" University of Timisoara, Laboratory of Steel Structures, in Romania, full details being found in Bordea (2010).

5.4.1 Introduction and scope

Buckling restrained braces (BRB) constitute a specific type of anti-seismic device that falls under the category of Displacement Dependent Devices (DDDs) and the sub-category of Nonlinear Devices (NLDs). Due to their strongly non linear behaviour, buckling restrained braces are normally used as main componentes of energy dissipating bracing systems with high energy dissipating capability and can be used to change favourably the dynamic characteristics of a structural system. They are based on the hysteretic behaviour of mild steel elements axially strained in tension and compression well beyond the elastic limit (Annex D of EN 15129). The elements are restrained (usually by an external concrete cylinder) so as to avoid buckling in compression. Their behaviour should be appropriately taken into account in the non linear analyses of the structural systems including these devices.

The design of buckling restrained braces (BRB) shall be based upon the results from qualifying cyclic tests in accordance with the procedures and acceptance criteria of EN 15129 (CEN, 2009). EN 15129 requires for qualifying that tests should be done on individual buckling-restrained braces and on buckling-restrained brace sub-assemblages.

5.4.2 Test specifications

The objective of the research is to evaluate the performance of buckling restrained braces (BRB) by Initial Type Test (ITT). Twelve specimens have been tested monotonically and cyclically, according to ECCS (1986) and AISC (2005) loading protocols.

According to EN 15129, a BRB shall be able to sustain the design displacement, d_{bd}, which is defined as the total displacement that the device will undergo when the structural system is subjected to the design seismic action alone according to EC8-1 multiplied by the partial factor γ_b, where γ_b is a partial factor not less than 1.1. It is therefore of interest to evaluate the maximum displacement that can be attained by the device. A first group of tests, called "Initial Type Test (ITT)" is used whenever new devices are designed. A second group of tests refers to the control of the production and it is necessary to achieve an appropriate level of confidence in the conformity of the product. Table 5.2 summarizes the main specifications of these two types of tests.

Table 5.2 – Tests for Displacement Dependant Devices (selected from EN 15129, Table 17)

Type of Devices		Subject of control	Frequency
Non linear	ITT (Initial Type Testing)	Evaluation of force vs. displacement cycle	1 prototype
		Ramp test	1 prototype
	FPC (Factory Production Control)	Evaluation of force vs. displacement cycle	2 %
		Ramp test	2 %

5. Design Assisted by Testing

5.4.3 Test specimens

In order to determine the effectiveness of the BRB elements, twelve specimens have been designed and executed (Bordea, 2010). Three debonding materials have been used, i.e. rubber with 3 mm thickness, asphaltic bitumen with 2 mm thickness and polyethylene film of 1 mm thickness. The materials used for the test specimens were as follows: for the steel core - S 275 steel (f_y = 275 N/mm², f_u = 400 N/mm², A % = 34 %); infill material - concrete C 40/50. Figure 5.17 shows a schematic of a tested BRB with polyethylene foil, 1mm thick debonding material.

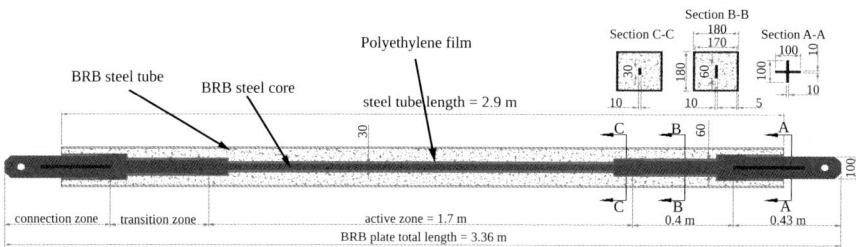

Figure 5.17 – Details and geometry of BRB (debonding material – polyethylene foil, 1mm thick)

5.4.4 Test setup and loading protocol for ITT

A schematic of the test setup is shown in Figure 5.18a while Figure 5.18b shows one specimen installed and the loading arrangement.

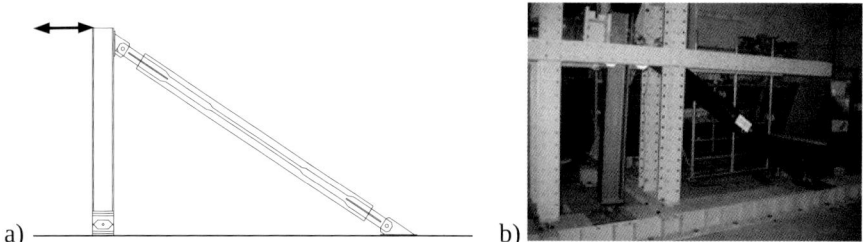

Figure 5.18 – Subassembly test setup: a) schematic of test setup; b) specimen installed and the loading arrangement

First, the monotonic tests were done according to AISC (2005) recommendations, one in tension and one in compression. After the monotonic tests, two cyclic tests for each type of BRB have been performed, one using AISC (2005) loading protocol and one using ECCS loading protocol, see Figure 5.19. The main difference between the two cyclic loading protocols is related to the different number of cycles per loading amplitude. Thus, specimens tested under ECCS loading protocol can be more prone to fail due to low cycle fatigue compared to AISC (2005), due to the larger number of cycles.

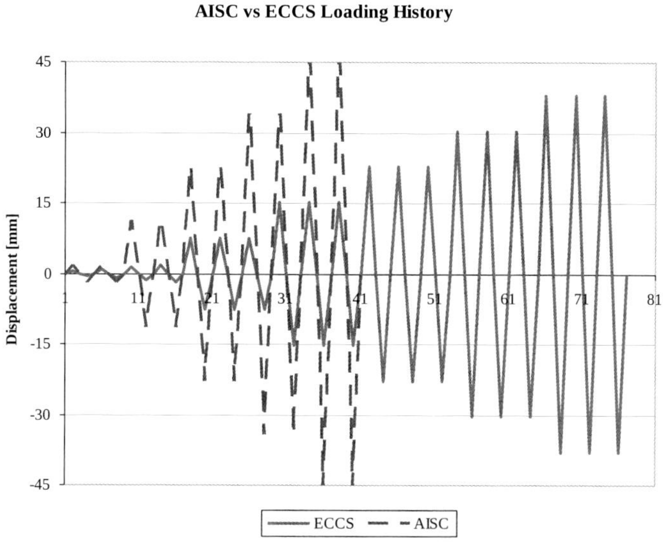

Figure 5.19 – ECCS vs. AISC cyclic loading protocols

5.4.5 Results

The results have shown that debonding material can significantly influence the deformation capacity. The capacity in compression is larger than the capacity in tension by approximately 30 %. The reference values of F_y and D_y amounted to 128 kN and 1.91 mm, respectively and were determined as the mean values of the individual tests.

Figure 5.20 plots the force - displacement curves for the monotonic tests. Due to the special characteristics of buckling restrained braces, the ultimate force under tension (curve T) considerably differs from the ultimate force under compression (curve C). The debonding material can have an important influence on the deformation capacity of the BRBs. Polyethylene film material have shown better behaviour compared to

5. Design Assisted by Testing

other debonding materials and therefore has been selected for cyclic tests. Figure 5.21 shows the hysteresis behaviour of braces. A reduction of the ultimate displacement due to low cycle fatigue can be observed with the ECCS procedure in comparison to the AISC loading protocol. There is a good agreement between envelopes of the cyclic curves and monotonic curves (Figure 5.22).

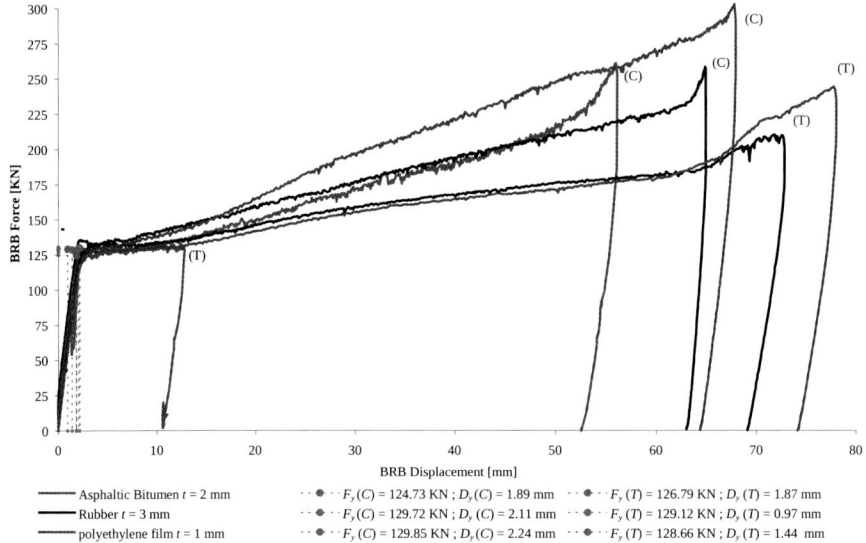

Figure 5.20 – The monotonic behaviour of the BRB specimens (compression vs. tension)

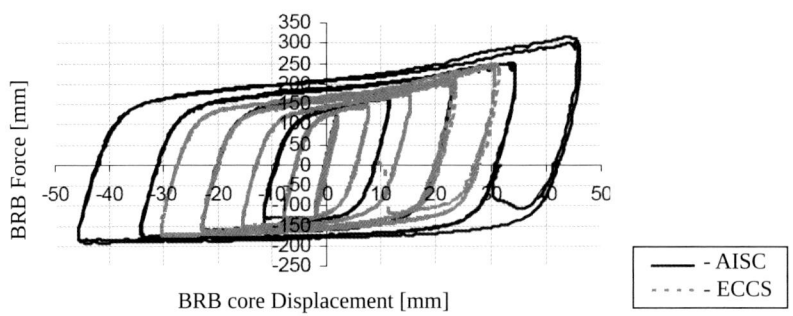

Figure 5.21 – Hysteretic behaviour of BRB with polyethylene film debonding material, AISC and ECCS loading protocols

5.4 APPLICATION: EXPERIMENTAL QUALIFICATION OF BUCKLING RESTRAINED BRACES

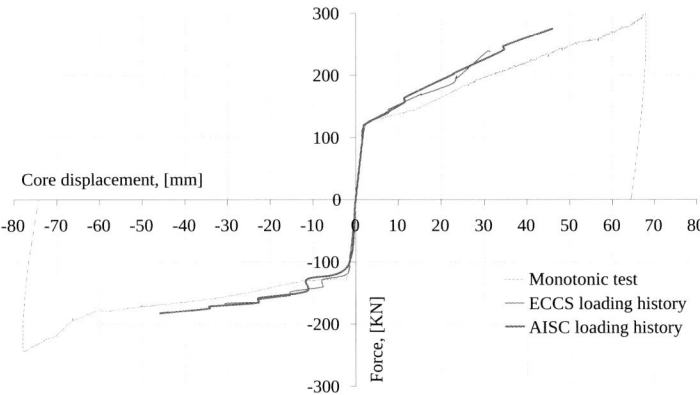

Figure 5.22 – Monotonic tests vs. the envelopes from AISC and ECCS loading protocols, specimens with polyethylene film debonding material

The experimental results have shown a good performance of the BRB, large deformation capacity and a very stable behaviour when loaded cyclically. Debonding material can have an important influence on the deformation capacity of the BRBs. Polyethylene film material was selected on the base of these observations. For the calibration of the numerical model, the compression-strength adjustment factor, β, that accounts for the compression overstrength (with respect to tension strength), and tension strength adjustment factor, ω, that accounts for strain hardening, were evaluated as a mean value of the two loading protocols, resulting a value of 1.3 for β, and a value of 2.2 for ω.

5.4.6 Fabrication Production Control tests

In order to validate the conformity of the selected BRBs, a system composed by a RC portal frame extracted from the initial structures and equipped with the seismic device (i.e. BRB) was tested (see Figure 5.23). The first test was the ramp test, and served for the static evaluation of the failure displacement. The second test served for the evaluation of force vs. displacement cycle. Both tests confirmed the proper behaviour of BRB and the capacity to sustain the maximum displacements demands (for details, see Dinu *et al*, 2012).

5. Design Assisted by Testing

R.C. Portal frame strengthened with BRB

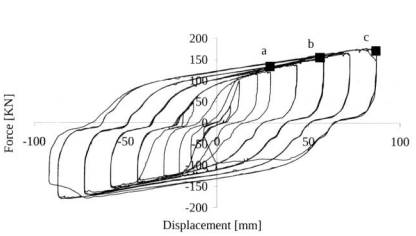
Envelope of cyclic horizontal displacement

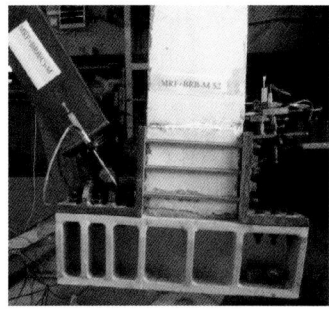
Column base-Brace connection detailing

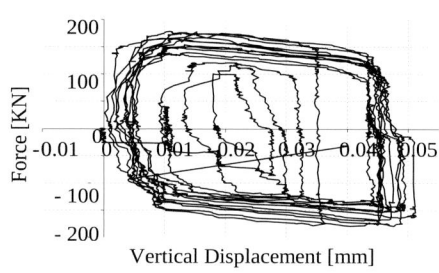
Cyclic loops of vertical displacement

Beam-Brace connection detailing

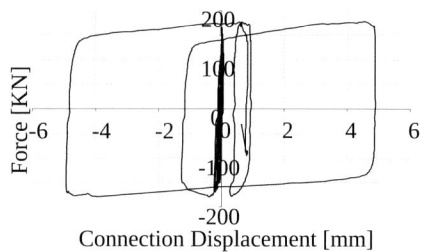
Horizontal slip

Figure 5.23 – General view of the specimen after the test (top) and connecting details (bottom)

Chapter 6

MULTI-STOREY BUILDING WITH MOMENT RESISTING FRAMES

6.1 BUILDING DESCRIPTION AND DESIGN ASSUMPTIONS

This chapter illustrates and discusses the main design steps and the verifications required by EC8-1 for a multistorey building equipped with Moment Resisting Frames (MRFs). Firstly, a general description of the geometric features and design assumptions (i.e. materials, loads, seismic action, etc.) is provided. In a second part of the chapter, the assumptions for structural modelling and the methods of analysis adopted to calculate the seismic induced effects are presented. Subsequently, the design and verifications for both ultimate limit state and damage limitation limit state are described and discussed. Lastly, an advanced assessment of the seismic behaviour of the building is carried out using advanced methods. In this case, a nonlinear static pushover analysis is performed for illustrative purposes only, as this multistorey building would not require such an advanced approach. Nevertheless, the nonlinear pushover analysis confirms the adequacy of the design.

6.1.1 Building description

The case study is a six storey office building with a rectangular plan layout and an area of 31.00 m × 24.00 m. The storey height is equal to 3.50 m with exception of the first floor, which is 4.00 m high. Figure 6.1 shows the architectural layout of the typical floor.

6. MULTI-STOREY BUILDING WITH MOMENT RESISTING FRAMES

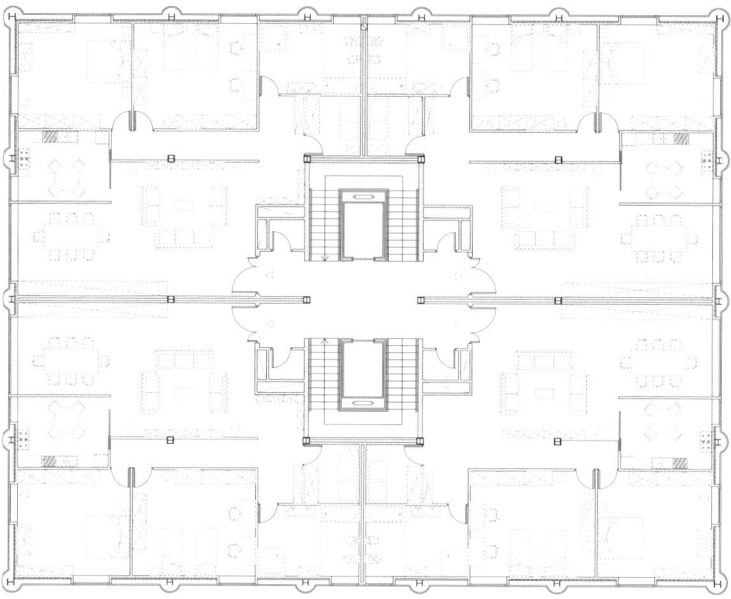

Figure 6.1 – Architectural plan of the typical floor

Composite slabs with profiled steel sheeting are adopted and designed in order to resist the vertical loads and to behave as horizontal rigid diaphragms able to transmit the seismic actions to the seismic resistant frames (see sub-chapter 6.2). The slabs are supported by hot rolled I-section beams belonging to the MRFs placed along the two main horizontal axes of the building. The connection between the slabs and the beams is provided by ductile headed shear studs that are directly welded through the metal deck to the beam flange. As already discussed in chapter 4, the composite slabs would act compositely with the steel beams, thus impairing the performance of the all steel beam-to-column joints. In order to avoid the composite action and to inhibit the load transfer from the slab to the column, in the worked example described within this chapter all moment resisting beam-to-column joints are assumed to be detailed as described in sub-chapter 4.3 (see Figure 4.7).

Figure 6.2 shows the structural layout of the typical floor, illustrating the location of the primary MRFs (indicated by thick bold lines), while Figure 6.3 shows the MRF vertical configuration in the two main plan directions. The primary MRFs are placed only in some spans, while the remaining members form secondary frames that are designed to resist gravity loads only. Thus, the corresponding beams and columns are designed with pinned joints, while the beam-to-column joints of the primary MRFs are designed with rigid full strength joints. This solution has gained increasing attention by designers, because the number of expensive connections is reduced and a more economical design may be achieved when compared to spatial MRFs, where all beam-to-column joints are made as rigid and full strength.

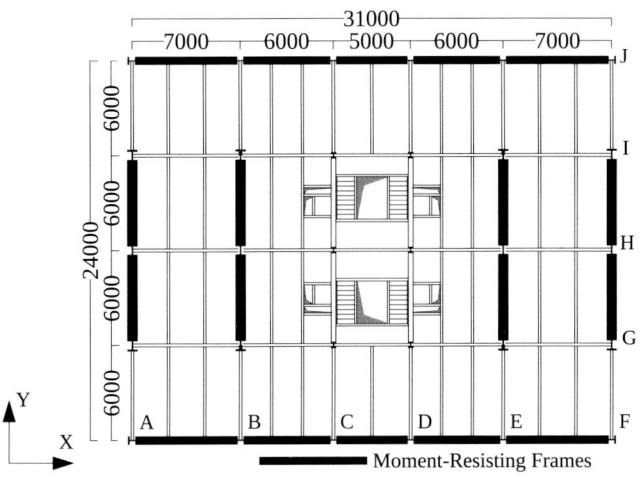

Figure 6.2 – Structural plan of the typical floor

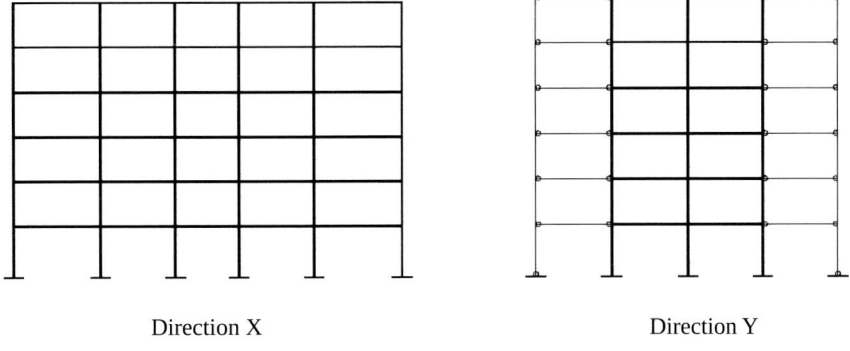

Figure 6.3 – Structural elevation of the primary MRFs

6.1.2 Normative references

Besides the seismic recommendations, the structural design was carried out according to the following European codes:

6. MULTI-STOREY BUILDING WITH MOMENT RESISTING FRAMES

- EN 1990 (2002) Eurocode 0: Basis of structural design;
- EN 1991-1-1 (2002) Eurocode 1: Actions on structures - Part 1-1: General actions - Densities, self-weight, imposed loads for buildings;
- EN 1993-1-1 (2005) Eurocode 3: Design of steel structures - Part 1-1: General rules and rules for buildings;
- EN 1993-1-8 (2005) Eurocode 3: Design of steel structures - Part 1-8: Design of joints;
- EN 1994-1-1 (2004) Eurocode 4: Design of composite steel and concrete structures - Part 1.1: General rules and rules for buildings.

For the sake of generality, a specific National Annex is not used. Hence, the calculation examples were carried out using the recommended values of the safety factors.

6.1.3 Materials

Since the measured yield stress of steel was not known a-priori, the material properties of steel were accounted for according to clause 6.2(3) a of EC8-1, as explained in section 3.2.1. S235 steel grade was specified for both dissipative and non-dissipative members, assuming $\gamma_{ov} = 1.25$, as recommended by EC8-1. It is noted that the material overstrength factor γ_{ov} for S235 can be substantially larger than the EC8 recommended value of 1.25. For example, the Romanian code (MRDPA, 2013) recommends $\gamma_{ov} = 1.40$ while the Italian code (NTC, 2008) recommends $\gamma_{ov} = 1.20$. In specific applications, the relevant National Annex of EC8-1 should be used, where the nationally determined parameters are reported.

Table 6.1 reports the material properties and the relevant partial safety factors for the verification checks.

Table 6.1 – Material properties and partial factors

Grade (-)	f_y (N/mm²)	f_u (N/mm²)	γ_M (-)	γ_{ov} (-)	E (N/mm²)
S235	235	360	$\gamma_{M0} = 1.00$ $\gamma_{M1} = 1.00$ $\gamma_{M2} = 1.25$	1.25	210000

6.1.4 Actions

6.1.4.1 Characteristic values of unit loads

Table 6.2 summarizes the characteristic values of both permanent (G_k) and variable actions (Q_k), corresponding to Category B for building occupancy and usage (i.e. corresponding to office buildings according to EN 1991-1-1). The value for the slab weight is inclusive of steel sheeting, infill concrete, finishing floor elements and partition walls. Analogously, the weight of the stairwells includes the contribution of the flights, steps and landings.

Table 6.2 – Characteristic values of vertical permanent and live loads

	G_k (kN/m²)	Q_k (kN/m²)
Storey slab	4.20	2.00
Roof slab	3.60	0.50
		1.00 (Snow)
Stairs	1.68	4.00
Claddings	2.00	(-)

6.1.4.2 Seismic action

A reference peak ground acceleration equal to $a_{gR} = 0.25g$ (where g is the acceleration of gravity), a type C soil, a type 1 spectral shape and importance factor γ_I equal to 1.0 were assumed.

The spectral parameters for soil type C (Table 2.2 or Table 3.2 of EC8-1) are $S = 1.15$, $T_B = 0.20$ s, $T_C = 0.60$ s and $T_D = 2.00$ s. In addition, the lower bound factor β for the design response spectrum (see section 2.3.5) was assumed equal to 0.2, as recommended in section 3.2.2.5 of EC8-1.

As described in section 2.3.5, the design response spectrum is obtained according to equations (2.23), namely by dividing the ordinates of elastic response spectrum by the behaviour factor q. Figure 6.4 depicts both the resulting elastic and the design spectra.

6. MULTI-STOREY BUILDING WITH MOMENT RESISTING FRAMES

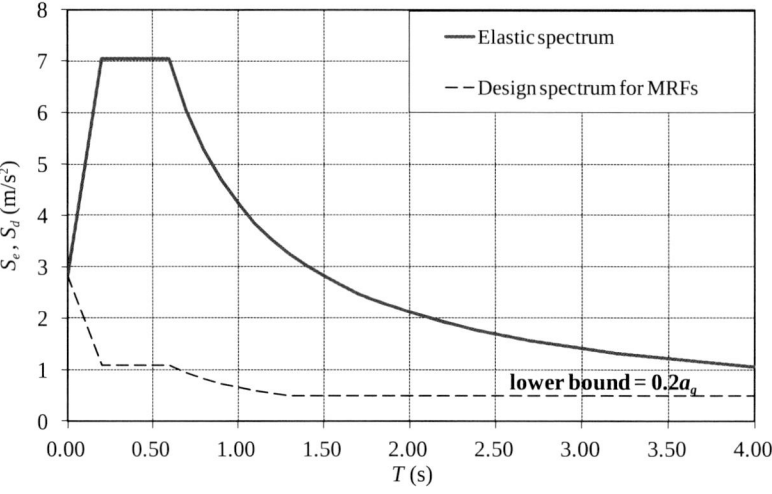

Figure 6.4 – Elastic and design response spectra

The behaviour factor was chosen on the basis of the intended design concept, corresponding to DCH structures (see concept b in sub-chapter 3.1). In addition, since multiple MRFs are considered, the behaviour factor according to clause 6.3.2 (see Figure 3.3 and equation (3.3)) is given by:

$$q = \frac{\alpha_u}{\alpha_1} \cdot q_0 = 1.3 \cdot 5 = 6.5 \qquad (6.1)$$

where q_0 is the reference value of the behaviour factor for systems regular in elevation (see Table 3.2), while α_u/α_1 is the plastic redistribution parameter (see Fig. 3.2a).

6.1.4.3 Combination of actions for seismic design situations

The load combination given by equation (2.24) (see section 2.3.6) can be rewritten as follows:

$$\sum_{j \geq 1} G_{k,j} + A_{Ed} + \sum_{i \geq 1} \psi_{2,i} Q_{k,i} \qquad (6.2)$$

where $G_{k,j}$ is the characteristic value of permanent action j (the self weight and all other dead loads), A_{Ed} is the design seismic action (corresponding to the reference return period multiplied by the importance factor), $Q_{k,i}$ is the characteristic value of variable action i and $\psi_{2,i}$ is the combination coefficient for the quasi-permanent value of the variable action i, which is a function of the destination of use of the building.

The values of $\psi_{2,i}$ are given in Annex A1 of EN 1990 (see Table 2.4). In this case, the values of $\psi_{2,i}$ are summarized in Table 6.3 for the main load categories.

Table 6.3 – Combination coefficients for both loads and masses in seismic design condition

Type of variable actions	ψ_{2i}	φ	ψ_{Ei}
Storey (see Category B – Office areas)	0.30	0.5	0.15
Roof	0.30	1	0.3
Snow (the site is located at altitude $H > 1000$ m a.s.l.)	0.20	1	0.2
Stairs (see Category C – congregation areas)	0.60	0.8	0.48

6.1.4.4 Masses

In accordance with clause 3.2.4(2)P, the inertial effects in the seismic design situation have to be evaluated by taking into account the presence of the masses corresponding to the combination of permanent and variable gravity loads as given by equation (2.25) in section 2.3.6.

$$\sum_{j\geq 1} G_{k,j} + \sum_{i\geq 1} \psi_{E,i} Q_{k,i}$$

The values of $\psi_{E,i}$ are calculated according to EC8-1 (see equation 2.25 and Table 2.5). In this case, the values of $\psi_{E,i}$ and the corresponding combination coefficients φ are summarized in Table 6.3.

Table 6.4 summarizes the seismic weights and the related masses at each floor that incorporate the weight of the pre-designed beams and columns of the whole structure (i.e. both seismic and gravity load resisting members). Seismic weights per unit floor area and the moment of inertia of masses are also reported.

6. MULTI-STOREY BUILDING WITH MOMENT RESISTING FRAMES

Table 6.4 – Seismic weights and masses

Storey (-)	G_k (kN)	Q_k (kN)	Seismic Weight (kN)	Seismic Weight (kN/m²)	Seismic Mass (kN s²/m)	Moment of inertia of masses (kN s²/m)
VI	3256.27	1326.00	3579.67	4.81	364.90	48139.74
V	3992.08	1608.00	4233.28	5.69	431.53	56929.95
IV	3994.08	1608.00	4235.28	5.69	431.73	56956.34
III	4020.54	1608.00	4261.74	5.73	434.43	57312.54
II	4034.87	1608.00	4276.07	5.75	435.89	57505.15
I	4092.99	1608.00	4334.19	5.83	441.81	58286.15

6.1.5 Pre-design

The building examined in this worked example satisfies the criteria for regularity both in plan and in elevation. In particular, as discussed in sub-section 2.4.3.1, the building is regular in plan because it complies with EC8-1 requirements (sub-section 4.2.3.2) as explained in 6.2.1.

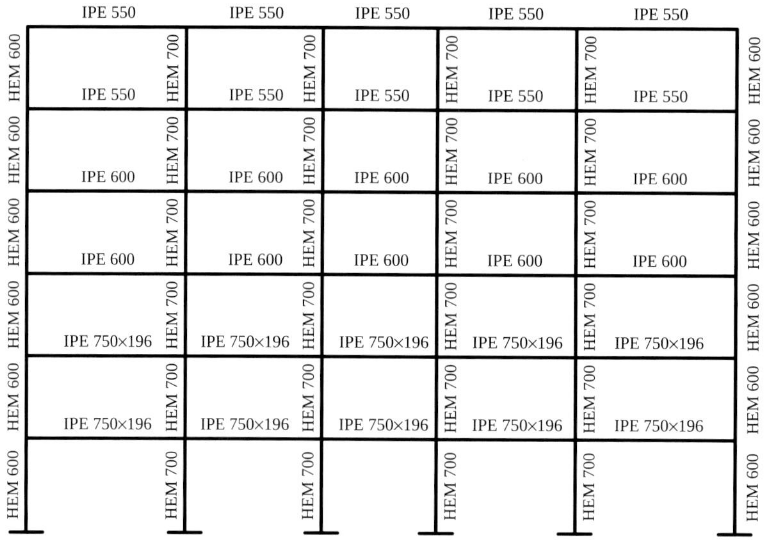

Figure 6.5 – Member cross sections of MRFs in X (a) and in Y (b) direction

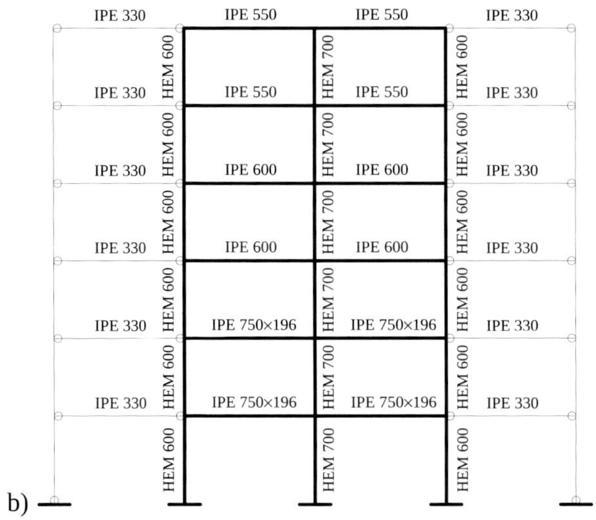

6.2 STRUCTURAL ANALYSIS AND CALCULATION MODELS

```
  IPE 330      IPE 550      IPE 550      IPE 330
┌─────────┬─────────────┬─────────────┬─────────┐
│         │             │             │         │
│ IPE 330 │   IPE 550   │   IPE 550   │ IPE 330 │
│ HEM 600 │   HEM 700   │   HEM 700   │ HEM 600 │
│         │             │             │         │
│ IPE 330 │   IPE 600   │   IPE 600   │ IPE 330 │
│         │             │             │         │
│ IPE 330 │   IPE 600   │   IPE 600   │ IPE 330 │
│         │             │             │         │
│ IPE 330 │  IPE 750×196│  IPE 750×196│ IPE 330 │
│         │             │             │         │
│ IPE 330 │  IPE 750×196│  IPE 750×196│ IPE 330 │
│         │             │             │         │
b)       HEM 700       HEM 700       HEM 600
```

Figure 6.5 – Member cross sections of MRFs in X (a) and in Y (b) direction (continuation)

6.2 STRUCTURAL ANALYSIS AND CALCULATION MODELS

6.2.1 General features

The building examined in this worked example satisfies the criteria for regularity both in plan and in elevation. In particular, as discussed in sub-section 2.4.3.1, the building is regular in plan because it complies with the following requirements (section 4.2.3.2 of EC8-1):

- The building structure is symmetrical in plan with respect to two orthogonal axes in terms of both lateral stiffness and mass distribution;
- The plan configuration is compact; in fact, each floor may be delimited by a polygonal convex line. Moreover, in plan set-backs or re-entrant corners or edge recesses do not exist;
- The structure has rigid in plan diaphragms. It was assumed to use composite slabs that in accordance to clause 9.2.1.2(P) of EN 1994-1-1 provide sufficient diaphragm behaviour (a slab thickness greater than 50 mm and total depth greater than 90 mm, as shown in sub-chapter 4.2);

6. MULTI-STOREY BUILDING WITH MOMENT RESISTING FRAMES

- The in plan slenderness ratio $\lambda = L_{max}/L_{min}$ of the building is lower than 4 (i.e. 31000/24000 = 1.29), where L_{max} and L_{min} are the larger and smaller plan dimensions of the building, measured in two orthogonal directions (see Figure 2.10 in sub-section 2.4.3.2);
- At each level and for both the X and Y directions, the structural eccentricity e_o (which is the nominal distance between the centre of stiffness and the centre of mass) is practically negligible and the torsional radius r is larger than the radius of gyration l_s of the floor mass in plan.

The position of the centre of stiffness and the torsional radius of each floor may be computed from equations (2.29) and (2.30) as the centre of the moments of inertia of the cross sections of the vertical elements of the MRFs as follows (sub-section 2.4.3.2):

$$x_{cs} = \frac{\sum(x \cdot K_x)}{\sum(K_x)} \simeq \frac{\sum(x \cdot EI_y)}{\sum(EI_y)}$$

$$y_{cs} = \frac{\sum(y \cdot K_y)}{\sum(K_y)} \simeq \frac{\sum(y \cdot EI_x)}{\sum(EI_x)}$$

(6.3)

$$r_x = \sqrt{\frac{\sum(\bar{y}_i^2 \cdot K_{x,i})}{\sum(K_{x,i})}} \simeq \sqrt{\frac{\sum(\bar{y}_i^2 \cdot EI_{y,i})}{\sum(EI_y)}}$$

$$r_y = \sqrt{\frac{\sum(\bar{x}_i^2 \cdot K_{y,i})}{\sum(K_{y,i})}} \simeq \sqrt{\frac{\sum(\bar{x}_i^2 \cdot EI_{x,i})}{\sum(EI_{x,i})}}$$

(6.4)

where $K_{x,i}$ and $K_{y,i}$ are the lateral stiffness at each storey of the seismic resisting systems in x and y direction of the plan (see Figure 6.2), while $\bar{y}_i$ and $\bar{x}_i$ are the corresponding distances from the centre of stiffness. Since the seismic resisting systems are planar, their stiffness in perpedicular direction is negligible as well as the contribution of secondary frames that are designed to resist solely the gravity loads. In case of a MRF, it can be reasonably consider for hand calculation the the bending stiffnesses of the vertical elements EI_y and EI_x of the MRFs in a vertical plan parallel to x and y axis, respectively for K_x and K_y.

6.2 Structural Analysis and Calculation Models

The eccentricity e_o and the torsional radius r should satisfy conditions given by equations (2.27) and (2.28), namely:

$$e_{ox} \leq 0.30 \cdot r_x \quad \text{and} \quad e_{oy} \leq 0.30 \cdot r_y \tag{6.5}$$

$$r_x \geq l_s \quad \text{and} \quad r_y \geq l_s \tag{6.6}$$

The radius of gyration l_s of the floor masses in plan is defined as the square root of the ratio between the polar moment of inertia of masses in plan with respect to the centre of mass of the floor and floor mass. In this worked example the floor area is rectangular with dimensions $l_x = 31$ m and $l_y = 24$ m, thus assuming the mass uniformly distributed over the floor, l_s is equal to:

$$l_s = \sqrt{\left(\frac{l_x^2 + l_y^2}{12}\right)} = 11.32 \text{ m} \tag{6.7}$$

Table 6.5 summarizes the calculation of the structural eccentricity for all floors:

Table 6.5 – Coordinates of the centre of stiffness (CS), centre of mass (CM), structural eccentricity e_o, torsional radius (r) and verifications for each floor

Storey	X_{cs} (m)	Y_{cs} (m)	r_x (m)	r_y (m)	X_{cm} (m)	Y_{cm} (m)	e_{ox} (m)	e_{oy} (m)	l_s (m)	e_{ox}/r_x (-)	e_{oy}/r_y (-)	r_x/l_s (-)	r_y/l_s (-)
VI	15.5	12	12	12.5	15.5	12	0	0	11.32	0	0	1.06	1.10
V	15.5	12	12	12.5	15.5	12	0	0	11.32	0	0	1.06	1.10
IV	15.5	12	12	12.5	15.5	12	0	0	11.32	0	0	1.06	1.10
III	15.5	12	12	12.5	15.5	12	0	0	11.32	0	0	1.06	1.10
II	15.5	12	12	12.5	15.5	12	0	0	11.32	0	0	1.06	1.10
I	15.5	12	12	12.5	15.5	12	0	0	11.32	0	0	1.06	1.10

It is worth noting that the conditions expressed by equations (6.5) and (6.6) can be indirectly verified by controlling that the period of the first torsional mode of vibration is lower than the periods of the translational modes in the two horizontal directions, as can be seen in Table 6.6.

The building also satisfies the criteria for regularity in elevation (see sub-section 2.4.3.3), since it complies with the following requirements (section 4.2.3.3 of EC8-1):

- All seismic resisting systems are distributed along the building height without interruption from the base to the top of the building;
- Both lateral stiffness and mass at every storey practically remain constant and/or reduce gradually, without abrupt changes, from the base to the top of the building;
- The ratio of the actual storey resistance to the resistance required by the analysis does not vary disproportionately between adjacent storeys;
- There are no setbacks.

6.2.2 Modelling assumptions

As discussed in sub-chapter 2.6, the structural model must represent the distribution of stiffness and mass so that all significant deformation shapes and inertia forces are properly accounted for under seismic actions (section 4.3.1 of EC8-1).

In this worked example a 3D calculation model was used. All connections for the gravity load designed parts of the frame (beam–to-columns connections, column bases) were assumed as perfectly pinned, but the columns are considered continuous through each floor beam. All connections of the members belonging to the MRFs have been considered full strength and full rigid. In addition, the flexibility of the column web panel zone were not taken into account.

The tributary area of masses was evaluated as described in section 2.6.4. Masses were considered as lumped in a selected master joint for each floor, because the floor diaphragms may be taken as rigid in their planes. This assumption implies that numerical model gives axial forces in the beams equal to zero. Such a simplification is considered acceptable for MRF structures.

Since the structure is characterized by full symmetry of in-plan distribution of stiffness and masses, in order to ensure a minimum of torsional strength and stiffness and to limit the consequences of unpredicted torsional response, EC8-1 imposes accidental torsional effects (see sub-chapter 2.7). These take into account the possible uncertainties in the distributions of stiffness and mass and/or a possible torsional component of the ground motion about a vertical axis. The calculation of these accidental torsional effects is detailed in section 6.2.3.

6.2.3 Numerical models and method of analysis

The calculation model was developed in SAP 2000 (see Figure 6.6) on the basis of the basic assumptions described in section 6.2.2.

The effects of seismic action were determined by means of a linear-elastic modal response spectrum analysis. The design spectra shown in section 6.1.4.2 were used to estimate the seismic actions in the two horizontal directions of the structure. As discussed in section 2.5.3, according to clause 4.3.3.3.1(3) it is required to take into account a number of vibration modes that satisfy either of the following conditions:

1) the sum of the effective modal masses for the modes taken into account amounts to at least 90 % of the total mass of the structure;
2) all modes with effective modal masses greater than 5 % of the total mass are taken into account.

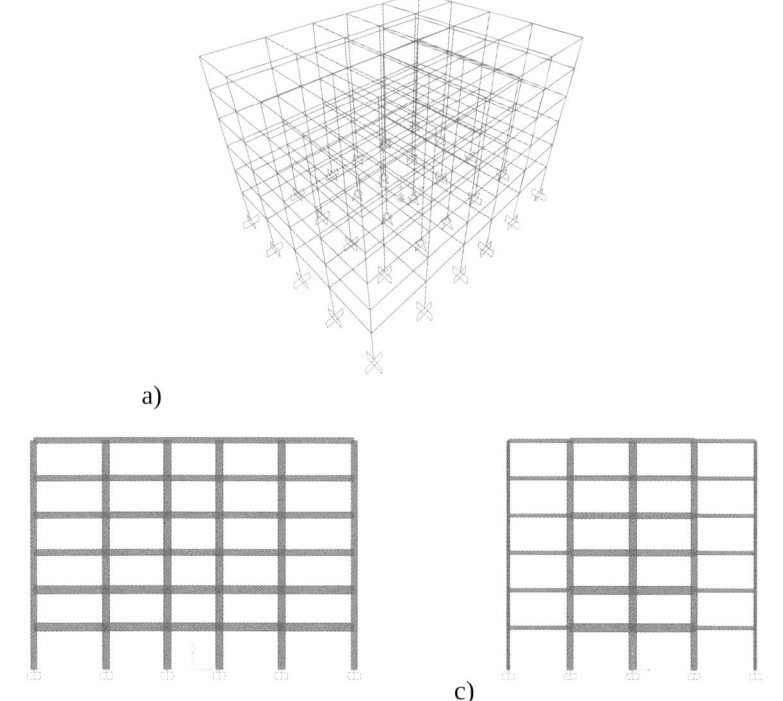

Figure 6.6 – Numerical model: 3D model (a) MRFs in X (b) and in Y (c) directions

In the present example, the first criterion was used, requiring six modes of vibration. The periods T_i and relevant participating mass ratios M_i are reported in Table 6.6, while the shapes of the first three modes (namely those translational in X and Y direction and the one torsional around Z direction) are shown in Figure 6.7.

The SRSS (Square Root of the Sum of the Squares) method is used to combine the modal maxima, since the first and the second modes of vibration in both X and Y direction (i.e. the 2nd and 5th for X direction, the 1st and the 4th for Y direction) are independent (as $T_2 \leq 0.9 T_1$, according to section 4.3.3.3.2 of EC8-1).

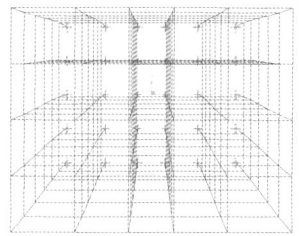
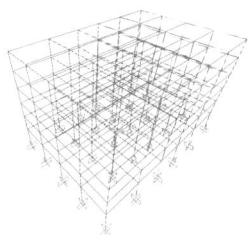

1st mode of vibration

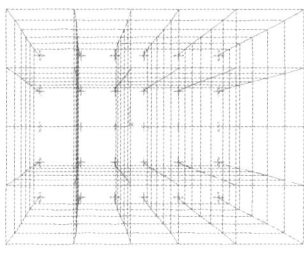
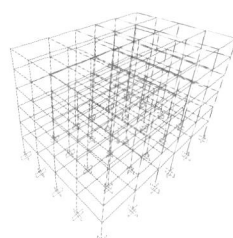

2nd mode of vibration

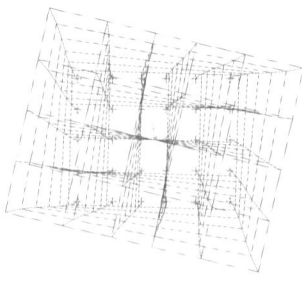
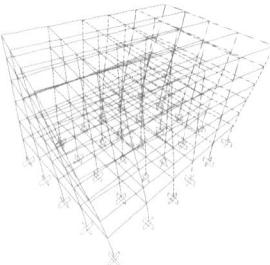

3rd mode of vibration

Figure 6.7 – Modal shapes of the main modes of vibrations

Table 6.6 – Periods and participating mass per mode of vibration

mode	T_i (s)	$M_{i,X}$ (%)	$M_{i,Y}$ (%)	$M_{i,RZ}$ (%)	$\Sigma(M_{i,X})$ (%)	$\Sigma(M_{i,Y})$ (%)	$\Sigma(M_{i,RZ})$ (%)
1	1.006	0.00	76.27	0.00	0.00	76.27	0.00
2	0.908	77.38	0.00	0.00	77.38	76.27	0.00
3	0.648	0.00	0.01	76.86	77.38	76.27	76.86
4	0.341	0.00	15.90	0.01	77.38	92.17	76.87
5	0.311	15.22	0.00	0.00	92.60	92.17	76.87
6	0.223	0.00	0.01	15.58	92.60	92.18	92.46

Accidental torsion effects were taken into account at every level by applying a torsional moment at the centre of mass of each storey, calculated according to equation (2.59) and given by the product of lateral force distribution and accidental eccentricity for each horizontal direction (see section 2.7.3). The value of accidental eccentricity was assumed as 5 % of the length in plan perpendicular to the direction of the considered seismic action (equation (2.56)). The lateral force distribution to estimate the torsional moment was derived using the lateral force method. Table 6.7 summarizes the accidental torsional moments.

Table 6.7 – Accidental torsional moments per floor

Storey (-)	Total seismic Mass (kN s²/m)	$S_d(T_{1,x}=T_1)$ (m/s²)	$S_d(T_{1,y}=T_2)$ (m/s²)	$F_{b,X}$ (kN)	$F_{b,Y}$ (kN)	$F_{X,i}$ (kN)	$F_{Y,i}$ (kN)	$M_{aX,i}$ (kNm)	$M_{aY,i}$ (kNm)
VI	2540.29	0.65	0.72	1396.98	1820.88	345.86	450.81	415.04	698.76
V						342.43	446.34	410.92	691.83
IV						275.98	359.72	331.17	557.57
III						210.67	274.60	252.81	425.63
II						144.12	187.86	172.95	291.18
I						77.91	101.55	93.49	157.40

The diagrams of internal forces (i.e. bending moments, shear and axial forces) due to both gravity and seismic loads are depicted qualitatively in Figure 6.8 and Figure 6.9 for the frames in the X and Y directions, respectively. As discussed in chapter 3, the internal forces obtained from the calculation

6. MULTI-STOREY BUILDING WITH MOMENT RESISTING FRAMES

models are used to design the dissipative members (i.e. the beams in case of MRFs), while the non-dissipative members (i.e. the columns of MRFs, with exception of the sections at their bases) should be designed for the combination given by gravity load induced effects and seismic induced effects, the latter magnified as given by equation 3.5 (see section 3.4.4).

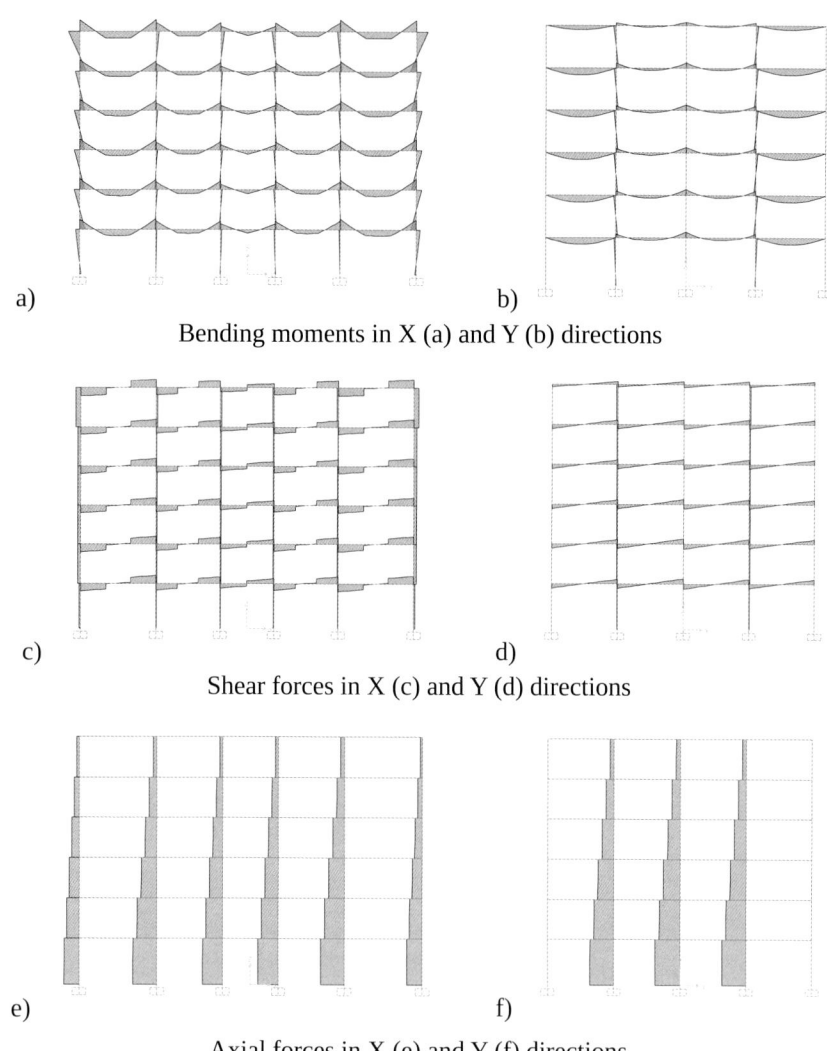

a) b)
Bending moments in X (a) and Y (b) directions

c) d)
Shear forces in X (c) and Y (d) directions

e) f)
Axial forces in X (e) and Y (f) directions

Figure 6.8 – Internal forces due to gravity loads.
(N.B. the Plots were scaled using different factors, but the same value was used for each type of internal force)

6.2 STRUCTURAL ANALYSIS AND CALCULATION MODELS

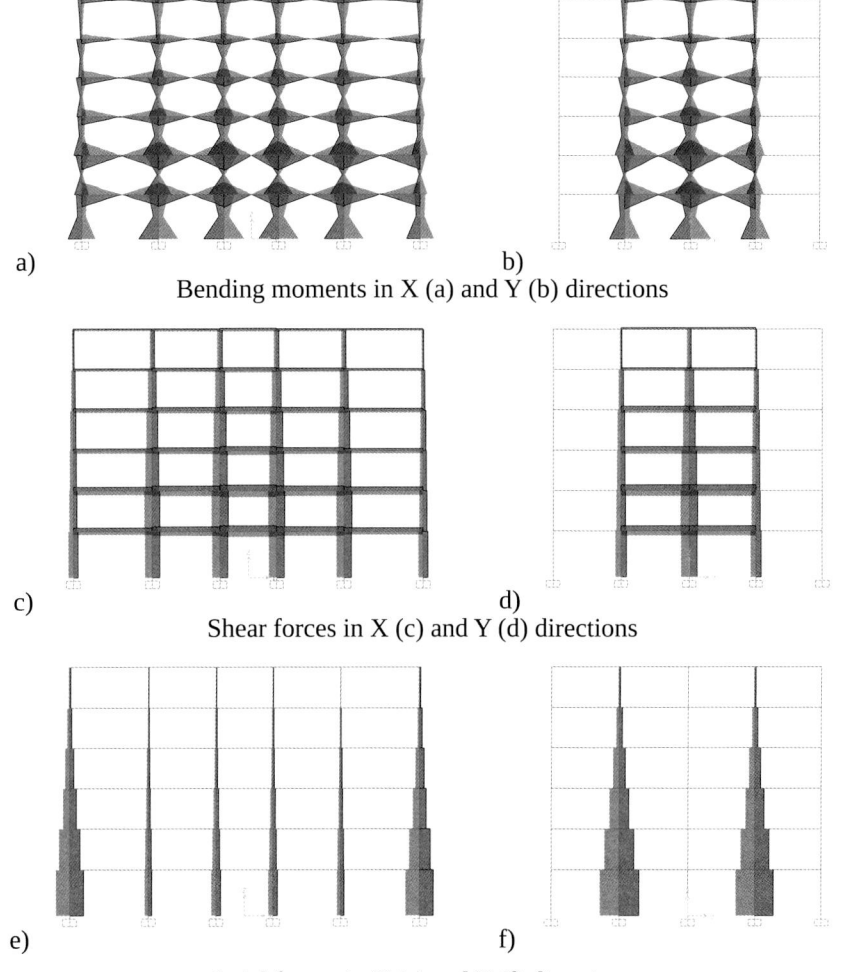

a) b)
Bending moments in X (a) and Y (b) directions

c) d)
Shear forces in X (c) and Y (d) directions

e) f)
Axial forces in X (e) and Y (f) directions

Figure 6.9 – Internal forces due to seismic actions
(N.B. the Plots were scaled using different factors, but the same value was used for each type of internal force; in addition, all action effects are defined positive due to the application of SRSS rule)

6.2.4 Imperfections for global analysis of frames

According to clause 5.3.2(4) of EN 1993-1-1, the effects of global imperfections for buildings sensitive to buckling in a sway mode can be disregarded if the following condition is satisfied:

$$H_{Ed} \geq 0.15 V_{Ed} \tag{6.8}$$

where H_{Ed} is the design value of the horizontal reaction at the bottom of the storey to the horizontal loads (namely the shear force at each storey); and V_{Ed} is the total design vertical load on the structure on the bottom of the storey.

The criterion given by equation (6.8) was controlled for both directions of the building, as shown in Table 6.8 where it can be observed that the condition is not satisfied. Therefore, the effects of global imperfections should be accounted for structural analysis. According to section 5.3.2 of EN 1993-1-1, the global imperfections can be modeled by means of a system of lateral forces $H_i = \phi \, V_{Ed,i}$, where the global sway imperfection ϕ is given as follows (Simões da Silva et al, 2016):

$$\phi = \phi_0 \cdot \alpha_h \cdot \alpha_m = 0.0024$$
$$\phi_0 = 1/200$$
$$\alpha_h = \frac{2}{\sqrt{h}} \text{ but } \frac{2}{3} \leq \alpha_h \leq 1 \Rightarrow \alpha_h = 0.667 \tag{6.9}$$
$$\alpha_m = \sqrt{0.5\left(1+\frac{1}{m}\right)} = 0.719$$

where h is the height of the building and m is the number of columns.

Table 6.8 – Global sway imperfections

Storey (-)	V_{Ed} (kN)	$H_{Ed,x}$ (kN)	$H_{Ed,y}$ (kN)	$0.15\,V_{Ed}$ (kN)	$H_{Ed,x}/V_{Ed}$ (-)	$H_{Ed,y}/V_{Ed}$ (-)	$H_i = \phi\,V_{Ed,i}$ (kN)
VI	3654.07	558.48	413.8	548.11	0.15	0.11	9.63
V	8128.55	1007.51	896.16	1219.28	0.12	0.11	21.42
IV	12605.03	1356.54	1259.76	1890.75	0.11	0.10	33.21
III	17107.97	1661.78	1532.51	2566.2	0.10	0.09	45.08
II	21625.24	1942.7	1800.51	3243.79	0.09	0.08	56.98
I	26200.63	2021.18	1878.63	3930.09	0.08	0.07	69.04

The calculated equivalent forces H_i are reported in Table 6.5. It is worth noting that the effects of induced by the lateral forces H_i should be considered for the calculation of the design effects at local (e.g. forces developing into both dissipative and non-dissipative members) and global level (e.g. displacements and second order effects).

6.2.5 Frame stability and second order effects

P-Δ effects were specified through the inter-storey drift sensitivity coefficient θ given by equation (2.68) (see sub-chapter 2.10), and are summarized in Tables 6.9 and 6.10 for calculation in the X and Y direction, respectively. As the coefficients θ are always smaller than 0.1, second order effects may be neglected, i.e. $\alpha = 1.00$.

Table 6.9 – Stability coefficients calculated in X direction

Storey (-)	P_{tot} (kN)	V_{tot} (kN)	h (mm)	$d_r = d_e \times q$ (mm)	$\theta = \dfrac{P_{tot} \cdot d_r}{V_{tot} \cdot h}$ (-)	$\alpha = \dfrac{1}{(1-\theta)}$ (-)
VI	3654.07	568.11	3500	18.85	0.03	1.04
V	8128.55	1028.93	3500	26.00	0.06	1.06
IV	12605.03	1389.75	3500	29.25	0.08	1.08
III	17107.97	1706.86	3500	27.95	0.08	1.09
II	21625.24	1999.68	3500	24.05	0.07	1.08
I	26200.63	2090.22	4000	19.50	0.06	1.07

Table 6.10 – Stability coefficients calculated in Y direction

Storey (-)	P_{tot} (kN)	V_{tot} (kN)	h (mm)	$d_r = d_e \times q$ (mm)	$\theta = \dfrac{P_{tot} \cdot d_r}{V_{tot} \cdot h}$ (-)	$\alpha = \dfrac{1}{(1-\theta)}$ (-)
VI	3654.07	423.43	3500	22.75	0.06	1.06
V	8128.55	917.58	3500	29.90	0.08	1.08
IV	12605.03	1292.97	3500	33.15	0.09	1.10
III	17107.97	1577.59	3500	30.55	0.09	1.10
II	21625.24	1857.49	3500	26.00	0.09	1.09
I	26200.63	1947.67	4000	20.15	0.07	1.07

6.3 DESIGN AND VERIFICATION OF STRUCTURAL MEMBERS

6.3.1 Design and verification of beams

The beam cross sections belong to Class 1, as defined by section 5.6 of EN 1993-1-1, which requires to satisfy the following conditions:

$$\frac{(d_b - 2r - 2t_f)}{t_w} \leq 72\varepsilon \quad \text{for web} \tag{6.10}$$

$$\frac{(b_f - 2r - t_w)}{t_f} \leq 9\varepsilon \quad \text{for flange} \tag{6.11}$$

where d_b is the beam depth, b_f is the flange width, t_f and t_w are the flange and web thickness, r is radius of root fillet and web thickness and ε is equal to $(235 / f_y)^{0.5}$.

Table 6.11 summarizes the local slenderness ratios of the adopted cross sections, showing that the latter are smaller than the limiting values for Class 1.

Table 6.11 – Classification of beam cross sections

Storey	cross section	$\dfrac{(d_b - 2r - 2t_f)}{t_w}$	72ε	$\dfrac{(b_f - 2r - t_w)}{t_f}$	9ε
VI	IPE 550	42.13		4.39	
V	IPE 550	42.13		4.39	
IV	IPE 600	42.83	72	4.21	9
III	IPE 600	42.83		4.21	
II	IPE 750×196	43.92		4.30	
I	IPE 750×196	43.92		4.30	

It was assumed that all beams are restrained against lateral-torsional buckling, as described in section 4.3.1 that illustrates examples of proper details (e.g. fly braces, etc.).

The stable length of beam between the section at a plastic hinge location and the adjacent lateral restraint can be assumed as the minimum given by the following equations (see EN 1993-1-1):

$$L_{stable} = 35 \cdot \varepsilon \cdot i_z \quad \text{for } 0.625 \leq \psi \leq 1 \tag{6.12}$$

$$L_{stable} = (60 - 40\psi) \cdot \varepsilon \cdot i_z \quad \text{for } -1 \leq \psi \leq 0.625 \tag{6.13}$$

$$L_{stable} = \frac{38 \cdot \varepsilon \cdot i_z}{\sqrt{\frac{1}{57.4} \cdot \left(\frac{N_{Ed}}{A}\right) + \frac{1}{756 \cdot C_1^2} \cdot \left(\frac{W_{pl.y}^2}{A \cdot I_t}\right) \cdot \left(\frac{f_y}{235}\right)^2}} \tag{6.14}$$

where $\varepsilon = (235/f_y)^{0.5}$ with f_y expressed in N/mm²; $\psi = M_{Ed,min}/M_{pl,Rd}$ is the ratio between the bending moments at both ends of the stable length among the lateral restraints, which can be conservatively assumed equal to 1 so that equations (6.11) and (6.12) give the same result; N_{Ed} is the design value of the compression force (to be expressed in [N]) in the member; A is the cross section area (to be expressed in [mm²]) of the member; $W_{pl,y}$ is the plastic section modulus of the member; I_t is the torsion constant of the member; f_y is the yield strength (to be expressed in [N/mm²]); and C_1 is a factor depending on the loading and end conditions that can be conservatively assumed equal to 1.

Table 6.12 summarizes the values of the calculated stable length per beam. The values given by equation (6.11) are the more conservative and should be used to detail the proper position of fly braces to restrain the beams.

Table 6.12 – Stable length of beam between the section at a plastic hinge location and the adjacent lateral restraint

Storey (-)	cross section (-)	L_{stable} Eq.(6.11) (mm)	L_{stable} Eq.(6.13) (mm)
VI	IPE 550	1557.50	2143.51
V	IPE 550	1557.50	2143.51
IV	IPE 600	1631.00	2226.93
III	IPE 600	1631.00	2226.93
II	IPE 750×196	1998.50	2663.13
I	IPE 750×196	1998.50	2663.13

6. MULTI-STOREY BUILDING WITH MOMENT RESISTING FRAMES

Figure 6.10 shows an example of detailing fly braces to restrain the beam of MRF against lateral torsional buckling. In addition, the shear connectors (e.g. Nelson studs) are located out of the beam segment where plastic hinge is expected to form, in order to avoid composite action at the beam-to-column joint (see chapter 4 for more details).

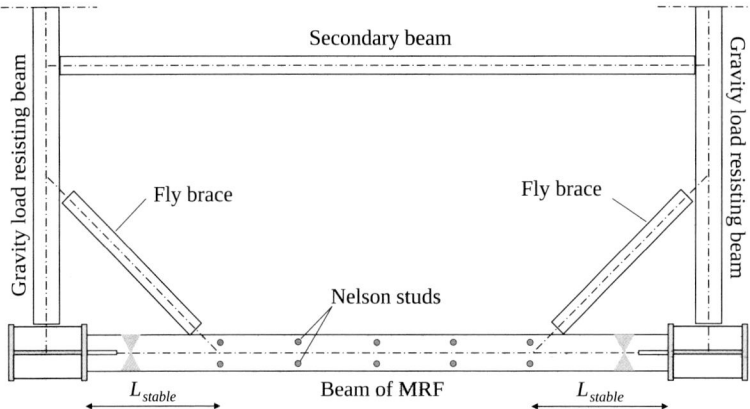

Figure 6.10 – Fly braces to avoid lateral torsional buckling of beams in MRFs and details of shear connectors to avoid composite action in the plastic hinge zone

Apart from the ultimate limit state verification under vertical loads, the bending and shear strengths are checked under the seismic load combination. All beams comply with the requirements given by equations (3.7), (3.8) and (3.9). With reference to the beam end sections, Table 6.13 reports the verification of the bending strength and the relevant overstrength factors Ω_i for the beams belonging to the A-B-C-D bays (see Figure 6.2) in the X direction, while the verifications for the beams belonging to the I-G-H-I bays (see Figure 6.2) are reported in Table 6.14. Because of symmetry, the seismic effects on the remaining bays of those allignements are simply mirrored.

The design overstrength ratio $\Omega = \min(M_{pl,Rd,i}/M_{Ed,i})$ (i.e. calculated according to equation (3.11)) is equal to 2.74 (highligthed in bold in Table 6.13) in the X direction and 2.82 (highligthed in bold in Table 6.14) in the Y direction, being similar because of structural symmetry.

Analysing the data reported in Tables 6.13 and 6.14, high values of beam overstrength can be recognized. This result is mainly due to the need to provide sufficient stiffness for both stability criteria and drift control, which imposes to oversize the cross sections of structural members.

6.3 DESIGN AND VERIFICATION OF STRUCTURAL MEMBERS

Table 6.13 – Flexural checks for beams belonging to MRF in X direction

	Storey	Left end					Right end					Ω
		$M_{Ed,G}$ (kNm)	$M_{Ed,E}$ (kNm)	$M_{Ed,I}$ (kNm)	M_{Ed} (kNm)	Ω_i (-)	$M_{Ed,G}$ (kNm)	$M_{Ed,E}$ (kNm)	$M_{Ed,I}$ (kNm)	M_{Ed} (kNm)	Ω_i (-)	
Span A-B	VI	103.47	74.25	0.19	177.91	3.68	105.60	71.12	2.09	178.81	3.66	
	V	99.44	106.28	3.70	209.42	3.13	88.34	104.28	3.63	196.25	3.34	
	IV	97.95	172.50	7.95	278.40	2.96	88.48	167.25	7.71	263.44	3.13	
	III	96.86	192.71	11.15	300.72	**2.74**	89.86	177.73	10.86	278.45	2.96	
	II	70.92	317.69	25.82	414.43	4.07	89.07	297.28	24.20	410.55	4.11	
	I	86.92	282.99	29.92	399.83	4.22	95.73	263.06	27.75	386.54	4.36	
Span B-C	VI	80.66	82.96	2.62	166.24	3.94	79.84	82.11	2.60	164.55	3.98	
	V	66.21	121.99	4.40	192.60	3.40	73.74	121.26	4.37	199.37	3.29	
	IV	66.93	190.95	8.93	266.81	3.09	73.37	189.36	8.86	271.59	3.04	**2.74**
	III	67.92	202.25	12.39	282.56	2.92	58.56	199.96	12.29	270.81	3.05	
	II	68.83	324.69	26.32	419.84	2.99	75.26	320.58	25.97	421.81	2.98	
	I	74.34	281.47	29.40	385.21	3.26	71.13	276.67	28.97	376.77	3.34	
Span C-D	VI	45.55	85.29	2.48	133.32	4.91	45.55	85.29	2.48	133.32	4.91	
	V	43.86	130.73	4.53	179.12	3.66	43.86	130.73	4.53	179.12	3.66	
	IV	44.97	204.88	9.45	259.30	3.18	44.97	204.88	9.45	259.30	3.18	
	III	44.85	218.89	13.40	277.15	2.98	44.85	218.89	13.40	277.15	2.98	
	II	48.57	338.71	27.68	414.96	3.03	48.57	338.71	27.68	414.96	3.03	
	I	48.73	297.95	31.42	378.10	3.33	48.73	297.95	31.42	378.10	3.33	

Table 6.14 – Flexural checks for beams belonging to MRF in Y direction

	Storey	Left end					Right end					Ω
		$M_{Ed,G}$ (kNm)	$M_{Ed,E}$ (kNm)	$M_{Ed,I}$ (kNm)	M_{Ed} (kNm)	Ω_i (-)	$M_{Ed,G}$ (kNm)	$M_{Ed,E}$ (kNm)	$M_{Ed,I}$ (kNm)	M_{Ed} (kNm)	Ω_i (-)	
Span G-H	VI	21.91	81.13	2.71	105.75	6.19	26.68	78.73	2.63	108.04	6.06	
	V	37.70	109.39	4.21	151.30	4.33	41.60	107.78	4.16	153.54	4.27	
	IV	36.60	218.60	11.00	266.20	3.10	43.59	212.09	10.69	266.37	3.10	
	III	36.69	225.48	15.08	277.25	2.98	43.45	220.17	14.69	278.31	2.97	
	II	34.06	378.27	33.44	445.77	3.78	47.85	353.71	31.31	432.87	3.89	
	I	32.65	329.40	37.45	399.50	4.22	47.83	305.42	34.63	387.88	4.35	**2.82**
Span H-I	VI	26.68	78.73	2.63	108.04	6.06	21.91	81.13	2.71	105.75	6.19	
	V	41.60	107.78	4.16	153.54	4.27	37.70	109.39	4.21	151.30	4.33	
	IV	43.59	212.09	10.69	266.37	3.10	36.60	218.60	11.00	266.20	3.10	
	III	43.45	220.17	14.69	278.31	2.97	36.69	225.48	15.08	277.25	2.98	
	II	47.85	353.71	31.31	432.87	2.90	34.06	378.27	33.44	445.77	**2.82**	
	I	47.83	305.42	34.63	387.88	3.24	32.65	329.40	37.45	399.50	3.15	

6. Multi-storey Building with Moment Resisting Frames

Tables 6.15 and 6.16 summarize the shear strength verifications (see equation (3.9)) for the beams of the MRFs in the X and Y directions, respectively. With this regard, the contribution due to capacity design (i.e. $V_{Ed,M} = (M_{pl,A} + M_{pl,B})/L_h$, being $M_{pl,A}$ and $M_{pl,B}$ the beam plastic moments with opposite signs at the end sections "A" and "B", while L_h is the length between the those sections) of shear force demand at both ends of each beam of MRF spans (see equation 3.9) was calculated considering that the flexural plastic hinges form at the column face, namely $L_h = L - (d_{c,A} + d_{c,B})/2$, where L is the theoretical node-to-node beam length, $d_{c,A}$ and $d_{c,B}$ are the depth of columns belonging to ends "A" and "B", respectively. This assumption is consistent with welded column-tree joints (see Figure 4.8).

Table 6.15 – shear checks for beams belonging to MRF in X direction

	storey	L_h (m)	$V_{pl,Rd}$ (kN)	Left end					Right end				
				$V_{Ed,G}$ (kN)	$V_{Ed,M}$ (kN)	$V_{Ed,I}$ (kN)	V_{Ed} (kN)	$\frac{V_{Ed}}{V_{pl,Rd}}$	$V_{Ed,G}$ (kN)	$V_{Ed,M}$ (kN)	$V_{Ed,I}$ (kN)	V_{Ed} (kN)	$\frac{V_{Ed}}{V_{pl,Rd}}$
Span A-B	VI	6.33	975.86	73.47	206.87	0.61	280.95	0.29	73.96	206.87	0.61	281.44	0.29
	V	6.33	975.86	68.42	206.87	1.05	276.34	0.28	65.16	206.87	1.05	273.08	0.28
	IV	6.33	1136.98	68.77	260.68	2.24	331.69	0.29	65.98	260.68	2.24	328.90	0.29
	III	6.33	1136.98	68.42	260.68	3.14	332.24	0.29	66.33	260.68	3.14	330.15	0.29
	II	6.33	1726.55	70.92	532.50	7.15	610.57	0.35	68.96	532.50	7.15	608.61	0.35
	I	6.33	1726.55	69.72	532.50	8.24	610.46	0.35	71.15	532.50	8.24	611.89	0.35
Span B-C	VI	5.28	975.86	63.37	247.90	0.87	312.14	0.32	63.00	247.90	0.87	311.77	0.32
	V	5.28	975.86	56.03	247.90	1.46	305.39	0.31	58.46	247.90	1.46	307.82	0.32
	IV	5.28	1136.98	56.63	312.38	2.96	371.98	0.33	58.82	312.38	2.96	374.17	0.33
	III	5.28	1136.98	56.95	312.38	4.11	373.44	0.33	72.98	312.38	4.11	389.47	0.34
	II	5.28	1726.55	58.83	475.87	8.72	543.42	0.31	61.04	475.87	8.72	545.63	0.32
	I	5.28	1726.55	60.52	475.87	9.73	546.12	0.32	59.37	475.87	9.73	544.97	0.32
Span C-D	VI	4.28	975.86	37.02	305.76	0.99	343.78	0.35	37.02	305.76	0.99	343.78	0.35
	V	4.28	975.86	39.70	305.76	1.81	347.28	0.36	39.70	305.76	1.81	347.28	0.36
	IV	4.28	1136.98	40.13	385.30	3.78	429.21	0.38	40.13	385.30	3.78	429.21	0.38
	III	4.28	1136.98	40.13	385.30	5.36	430.79	0.38	40.13	385.30	5.36	430.79	0.38
	II	4.28	1726.55	41.96	586.95	11.07	639.98	0.37	41.96	586.95	11.07	639.98	0.37
	I	4.28	1726.55	41.95	586.95	12.57	641.47	0.37	41.95	586.95	12.57	641.47	0.37

6.3 Design and Verification of Structural Members

Table 6.16 – shear checks for beams belonging to MRF in Y direction

	storey	L_h (m)	$V_{pl,Rd}$ (kN)	Left end					Right end			
				$V_{Ed,G}$ (kN)	$V_{Ed,M}$ (kN)	$V_{Ed,I}$ (kN)	V_{Ed} (kN)	$\dfrac{V_{Ed}}{V_{pl,Rd}}$	$V_{Ed,G}$ (kN)	$V_{Ed,M}$ (kN)	$V_{Ed,I}$ (kN)	V_{Ed} (kN)	$\dfrac{V_{Ed}}{V_{pl,Rd}}$
Span G-H	VI	5.33	975.86	24.17	245.67	0.89	270.73	0.28	25.76	245.67	0.89	272.32	0.28
	V	5.33	975.86	39.53	245.67	1.40	286.59	0.29	40.83	245.67	1.40	287.89	0.30
	IV	5.33	1136.98	39.94	309.57	3.62	353.13	0.31	42.27	309.57	3.62	355.46	0.31
	III	5.33	1136.98	39.98	309.57	4.96	354.51	0.31	42.23	309.57	4.96	356.76	0.31
	II	5.33	1726.55	41.00	632.37	10.79	684.16	0.40	45.59	632.37	10.79	688.75	0.40
	I	5.64	1726.55	40.71	597.62	12.01	650.34	0.38	45.76	597.62	12.01	655.39	0.38
Span H-I	VI	5.28	975.86	25.76	245.67	0.89	272.32	0.28	24.17	245.67	0.89	270.73	0.28
	V	5.28	975.86	40.83	245.67	1.40	287.89	0.30	39.53	245.67	1.40	286.59	0.29
	IV	5.28	1136.98	42.27	309.57	3.62	355.46	0.31	39.94	309.57	3.62	353.13	0.31
	III	5.28	1136.98	42.23	309.57	4.96	356.76	0.31	39.98	309.57	4.96	354.51	0.31
	II	5.28	1726.55	45.59	632.37	10.79	688.75	0.40	41.00	632.37	10.79	684.16	0.40
	I	5.28	1726.55	45.76	597.62	12.01	655.39	0.38	40.71	597.62	12.01	650.34	0.38

For illustration, based on the values from Tables 6.13 and 6.15, the seismic demands on left end of the beam in the bay A-B at first storey (IPE 750×196) are (equations (3.7), (3.9) and (3.10)):

$$M_{Ed} = M_{Ed,G} + \alpha \cdot (M_{Ed,I} + M_{Ed,E}) =$$
$$= 86.92 \text{ kNm} + 1.0 \cdot (29.92 \text{ kNm} + 382.99 \text{ kNm}) = \quad (6.15)$$
$$= 399.83 \text{ kNm} < 1257.25 = M_{pl,Rd}$$

$$V_{Ed} = V_{Ed,G} + \alpha \cdot (V_{Ed,M} + V_{Ed,I}) = V_{Ed,G} + \alpha \cdot \left(V_{Ed,I} + \left(\frac{2M_{pl,Rd}}{L_h}\right)\right) =$$
$$= 69.72 \text{ kN} + \alpha \cdot \left(8.24 \text{ kN} + \left(\frac{2 \cdot 1257.25}{6.33}\right) \text{kN}\right) = \quad (6.16)$$
$$= 610.46 \text{ kN} < 823.28 = 0.5 \cdot V_{pl,Rd}$$

6. Multi-storey Building with Moment Resisting Frames

where $M_{Ed,G}$, $M_{Ed,E}$ and $M_{Ed,I}$ are the bending moments given by the numerical model for gravity (i.e. due to $G_k + 0.3Q_k$), the seismic loads and the forces H_i induced by the overall sway imperfections, respectively; α is the stability coefficient for the first storey in X direction, taken as 1.00 because the stability coefficient $\theta \leq 0.10$; $V_{Ed,G}$ is the shear force given by numerical model for gravity loads in seismic conditions, while $V_{Ed,M}$ is the shear force due to the hierarchy criterion corresponding to plastic hinges formed at both beam ends.

6.3.2 Design and verification of columns

A wide flange HEM 600 was used for all external columns of MRF, while a HEM 700 was adopted for the internal columns. This choice was mainly conditioned by the need to satisfy stability and drift limits. Indeed, as can be seen from the verification check ratios, smaller profiles could satisfy the hierarchy requirements, which are given by equation (3.12), (3.13) and (3.14).

Tables 6.17 and 6.18 summarize the M-N strength and buckling verifications for columns belonging to vertical "A" (see Figure 6.2) of the MRF in the X directions and vertical "G" (see Figure 6.2) of the MRF in the Y direction, respectively. Since a 3D structural analysis was carried out, the bending moments were obtained for both directions (i.e. y-y and z-z) of each column section.

The strength checks of column cross section against biaxial bending and axial force were carried out according to clause 6.2.9.1(6) of EN 1993-1-1, namely:

$$\left(\frac{M_{y,Ed}}{M_{N,y,Rd}}\right)^2 + \left(\frac{M_{z,Ed}}{M_{N,z,Rd}}\right)^{5N_{Ed}/N_{pl,Rd}} \leq 1 \tag{6.17}$$

where $M_{y,Ed}$ and $M_{z,Ed}$ are the design bending moments in the two main axis of the column, while $M_{N,y,Rd}$ and $M_{N,z,Rd}$ are the corresponding bending strength with axial interaction according to clause 6.2.9.1(5) of EN 1993-1-1.

The stability verifications of the columns in bending and axial compression were carried out according to clause 6.3.3(4) of EN 1993-1-1, namely:

$$\frac{N_{Ed}}{\chi_y \frac{N_{Rk}}{\gamma_{M1}}} + k_{yy} \frac{M_{y,Ed} + \Delta M_{y,Ed}}{\chi_{LT} \frac{M_{y,Rk}}{\gamma_{M1}}} + k_{yz} \frac{M_{z,Ed} + \Delta M_{z,Ed}}{\frac{M_{z,Rk}}{\gamma_{M1}}} \leq 1 \tag{6.18}$$

$$\frac{N_{Ed}}{\chi_z N_{Rk}} + k_{zy}\frac{M_{y,Ed}+\Delta M_{y,Ed}}{\chi_{LT}M_{y,Rk}} + k_{zz}\frac{M_{z,Ed}+\Delta M_{z,Ed}}{M_{z,Rk}} \leq 1 \qquad (6.19)$$
$$\gamma_{M1}\gamma_{M1}\gamma_{M1}$$

where: N_{Ed}, $M_{y,Ed}$ and $M_{z,Ed}$ are the design values of the compression force and the maximum moments about the y-y and z-z axis along the member, respectively; $\Delta M_{y,Ed}$, $\Delta M_{z,Ed}$ are the moments due to second order effects; χ_y and χ_z are the reduction factors due to flexural buckling; χ_{LT} is the reduction factor due to lateral torsional buckling; k_{yy}, k_{yz}, k_{zy}, k_{zz} are the interaction factors evaluated according to Annex B of EN 1993:1-1.

In addition, the shear strength verifications for columns "A" and "G" are reported in Table 6.19 and 6.20, respectively.

The ECCS Eurocode Design Manual on EC3-1-1 (Simões da Silva et al, 2016) provides for more details and examples on both strength and stability checks.

Table 6.17 – Flexural checks for columns "A" in X direction

	storey	$M_{y,Ed}$ (kNm)	$M_{z,Ed}$ (kNm)	N_{Ed} (kN)	$M_{NRd,y}$ (kNm)	$M_{NRd,z}$ (kNm)	$N_{pl,Rd}$ (kN)	Eq. (6.16)	Eq. (6.17)**	Eq. (6.18)**
top end	VI	385.85	0.00	176.30	2061.42	453.55	8546.95	0.04	0.182	0.139
	V	467.05	19.85	408.07	2061.42	453.55	8546.95	0.53	0.277	0.221
	IV	532.58	14.38	710.00	2061.42	453.55	8546.95	0.31	0.367	0.292
	III	392.38	13.70	1021.34	2061.42	453.55	8546.95	0.16	0.459	0.372
	II	588.72	18.04	1469.55	2053.88	453.55	8546.95	0.14	0.534	0.448
	I	501.33	11.69	1878.40	2061.42	453.55	8546.95	0.17	0.453	0.452
bottom end	VI	124.50	19.85	186.10	2061.42	453.55	8546.95	0.73	0.182	0.139
	V	283.82	14.38	417.93	2061.42	453.55	8546.95	0.46	0.277	0.221
	IV	458.02	13.70	719.88	2061.42	453.55	8546.95	0.28	0.367	0.292
	III	757.93	18.08	1031.21	2061.42	453.55	8546.95	0.28	0.459	0.372
	II	689.99	11.65	1479.43	2053.88	453.55	8546.95	0.16	0.534	0.448
	I*	368.59	74.30	1570.65	2061.42	453.55	8546.95	0.44	0.453	0.452

* According to EN 1998-1 Clause 6.3.1(2) plastic hinges may form at bases of columns. Therefore, for that section the seismic induced effects are taken directly from calculation models, as for the beams.
** The buckling check are representative of the whole member. Therefore the maxima induced effects were accounted for these verification checks.

6. MULTI-STOREY BUILDING WITH MOMENT RESISTING FRAMES

Table 6.18 – Flexural checks for columns "G" in Y direction

	storey	$M_{y,Ed}$ (kNm)	$M_{z,Ed}$ (kNm)	N_{Ed} (kN)	$M_{NRd,y}$ (kNm)	$M_{NRd,z}$ (kNm)	$N_{pl,Rd}$ (kN)	Eq. (6.16)	Eq. (6.17)**	Eq. (6.18)**
top end	VI	339.25	0.00	275.92	2061.42	453.55	8546.95	0.03	0.174	0.141
	V	463.24	19.22	597.67	2061.42	453.55	8546.95	0.38	0.317	0.257
	IV	585.12	13.28	1058.26	2061.42	453.55	8546.95	0.19	0.443	0.359
	III	447.68	11.54	1526.45	2037.37	453.55	8546.95	0.09	0.566	0.469
	II	679.92	15.94	2174.86	1848.93	416.53	8546.95	0.15	0.673	0.579
	I	535.11	10.21	2759.64	2034.91	453.55	8546.95	0.10	0.526	0.535
bottom end	VI	100.85	19.22	285.71	2061.42	453.55	8546.95	0.60	0.174	0.141
	V	370.61	13.28	607.47	2061.42	453.55	8546.95	0.32	0.317	0.257
	IV	550.71	11.54	1068.06	2061.42	453.55	8546.95	0.17	0.443	0.359
	III	889.20	15.94	1536.24	2037.37	453.55	8546.95	0.24	0.566	0.469
	II	821.70	10.21	2184.67	1848.93	416.53	8546.95	0.21	0.673	0.579
	I*	387.79	96.59	2320.25	2034.91	453.55	8546.95	0.29	0.526	0.535

* According to EN 1998-1 Clause 6.3.1(2) plastic hinges may form at bases of columns. Therefore, for that section the seismic induced effects are taken directly from calculation models, as for the beams.

** The buckling check are representative of the whole member. Therefore the maxima induced effects were accounted for these verification checks.

Table 6.19 – Shear checks for columms belonging to vertical "A" in X direction

Storey (-)	$V_{Ed,G}$ (kN)	$V_{Ed,E}$ (kN)	$V_{Ed,I}$ (kN)	V_{Ed} (kN)	$V_{pl,Rd}$ (kN)	$\dfrac{V_{pl,Rd}}{V_{Ed}}$
VI	46.52	20.87	0.19	125.28	2085.36	16.65
V	25.01	47.97	1.36	206.03	2085.36	10.12
IV	28.43	64.36	3.08	271.30	2085.36	7.69
III	27.18	76.83	5.33	317.11	2085.36	6.58
II	28.38	86.65	8.19	355.37	2085.36	5.87
I***	12.38	108.45	13.82	421.63	2085.36	4.95

*** Differently from flexural actions the shear demands are computed according to hierarchy criterion given by equation (3.4) in section 3.4.4, in order to avoid shear forces could impair the rotation capacity of columns at their base.

Table 6.20 – Shear checks for columns belonging to vertical "G" in Y direction

Storey (-)	$V_{Ed,G}$ (kN)	$V_{Ed,E}$ (kN)	$V_{Ed,I}$ (kN)	V_{Ed} (kN)	$V_{pl,Rd}$ (kN)	$\dfrac{V_{pl,Rd}}{V_{Ed}}$
VI	11.83	23.67	0.28	103.62	2085.36	20.13
V	10.42	56.43	1.84	229.26	2085.36	9.10
IV	10.55	77.79	4.12	312.22	2085.36	6.68
III	9.99	92.07	7.02	367.05	2085.36	5.68
II	10.33	104.63	10.74	416.09	2085.36	5.01
I***	4.71	116.96	15.89	458.28	2085.36	4.55

*** Differently from flexural actions the shear demands are computed according to hierarchy criterion given by equation (3.4) in section 3.4.4, in order to avoid shear forces could impair the rotation capacity of columns at their base.

In order to exemplify the application of capacity design requirements at global level (see equation (3.5)) for all columns, on the basis of the values reported in Tables 6.17 and 6.19, the seismic demands at the top end of the column at first storey in vertical "A" are computed as follows (equation (3.12)):

$$M_{y,Ed} = M_{y,Ed,G} + \alpha \cdot \left(M_{y,Ed,I} + 1.1 \cdot \gamma_{ov} \cdot \Omega \cdot M_{y,Ed,E} \right) =$$
$$= 35.95 \text{ kNm} + 1.00 \cdot (17.18 \text{ kNm} + 1.1 \cdot 1.25 \cdot 2.74 \cdot 118.77 \text{ kNm}) = \quad (6.20)$$
$$= 484.15 \text{ kNm}$$

$$M_{z,Ed} = M_{z,Ed,G} + \alpha \cdot \left(M_{z,Ed,I} + 1.1 \cdot \gamma_{ov} \cdot \Omega \cdot M_{z,Ed,E} \right) =$$
$$= 0.00 \text{ kNm} + 1.00 \cdot (0.63 \text{ kNm} + 1.1 \cdot 1.25 \cdot 2.74 \cdot 2.93 \text{ kNm}) = \quad (6.21)$$
$$= 11.69 \text{ kNm}$$

$$N_{Ed} = N_{Ed,G} + \alpha \cdot \left(N_{Ed,I} + 1.1 \cdot \gamma_{ov} \cdot \Omega \cdot N_{Ed,E} \right) =$$
$$= 686.22 \text{ kN} + 1.00 \cdot (22.49 \text{ kN} + 1.1 \cdot 1.25 \cdot 2.74 \cdot 309.98 \text{ kN}) = \quad (6.22)$$
$$= 1878.40 \text{ kN}$$

$$V_{Ed} = V_{Ed,G} + \alpha \cdot \left(V_{Ed,I} + 1.1 \cdot \gamma_{ov} \cdot \Omega \cdot V_{Ed,E} \right) =$$
$$= 12.38 \text{ kN} + 1.00 \cdot (13.82 \text{ kN} + 1.1 \cdot 1.25 \cdot 2.74 \cdot 108.45 \text{ kN}) = \quad (6.23)$$
$$= 421.63 \text{ kN}$$

6. Multi-storey Building with Moment Resisting Frames

The seismic effects in the columns of the MRFs in the X direction are magnified by $\alpha\,1.1\gamma_{ov}\Omega = 1.00 \times 1.1 \times 1.25 \times 2.74 = 3.77$ and $1.17 \times 1.1 \times 1.25 \times 2.82 = 3.88$ in the Y direction, considerably increasing the design forces. Notwithstanding so large design forces, it is worth noting that the need to satisfy both stability criterion and drift limits led to constant column cross sections along the building height and this explains the so large overstrength factors reported in Tables 6.13 and 6.14.

Another aspect to be accounted for is related to shear strength of columns, which should be at least larger than twice the relevant seismic induced effect (i.e. $V_{Ed}/V_{pl,Rd} \leq 0.5$, according to clause 6.6.3(4)). This requirement could influence the choice of column cross sections in case of stocky frames with reduced number of moment resisting spans.

It is noted that the seismic demand of M-N effects at the column bases reported in Tables 6.17 and 6.18 are computed differently from those determined for the other column sections. Indeed, according to clauses 6.3.1(2) and 6.6.1(1)P, plastic hinges may form at the base of the columns in order to induce a very ductile overall mechanism. Hence, seismic induced effects are taken directly from calculation models without magnification, as done for beams. For illustration, the calculation of seismic demands on the bottom end of the column "A" at first storey is reported as follows:

$$M_{y,Ed} = M_{y,Ed,G} + \alpha \cdot \left(M_{y,Ed,I} + M_{y,Ed,E}\right) =$$
$$= 13.56 \text{ kNm} + 1.00 \cdot \left(38.11 \text{ kNm} + 316.92 \text{ kNm}\right) = \quad (6.24)$$
$$= 368.59 \text{ kNm}$$

$$N_{Ed} = N_{Ed,G} + \alpha \cdot \left(N_{Ed,I} + N_{Ed,E}\right) =$$
$$= 697.43 \text{ kN} + 1.00 \cdot \left(22.49 \text{ kN} + 309.98 \text{ kN}\right) = 1029.90 \text{ kN} \quad (6.25)$$

$$V_{Ed} = V_{Ed,G} + \alpha \cdot \left(V_{Ed,I} + 1.1 \cdot \gamma_{ov} \cdot \Omega \cdot V_{Ed,E}\right) =$$
$$= 10.99 \text{ kN} + 1.00 \cdot \left(13.82 \text{ kN} + 1.1 \cdot 1.25 \cdot 2.74 \cdot 108.45 \text{ kN} +\right) = \quad (6.26)$$
$$= 421.63 \text{ kN}$$

The large safety margins for the column sections at their base (see Tables 6.17 and 6.18) contrast with the possibility to activate plastic deformations. Since the columns are designed to satisfy both capacity design and stability criteria, their cross sections are generally very large and this is more evident at the first storey. Therefore, if designers aims at exploiting the contribution of plastic rotation capacity of column bases to overall plastic mechanisms, dog-bone shaped columns could be envisaged at the bases.

EC8-1 does not clearly specify how the capacity design for the shear forces should be applied for column bases. Anyway, in order to avoid that shear forces could impair the rotation capacity of columns at their base, the seismic induced shear effects are computed according to hierarchy criterion given by equation (3.5), as for all other columns.

In addition to the member checks, EN 1998-1 clause 4.4.2.3(4) (CEN, 2004a) requires that the local hierarchy criterion is verified for all beam-to-column joints (equation (2.71)). as follows:

$$\sum M_{Rc} \geq 1.3 \sum M_{Rb} \qquad (6.27)$$

where $M_{Rc,1}$ and $M_{Rc,2}$ are the flexural strengths $M_{N,Rd}$ (i.e. accounting for M-N interaction) of the column sections below and above the joint, respectively; $M_{Rb,1}$ and $M_{Rb,2}$ are the plastic moments of beams at both sides of the joint.

For illustration, this check is reported in detail for both external and internal joints of the MRF span A-B at the first floor (see Figure 6.2), as follows:

$$\sum M_{Rc} = M_{Rc,1} + M_{Rc,2} = 2061.42 + 2061.42 \text{ kNm} = 4122.84 \text{ kNm}$$
$$\sum M_{Rb} = M_{Rb,1} = 1685.89 \text{ kNm} \qquad (6.28)$$
$$\frac{\sum M_{Rc}}{\sum M_{Rb}} = 3.2 > 1.3$$

$$\sum M_{Rc} = M_{Rc,1} + M_{Rc,2} = 2476.9 \text{ kNm} + 2476.9 \text{ kNm} = 4953.8 \text{ kNm}$$
$$\sum M_{Rb} = M_{Rb,1} + M_{Rb,2} = 1685.89 \text{ kNm} + 1685.89 \text{ kNm} = 3371.78 \text{ kNm} \qquad (6.29)$$
$$\frac{\sum M_{Rc}}{\sum M_{Rb}} = 1.47 > 1.3$$

The verification checks of the local hierarchy criterion for external and internal joints belonging to columns "A" and "B" in the X direction are summarized in Table 6.21, showing that the local capacity design criterion is widely satisfied.

6. Multi-storey Building with Moment Resisting Frames

Table 6.21 – Local hierarchy criterion for external and inner columns in X direction

storey	$M_{Rc,1}$ (kNm)	$M_{Rc,2}$ (kNm)	M_{Rb} (kN)	$\dfrac{\sum M_{Rc}}{\sum M_{Rb}}$	$M_{Rc,1}$ (kNm)	$M_{Rc,2}$ (kNm)	$M_{Rb,1}$ (kNm)	$M_{Rb,2}$ (kNm)	$\dfrac{\sum M_{Rc}}{\sum M_{Rb}}$
VI	2061.42	-	654.95	3.15	2476.9	-	654.95	654.95	1.89
V	2061.42	2061.42	654.95	6.29	2476.9	2476.9	654.95	654.95	3.78
IV	2061.42	2061.42	825.32	5.00	2476.9	2476.9	825.32	825.32	3.00
III	2061.42	2061.42	825.32	5.00	2476.9	2476.9	825.32	825.32	3.00
II	2061.42	2061.42	1685.89	2.45	2476.9	2476.9	1685.89	1685.89	1.47
I	2061.42	2061.42	1685.89	2.45	2476.9	2476.9	1685.89	1685.89	1.47

6.3.3 Panel zone of beam-to-column joints

As described in section 3.5.3, the column web panel zones of beam-to-column joints should resist the forces developing in the adjacent dissipative elements (i.e. either beams or connections), in this case the connected beams.

The design shear forces acting into the column web panel (see Figure 3.7) should be calculated according to equation (3.17), namely as follows:

$$V_{wp,Ed} = \dfrac{\sum M_{pl,Rd,i}}{z} - \left(\dfrac{V_{c1,Ed} - V_{c2,Ed}}{2}\right)$$

where $M_{pl,Rd,i}$ is the plastic bending moment of beam "i" (or its projected value at column face if some stiffeners are used to strengthen the connection) and z is the beam lever arm, which can be calculated as $z = d_b - t_f$; d_b is the beam depth and t_f the relevant flange thickness; $V_{c1,Ed}$ and $V_{c2,Ed}$ are the shear forces in the lower and upper portions of the column, respectively.

The resistance of the column web panel is given by equation (3.18), namely as follows:

$$V_{wp,Rd} = V_{wc,Rd} + \Delta V_{wp,add,Rd} = \dfrac{0.9 f_{y,wc} \cdot A_{vc}}{\sqrt{3} \cdot \gamma_{M0}} + \dfrac{4 M_{pl,fc,Rd}}{d_s}$$

where $V_{wc,Rd}$ is the design plastic shear resistance of the unstiffened column web panel and $\Delta V_{wc,add,Rd}$ the overstrength contribution due to a mechanism involving the plastic moment capacity of the column flanges $M_{pl,fc,Rd}$.

Double-stiffened welded beam-to-column joints with continuity plates (CP) are considered (see Figure 6.11). The comparison between the web panel shear strength and demand (see Table 6.22) shows that the column web panels belonging to the external joints satisfy the capacity design requirements except for the joints at the first and second storey, where there are the deeper beams (i.e. IPE 750 ×196). Therefore, supplementary web plates are designed to fulfil the strength requirement. Steel grade S235 is used for these additional plates and the relevant thickness is reported in Table 6.22. It should be noted that, since no plug welds are considered (see Figure 4.17), the design of the minimum thickness of the strengthening plates is conditioned by the stability requirement of EN 1993-1-8 to prevent the buckling or separation of the lapped parts, namely $b_s \le 40\varepsilon \cdot t_s$, where b_s is the depth of the strengthening plate, t_s is its thickness and $\varepsilon = (235/f_y)^{0.5}$.

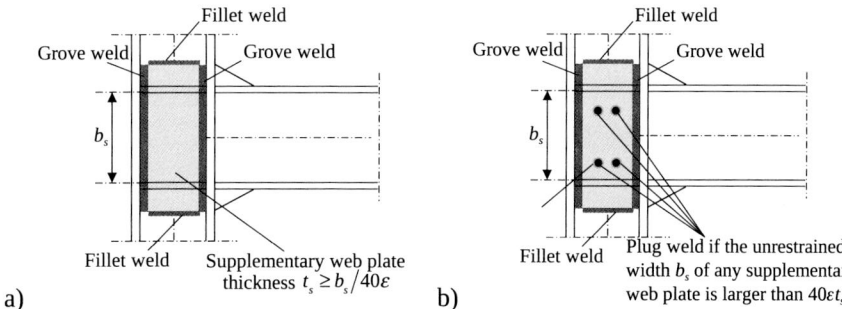

Figure 6.11 – Detail of beam-to-column joint:
a) stocky and b) supplementary web plate

Table 6.22 – Verification of shear panel zone and design of additional web plates for external joints of column A in X direction

storey	$V_{Wp,Ed}$ (kN)	CP (mm)	t_s (mm)	$V_{Wp,Rd}$ (kN)	$V_{Wp,Rd}$ (kN)	$\dfrac{V_{wp,Ed}}{\left(V_{wp,Rd}+\Delta V_{wp,Rd}\right)}$
VI	1103.97	20	-	1947.39	-	0.57
V	1269.63	20	-	1947.39	-	0.65
IV	1453.15	20	-	1941.54	-	0.75
III	1443.42	20	-	1941.54	-	0.74
II	2283.28	25	485×12	1955.73	710.68	0.86
I	2297.29	25	485×12	1955.73	710.68	0.86

6. MULTI-STOREY BUILDING WITH MOMENT RESISTING FRAMES

All internal joints also need to be strengthened by supplementary plates welded to the column panel zone. The designed details of the stiffeners and the relevant verification checks are shown in Table 6.23 where the number and the thickness of designed strengthening plates (t_s) for the panel zone of column "B" are reported.

Table 6.23 – Verification of shear panel zone and design of additional web plates for internal joints of column B in X direction

storey	$V_{Wp,Ed}$ (kN)	CP (mm)	t_s (mm)	$V_{Wp,Rd}$ (kN)	$\Delta V_{Wp,add,Rd}$ (kN)	$\dfrac{V_{wp,Ed}}{\left(V_{wp,Rd}+\Delta V_{wp,add,Rd}\right)}$
VI	2242.08	20	580×5	2183.07	354.12	0.88
V	2170.66	20	580×5	2183.07	354.12	0.86
IV	2424.15	20	580×5	2177.21	354.12	0.96
III	2317.07	20	580×5	2177.21	354.12	0.92
II	3910.86	25	2×580×15	2191.40	2124.71	0.91
I	3878.08	25	2×580×15	2191.40	2124.71	0.90

To illustrate this check, the verification of an internal joint at the first storey in the X direction is performed. The shear force demand into column web panel of the internal joint is calculated as follows (equation (3.17)):

$$V_{wp,Ed} = \frac{\sum M_{pl,Rd,i}}{z} - \left(\frac{V_{c1,Ed}-V_{c2,Ed}}{2}\right) = \\ = \frac{2 \cdot 1685.89}{(0.770-0.0254)} - \left(\frac{661.49+638.98}{2}\right) = 3878.08 \text{ kN} \quad (6.30)$$

It is clear that the web panel shear strength $V_{wp,Rd}$ given by equation (3.18) is smaller than the demand $V_{wp,Ed}$ obtained by equation (6.30). Hence, additional web plates are necessary to satisfy the EC8-1 capacity design criterion. Assuming that the width of the required plate is identical to the column web depth d_{wc}, the minimum thickness t_s can be obtained by imposing that the shear strength of the additional area should balance the force given by ($V_{wp,Ed} - V_{wp,Rd}$), as follows:

6.4 DAMAGE LIMITATION

$$t_s \cdot d_{wc} \cdot 0.9 \cdot \frac{f_y}{\sqrt{3}} = t_s \cdot 580 \cdot 0.9 \cdot \frac{235}{\sqrt{3}} \geq 1686.68 \text{ kN} \tag{6.31}$$

$t_s \geq 23.82$ mm → 2 plates 15 mm thick, one per web side

The thickness t_s obtained in equation (6.31) does not satisfy the slenderness limit (i.e. $b_s \leq 40\varepsilon \cdot t_s$). Hence, four plug welds per supplementary web plate are needed to prevent the buckling or separation of the lapped parts (see Figure 6.11b).

6.4 DAMAGE LIMITATION

As illustrated in section 2.11.2, the damage limitation requirement is satisfied if, under a seismic action having a larger probability of occurrence than the design seismic action, the interstorey drift demand is less than the upper limit given by equation (2.73) (see section 4.4.3.2 of EC8-1). Assuming brittle non-structural elements, the interstorey drift limit is equal to 0.50 % of h. Moreover, concerning the displacement reduction factor ν, the recommended value is assumed, i.e. ν = 0.5 (the structure belongs to class II). The displacements d_s induced by the design seismic actions were computed as follows:

$$\nu d_s = \nu q \cdot d_e \tag{6.32}$$

where d_s is the displacement of the structural system induced by the design seismic action, q is the behaviour factor and d_e is the displacement of the structural system, obtained from a linear elastic analysis under the design seismic forces.

Figures 6.12a,b show the lateral storey displacements subject to the seismic design loads in both X and Y directions, while Tables 6.24 and 6.25 report the relevant drift verification checks that are satisfied for each storey.

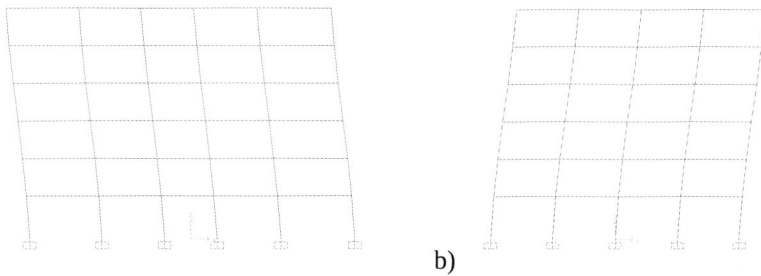

Figure 6.12 – Lateral displacement shapes: X (a) and Y (b) directions

Table 6.24 – Damage limitation check for MRFs in X direction

Storey (-)	h (mm)	$vd_r = v(d_e \times q)$ (mm)	vd_r/h (%)	Performance Limit (%)
VI	3500	9.43	0.27	0.50
V	3500	13.00	0.37	
IV	3500	14.63	0.42	
III	3500	13.98	0.40	
II	3500	12.03	0.34	
I	4000	9.75	0.24	

Table 6.25 – Damage limitation check for MRFs in Y direction

Storey (-)	h (mm)	$vd_r = v(d_e \times q)$ (mm)	vd_r/h (%)	Performance Limit (%)
VI	3500	11.38	0.33	0.50
V	3500	14.95	0.43	
IV	3500	16.58	0.47	
III	3500	15.28	0.44	
II	3500	13.00	0.37	
I	4000	10.08	0.25	

6.5 PUSHOVER ANALYSIS AND ASSESSMENT OF SEISMIC PERFORMANCE

6.5.1 Introduction

According to Eurocode 8, the seismic performance of structural systems can be evaluated by comparing the displacement capacity, provided by nonlinear static analyses, with the seismic displacement demand. The EC8-1 analysis procedure (see Annex B of EC8-1) is based on the N2 method (Fajfar,

2000), which evaluates the displacement demand by defining a single degree-of-freedom (SDOF) system equivalent to the examined structure (section 2.5.4).

The EC8-1 procedure can be summarized according to the following steps:

1. Modelling the structural members and secondary elements (if relevant) using non-linear techniques for considering material and geometrical sources of non-linearity;
2. Execution of pushover analyses by applying, alternatively, two sets of horizontal forces to the structural model and identification of the displacements at the formation of a plastic mechanism to be used as the initial approximation for target displacement;
3. Transformation of the structural response into an equivalent SDOF system (equivalent bilinear SDOF model - base shear versus displacement of the participant mass based on the equal energy principle);
4. Evaluation of the seismic demand for SDOF and MDOF systems;
5. Comparison between the structural capacity and the seismic demand (also accounting for torsional effects).

Hereinafter, each step is discussed and detailed with reference to the assessment of the seismic response of the MRF in the X direction designed in the previous sections.

6.5.2 Modelling assumptions

6.5.2.1 Plastic hinges

The non-linear modelling of the structural behaviour is a crucial issue that significantly influences the accuracy and the effectiveness of the structural assessment. EC8 does not provide exhaustive data and recommendations about the modelling of non-linear response of structural members. On the contrary, detailed and accurate assumptions are provided by FEMA 356. Therefore, the FEMA 356 assumptions for modelling the structural system are adopted, while the assessment procedure is entirely compliant with EC8.

In particular, structural members were modelled using lumped plasticity (see section 2.6.4), namely made of an elastic part and plastic hinges fixed a-priori at both ends of the element length (see Figure 6.13).

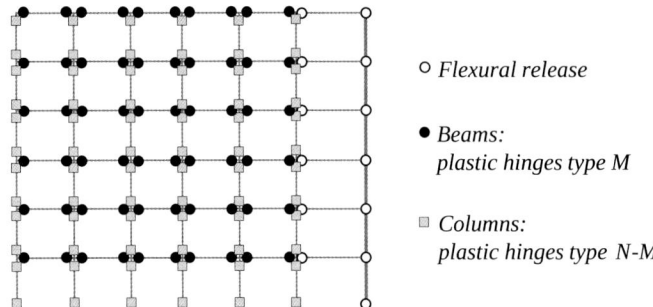

Figure 6.13 – Position and types of plastic hinges

The force-deformation relationship characterizing the non-linear behaviour of the plastic hinge model were derived as recommended by FEMA 356. The criteria and the modelling parameters differ for primary and secondary components, as well as for ductile and brittle elements. With this regard, both EC8-1 and FEMA 356 define ductile elements as "deformation-controlled", while brittle elements are "force-controlled". In the first case, the plastic behaviour includes strain hardening and a strength-degradation with residual strength capacity. In the second case, the behaviour is characterized by an elastic range followed by loss of strength.

According to this definition, FEMA 356 provides a generalized force - deformation response curve for the plastic hinges (see Figure 6.14a), which allows covering all non-linear response curves by particularizing the proper modelling parameters. The point A corresponds to the unloaded condition. The point B corresponds to achievement of the plastic strength. The slope of line BC depends on the type of structural member and it is usually within the range of 0 % and 10 % of the initial slope (line AB). The point C corresponds to the peak strength of the component prior to its failure. The branch CD simulate the initial failure of the member and the segment DE corresponds to the residual strength, which can be zero in case of low-ductile systems. The response parameters (e.g. the plastic range "a", the ultimate deformation capacity "b" and the residual strength "c") are also provided by FEMA 356 in tabular form for the most representative structural members and relevant details.

Both FEMA 356 (chapter 5) and EN 1998-3 (Annex B) (CEN, 2005c) define criteria for the acceptable damage state condition of the plastic hinges associated to three limit states that are similar in both codes, see the three points on generalized force-displacement curve in Figure 6.14b. In particular, it can be roughly assumed that Immediate Occupancy (IO) corresponds to Damage Limitation (DL), Life Safety (LS) stands for Significant Damage (SD), while Collapse Prevention (CP) relates to Near Collapse (NC).

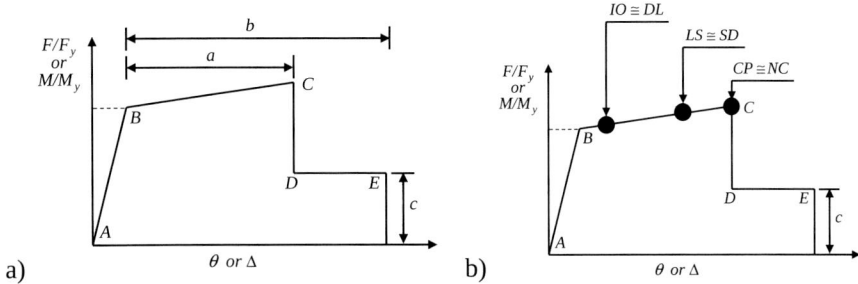

Figure 6.14 – Generalized force-deformation relationship of plastic hinge (a) and relevant acceptance criteria (b)

In this case, the response parameters were obtained from Table 5.6 of FEMA 356. Indeed, all beams satisfy the following conditions (where f_y is expressed in ksi according to US Units):

$$\frac{b_f}{2t_f} \leq \frac{52}{\sqrt{f_y}} \quad \text{and} \quad \frac{h}{t_w} \leq \frac{418}{\sqrt{f_y}} \tag{6.33}$$

In addition, the columns comply with the following requirements:

$$\frac{N_{Ed}}{N_{pl}} \leq 0.20 \quad \text{and} \quad \frac{b_f}{2t_f} \leq \frac{52}{\sqrt{f_y}} \quad \text{and} \quad \frac{h}{t_w} \leq \frac{300}{\sqrt{f_y}} \tag{6.34}$$

The acceptance criteria given by FEMA 356 for these types of members are the same of EN 1998-3 (see Table B.1) for class 1 members.

In line with the seismic assessment philosophy, it should be also noted that the plastic strength of structural members was calculated assuming the average yield stress of steel, namely $\gamma_{ov} \cdot f_y = 1.25 \cdot 235 = 294$ MPa. The hardening of plastic hinges was assumed according to AISC 341-10 (see chapter 4), as follows:

$$\gamma_{sh} = \frac{f_y + f_u}{2 \cdot f_y} = \frac{235 + 360}{2 \cdot 235} = 1.27 \tag{6.35}$$

6. MULTI-STOREY BUILDING WITH MOMENT RESISTING FRAMES

Table 6.26 reports the response parameters and the relevant acceptance criteria adopted for both beams and columns, where θ_y is the yield rotation of the member and M_p is its plastic bending strength as reported in Tables 6.27-6.28 for the members belonging to the perimeter MRF in the X direction (N.B. characterized by five moment resisting spans). The yield rotation θ_y was calculated as follows:

$$\theta_y = \frac{M_p L_b}{6EI_b} \qquad \text{for beams} \tag{6.36}$$

$$\theta_y = \frac{M_p L_c}{6EI_c}\left(1 - \frac{N}{N_p}\right) \qquad \text{for columns} \tag{6.37}$$

where L_b and L_c are respectively the lengths of the beam and the column; I_b and I_c are the second moment of area of the beam and the column cross sections.

Table 6.26 – Response parameters and acceptance criteria for plastic hinges

Type	a	b	c	DL	SD	NC
Beam	$9\theta_y$	$11\theta_y$	$0.6 M_p$	$1\theta_y$	$6\theta_y$	$8\theta_y$
Column	$9\theta_y$	$11\theta_y$	$0.6 M_p$	$1\theta_y$	$6\theta_y$	$8\theta_y$

Table 6.27 – Yield rotation for plastic hinges of beams (MRF in X direction)

beam profile (-)	L_b (mm)	$\gamma_{ov} f_y$ (MPa)	I_b (cm⁴)	M_p (kNm)	θ_y (rad)
IPE550	7000		67120	818.68	0.00678
IPE550	6000		67120	818.68	0.00588
IPE550	5000		67120	818.68	0.00484
IPE600	7000		92080	1031.65	0.00622
IPE600	6000	293.75	92080	1031.65	0.00534
IPE600	5000		92080	1031.65	0.00445
IPE750×196	7000		240300	2107.36	0.00487
IPE750×196	6000		240300	2107.36	0.00418
IPE750×196	5000		240300	2107.36	0.00348

Table 6.28 – Yield rotation for plastic hinges of columns (MRF in X direction)

storey (-)	column profile (-)	L_c (mm)	$\gamma_{ov}f_y$ (MPa)	I_c (cm^4)	M_p (kNm)	N/N_p (-)	θ_y (rad)
VI	HEM600	3500	293.75	237400	2576.78	0.012	0.00298
VI	HEM600	3500	293.75	237401	2576.78	0.030	0.00293
IV	HEM600	3500	293.75	237402	2576.78	0.047	0.00287
III	HEM600	3500	293.75	237403	2576.78	0.064	0.00282
II	HEM600	3500	293.75	237404	2576.78	0.082	0.00277
I	HEM600	4000	293.75	237405	2576.78	0.101	0.00310
Column "A" (see Figure 6.2)							
VI	HEM700	3500	293.75	329300	3096.13	0.016	0.00257
VI	HEM700	3500	293.75	329301	3096.13	0.037	0.00252
IV	HEM700	3500	293.75	329302	3096.13	0.061	0.00245
III	HEM700	3500	293.75	329303	3096.13	0.086	0.00239
II	HEM700	3500	293.75	329304	3096.13	0.114	0.00231
I	HEM700	4000	293.75	329305	3096.13	0.142	0.00256
Column "B" (see Figure 6.2)							
VI	HEM700	3500	293.75	329300	3096.13	0.014	0.00257
VI	HEM700	3500	293.75	329301	3096.13	0.036	0.00252
IV	HEM700	3500	293.75	329302	3096.13	0.060	0.00246
III	HEM700	3500	293.75	329303	3096.13	0.085	0.00239
II	HEM700	3500	293.75	329304	3096.13	0.113	0.00232
I	HEM700	4000	293.75	329305	3096.13	0.141	0.00256
Column "C" (see Figure 6.2)							

The bending moment – axial force domain of the columns were estimated according to section 6.2.9.1 of EN 1993-1-1 and shown in Figure 6.15.

6. MULTI-STOREY BUILDING WITH MOMENT RESISTING FRAMES

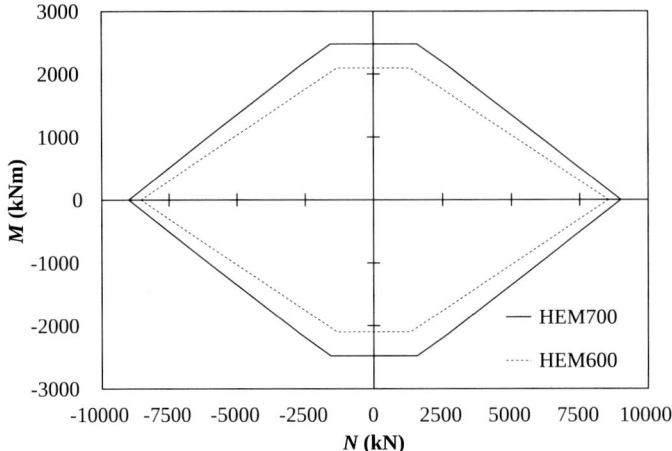

Figure 6.15 – M-N domain for plastic hinges of columns

It should be noted that the plastic contribution of column web panel was not accounted for because supplementary web plates were designed to guarantee larger resistance than that of the connected beams (see section 6.3.3). In case of weak column web panels it is important to model properly its plastic behaviour. A possible modelling strategy to simulate the shear response of panel zone consists of using a fictitious frame made of rigid elements reproducing the dimensions of the column web panel zone (Gupta and Krawinkler, 1999). Each rigid element is pinned at both ends, while at the upper corner two rotational springs (see Figure 6.16) reproduce the shear-distortion behaviour of the column web panel.

The nonlinear springs can be characterized by means of Ramberg-Osgood relationships with exponents ranging within 9-10, while the yielding parameters can be defined as follows:

$$V_y = \frac{f_y}{\sqrt{3}} A_v \qquad (6.38)$$

$$M_y = V_y \cdot \left(h_b - t_{f.b}\right) \qquad (6.39)$$

$$\gamma_y = \frac{f_y}{\sqrt{3} \cdot G} \qquad (6.40)$$

where M_y is the moment corresponding to the plastic shear V_y; A_v is the shear area of the column given according to EN-1993 and accounting for additional web plates, if relevant; h_b is the beam height and $t_{f,b}$ is the beam flange thickness; γ_y is the distortion corresponding to the yielding of the panel in shear.

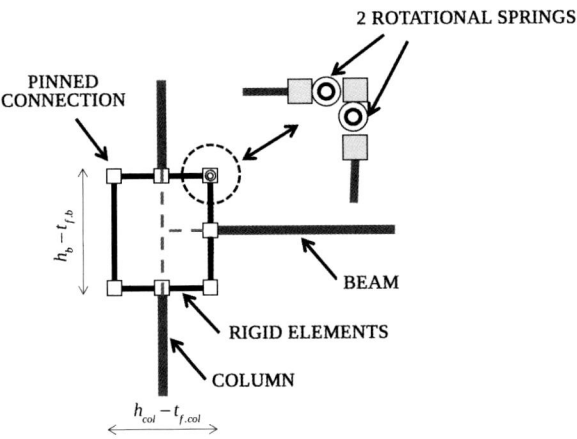

Figure 6.16 – Refined model for the column web panel zone

6.5.2.2 P-Delta effects

Since the building is regular both in plan and in elevation, 2D planar nonlinear models were considered. This approach is allowed by EC8-1 clause 4.3.1(5)), since the examined building satisfies the conditions given by clauses 4.2.3.2 and 4.3.3.1(8).

As discussed in section 2.6.4, in case of 2D models, the gravity loads directly applied by the tributary area of loads acting on the planar frame do not reflect the actual amount of vertical forces producing overall overturning effects. Hence, in order to account for the influence given by the complement of vertical loads, a leaning column was modelled (see Figure 2.33). This type of column is a fictitious zero-stiffness vertical element supporting the amount of vertical loads that are not directly applied to the 2D frame model. Therefore, the leaning column is made of truss elements with flexural releases at both ends. In addition, the leaning column is also connected at each storey level by a rigid diaphragm, as shown in Figure 6.17 where the additional vertical loads are also reported.

6. MULTI-STOREY BUILDING WITH MOMENT RESISTING FRAMES

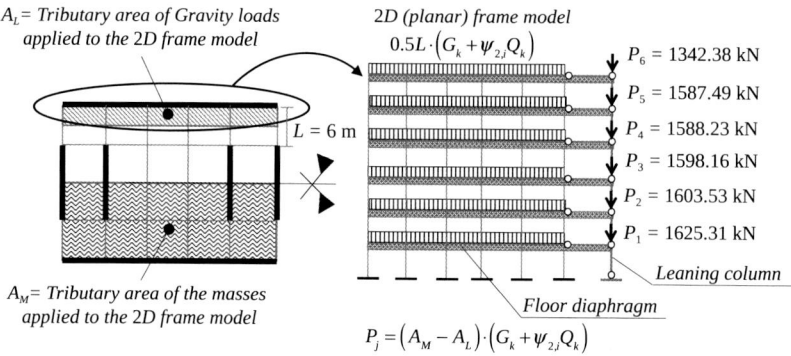

Figure 6.17 – Modelling of P-Δ effects in 2D model

6.5.3 Pushover analysis

The nonlinear static analysis should account for both gravity loads and horizontal increasing loads. Therefore, this type of analysis was carried out in two steps. First the entire value of the gravitational loads in the seismic condition (see equation (2.24)) are applied. Aftewards, the vertical loads remain constant and the lateral distributions of horizontal forces are applied in a second step. As mentioned in chapter 2, the structural analysis softwares allow to control the application of forces to the numerical model of the structure by means of two different strategies: 1) force control; 2) displacement control. The first case was adopted for gravity loads, because the magnitude of this load distribution is known a-priori and the structure is expected to resist elastically to these forces. Displacement control was used for the lateral forces, because it is imposed how much the structure should laterally deform, but the corresponding lateral reaction is unknown. In order to use displacement control, the horizontal displacement of a node at the roof in the direction of the applied forces was selected. Moreover, it is necessary to give the magnitude of the displacement that is the target for the analysis. Due to this assumption, it should be noted that the magnitude of the applied lateral forces is not important, only their distribution with the height of the structure.

For what concern the type of lateral forces, according to clause 4.3.3.4.2.2(1) two horizontal distributions were considered:

- a "uniform" distribution, based on mass proportional lateral forces, regardless of height (uniform response acceleration);

- a "modal" distribution, where the lateral forces are proportional to the first mode of vibration weighted with the masses at each storey.

For both cases, the lateral forces $\bar{F}_i$ depend on the normalized displacements ϕ_i and the masses m_i at the i-th storey, as follows:

$$\bar{F}_i = m_i \cdot \phi_i \qquad (6.41)$$

The profiles of displacements are both the first mode of vibration and the uniform lateral drifts that should be normalized at the roof displacement, namely by imposing $\phi_n = 1$, where n is the control node at the roof level. Table 6.29 reports the values of both distributions of lateral forces.

Table 6.29 – Lateral force distributions (pushover in X direction)

storey (-)	m_i (kN s²/m)	Modal pattern		Uniform pattern	
		ϕ_i (m/s²)	$\bar{F}_i$ (kN)	ϕ_i (m/s²)	$\bar{F}_i$ (kN)
VI	182.45	1.00	182.45	1.00	182.45
VI	215.77	0.88	188.92	1.00	215.77
IV	215.87	0.70	151.00	1.00	215.87
III	217.22	0.49	106.57	1.00	217.22
II	217.95	0.30	64.46	1.00	217.95
I	220.91	0.13	29.56	1.00	220.91

The capacity curves F_b-d_n (where F_b is the base shear and d_n is the roof displacement), determined for both force distributions given in Table 6.29 are depicted in Figure 6.18, where the points corresponding to the first plastic hinge and the formation of plastic mechanism are highlighted. The capacity curve obtained with the uniform load pattern is characterized by larger strength and stiffness than the modal load pattern. On the contrary, the displacement at the formation of the plastic mechanism is smaller. This result depends on the type of induced damage distribution. Indeed, the uniform load pattern concentrates the damage at the first storey, inducing a soft storey mechanism with poor plastic redistribution along the building height. The "modal" distribution is more demanding for the upper storeys, with more uniform plastic distribution from storey to storey.

According to EC8-1 (section 4.3.3.4.2.1), the obtained capacity curves allow to verify or revise the overstrength ratio α_u/α_1 (i.e. the structural redundancy), used to estimate the behaviour factor q, as well as the plastic mechanisms and the distribution of damage.

The calculated overstrength factors are reported in Table 6.30, while the relevant damage distributions are depicted in Figure 6.19. The minimum value of the ratio α_u/α_1 is equal to 1.34 and it is slightly larger than the value of 1.30 assumed for design.

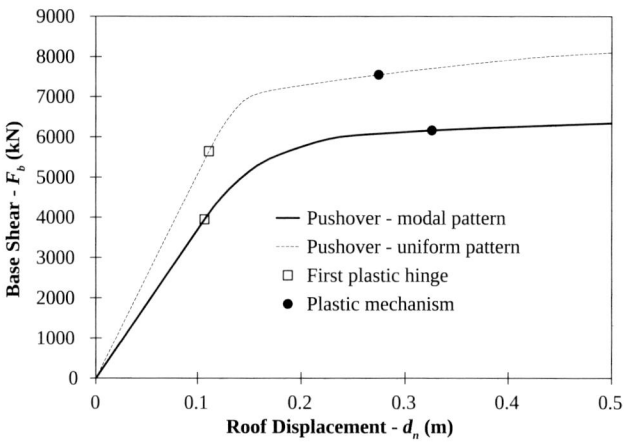

Figure 6.18 – Capacity curves for both modal and uniform pushover

Table 6.30 – Overstrength ratios α_u/α_1

Force distribution	$\alpha_1 \cdot F = F_1$ (kN)	$\alpha_u \cdot F = F_u$ (kN)	α_u/α_1 (-)
Modal	3946.16	6161.20	1.56
Uniform	5644.12	7540.85	1.34

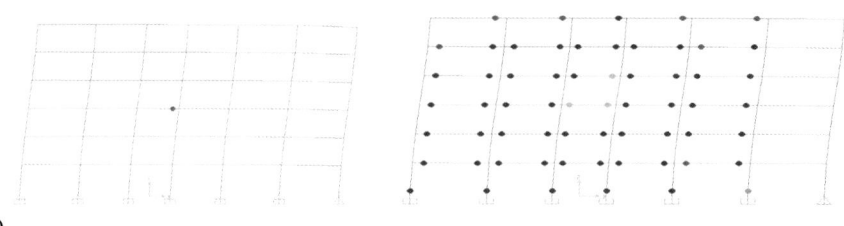

a)

First plastic hinge   Plastic mechanism

Figure 6.19 – Damage pattern for both modal (a) and uniform (b) force distribution

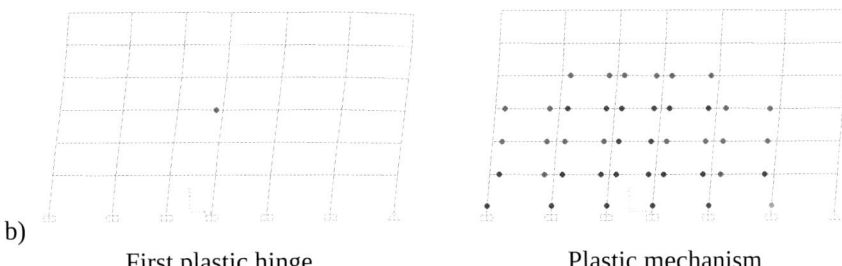

b)
           First plastic hinge   Plastic mechanism

Figure 6.19 – Damage pattern for both modal (a) and uniform (b) force distribution (continuation)

6.5.4 Transformation to an equivalent SDOF system

Once the capacity curves is obtained, it is possible to convert the response of the MDOF system into an equivalent elastic-perfectly plastic SDOF. The mass of the equivalent SDOF system is determined as:

$$m^* = \sum m_i \cdot \phi_i = \sum \overline{F}_i = 722.96 \text{ kN} \tag{6.42}$$

and the transformation factor is given by:

$$\Gamma = \frac{m^*}{\sum m_i \phi_i^2} = \frac{722.96}{528.80} = 1.37 \tag{6.43}$$

Hence, the capacity curve F_b-d_n of the MDOF (see Figure 6.20a) was converted into the response curve F^*-d^* of the equivalent SDOF system (see Figure 6.20b) as follows:

$$F^* = \frac{F_b}{\Gamma} \tag{6.44}$$

$$d^* = \frac{d_n}{\Gamma} \tag{6.45}$$

An equivalent bilinear force-displacement relationship was determined, with the yield force F_y^* and the yield displacement d_y^* of the SDOF. The yield force F_y^* represents the ultimate strength of the idealized system and it is equal to the base shear force at the formation of the plastic mechanism, as follows:

6. Multi-storey Building with Moment Resisting Frames

$$F_y^* = \frac{\alpha_u \cdot F}{\Gamma} = \frac{6161.20}{1.37} = 4506.54 \text{ kN} \quad (6.46)$$

The initial stiffness of the idealized system was determined in such a way that the areas under the actual and the idealized force–deformation curves are equal (see Figure 6.21). Based on this assumption, the yield displacement of the idealised SDOF system d_y^* is given by:

$$d_y^* = 2\left(d_m^* - \frac{E_m^*}{F_y^*}\right) = 2\left(0.238 - \frac{768.32}{4506.54}\right) = 0.135 \text{ m} \quad (6.47)$$

where E_m^* is the actual deformation energy up to the formation of the plastic mechanism; d_m^* is the displacement at the formation of the plastic mechanism.

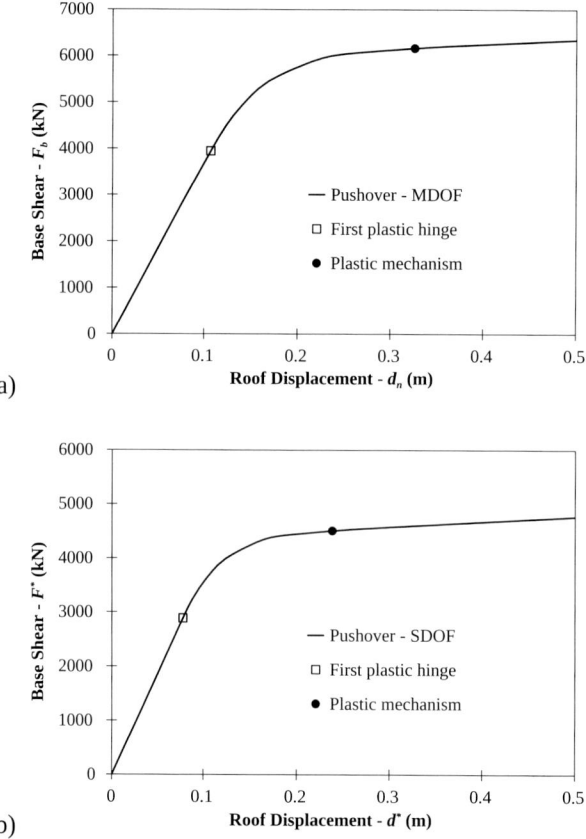

Figure 6.20 – Capacity curve of MDOF (a) and SDOF (b) system

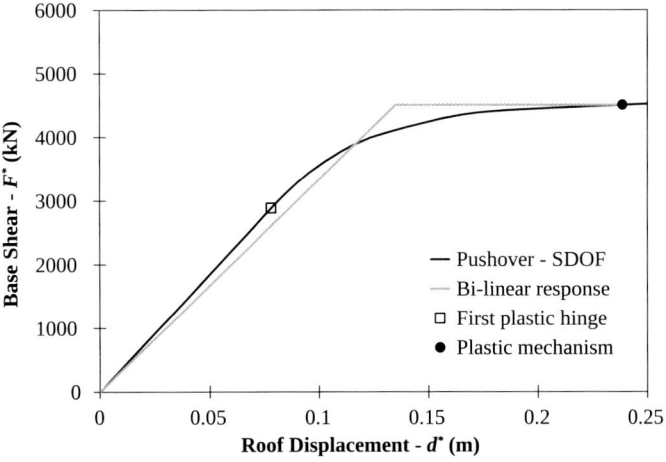

Figure 6.21 – Idealized elastic-perfectly plastic force vs displacement behaviour

6.5.5 Evaluation of the seismic demand

The seismic demand of the structure was obtained converting the target displacement d_t^* estimated for the idealized elastic-perfectly plastic response curve of the SDOF into the roof displacement d_n of the MDOF.

The target displacement d_t^* depends on the period of vibration T^* of the equivalent SDOF system, which was computed as follows:

$$T^* = 2\pi \sqrt{\frac{m^* d_t^*}{F_y^*}} = 2 \cdot 3.14 \sqrt{\frac{722.96 \cdot 0.24}{4506.54}} = 1.23 \text{ s} \qquad (6.48)$$

It should be noted that T^* is larger than the corner period T_C of the elastic response spectrum (i.e. $T_C = 0.6$ s as shown in sub-section 6.1.4.2).

The target displacement d_{et}^* of the elastic SDOF with period T^* is given by:

$$d_{et}^* = S_e(T^*) \left[\frac{T^*}{2\pi}\right]^2 = 3.44 \left[\frac{1.23}{2 \cdot 3.14}\right]^2 = 0.132 \text{ m} \qquad (6.49)$$

where $S_e(T^*)$ is the acceleration obtained from the elastic response spectrum at the period T^*.

Since the period $T^* > T_C$, the demand of the elastic-perfectly plastic SDOF d_t^* is equal to the demand of the unlimited elastic system d_{et}^*.

6. MULTI-STOREY BUILDING WITH MOMENT RESISTING FRAMES

Once known d_t^*, the target displacement d_t of the MDOF system was estimated as follows:

$$d_t = \Gamma d_t^* = 1.37 \cdot 0.132 \text{ m} = 0.181 \text{m} \tag{6.50}$$

6.5.6 Evaluation of the structural performance

The demand d_t was obtained with reference to a 2D model. Since the effects of accidental eccentricity should be also accounted for, the displacement d_t was magnified by the δ factor given by equation (2.58), which is equal to 1.6 due to the symmetry of the structural plan.

Therefore, the target displacement of the 2D MRF is given by:

$$d_t = 1.6 \cdot 0.181 \text{ m} = 0.290 \text{ m} \tag{6.51}$$

The capacity of the structure was verified as follows: i) the modal pushover was performed imposing $d_t = 0.290$ m as maximum roof displacement (see Figure 6.22); ii) it was checked that the local demand of plastic hinges does not exceed the limits imposed by the acceptance criteria at ultimate limit state (i.e. SD limit state according to EN 1998-3, see Table 6.26). These verifications are satisfied as shown in Table 6.31, where the ductility verification checks are reported for the elements most engaged in the plastic range.

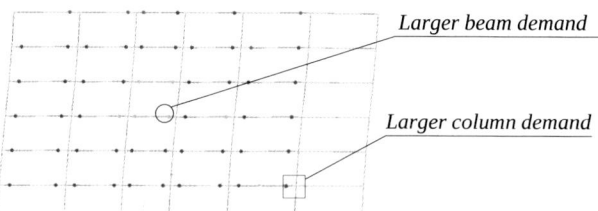

Figure 6.22 – Pushover response at $d_t = 0.290$ m

Table 6.31 – Yield rotation for plastic hinges of columns (MRF in X direction)

element (-)	profile (-)	Plastic rotation demand (rad)	Plastic rotation capacity (rad)
Beam	IPE 750×196	0.00634	0.02904
Column	HEM600	0.00275	0.01787

Chapter 7

MULTI-STOREY BUILDING WITH CONCENTRICALLY BRACED FRAMES

7.1 BUILDING DESCRIPTION AND DESIGN ASSUMPTIONS

This chapter illustrates and discusses the design steps and the verifications required by EC8-1 for a multistorey building equipped with Concentrically Braced Frames (CBFs). Both diagonal X and inverted-V CBF configurations are dealt with. Firstly, a general description of the geometric features and design assumptions (i.e. materials, loads, seismic action, etc.) is provided. In the second part of the chapter, the assumptions for structural modelling and the methods of analysis adopted to calculate the seismic induced effects are presented. Finally, the design and verifications for both ultimate limit state and damage limitation limit state are described and discussed.

This worked example uses the same building architectural definition as in chapter 6, all details being found in section 6.1.1 concerning architectural plan, geometrical properties, diaphragms and gravity load resisting systems. However, in order to allow for an independent reading of this chapter and easy comparison with the other case studies, a similar presentation is used that may include some duplicate information.

7.1.1 Building description

The worked example is a six storey office building with a rectangular plan layout and an area of 31.00 m $\times$ 24.00 m. The storey height is equal to 3.50 m with exception of the first floor, which is 4.00 m high.

7. MULTI-STOREY BUILDING WITH CONCENTRICALLY BRACED FRAMES

As discussed in chapter 2, the plan location of the braced bays plays a key role for building regularity. It is generally a compromise between architectural needs (i.e. the necessity of placing windows and doorways) and structural requirements (i.e. the necessity of distributing the seismic resistant elements along the perimeter of the building to guarantee adequate torsional strength and stiffness). In light of these considerations, the braces were distributed in order to guarantee large torsional rigidity without filling all spans around the perimeter of the building.

Figure 7.1 shows the structural layout of the typical floor, illustrating the location of the braces, while Figure 7.2 shows the CBF vertical configuration in the two main plan directions.

It is important to note that the choice of using two different CBF typologies does not have any structural motivations, but it derives only by the opportunity to deal with two different case studies in one example building.

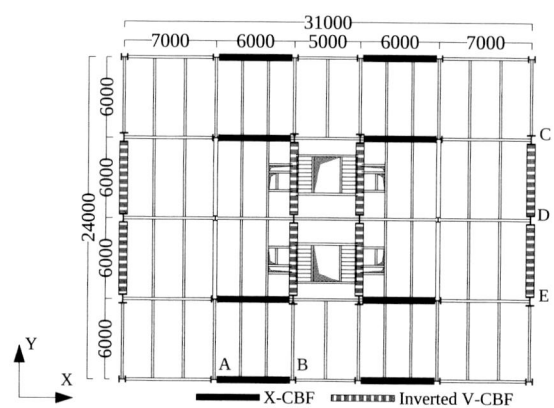

Figure 7.1 – Structural plan of the typical floor

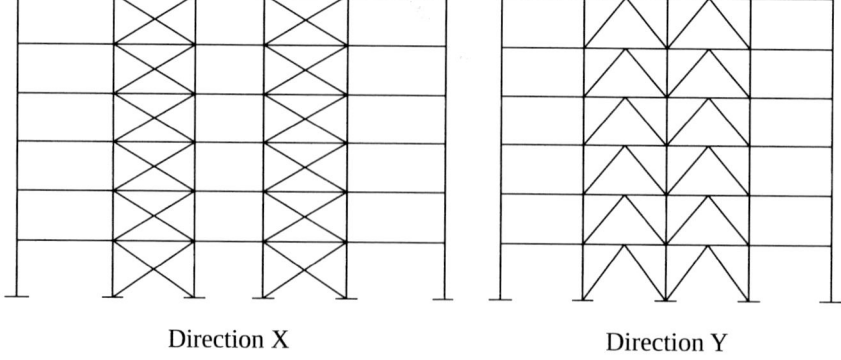

Figure 7.2 – Vertical layout of CBFs

7.1.2 Normative references

Besides the seismic recommendations, the structural design was carried out according to the following European codes:

- EN 1990 (2002) Eurocode 0: Basis of structural design;
- EN 1991-1-1 (2002) Eurocode 1: Actions on structures - Part 1-1: General actions - Densities, self-weight, imposed loads for buildings;
- EN 1993-1-1 (2005) Eurocode 3: Design of steel structures - Part 1-1: General rules and rules for buildings;
- EN 1993-1-8 (2005) Eurocode 3: Design of steel structures - Part 1-8: Design of joints;
- EN 1994-1-1 (2004) Eurocode 4: Design of composite steel and concrete structures - Part 1.1: General rules and rules for buildings.

For the sake of generality, a specific National Annex is not used. Hence, the calculation examples were carried out using the recommended values of the safety factors.

7.1.3 Materials

Similarly to chapter 6, the material properties of steel were accounted for according to clause 6.2(3)a. The mechanical properties of the materials used for the structural members are summarized in Table 7.1, where the partial factors are also indicated.

Table 7.1 – Material properties and partial factors

Grade	f_y (N/mm²)	f_t (N/mm²)	γ_M (-)	γ_{ov} (-)	E (N/mm²)
S235	235	360	$\gamma_{M0} = 1.00$	1.00	210000
S355	355	510	$\gamma_{M1} = 1.00$		
			$\gamma_{M2} = 1.25$		

Two different steel grades were used, namely S235 and S355. S235 was adopted for the dissipative members, while S355 was selected for the non

7. MULTI-STOREY BUILDING WITH CONCENTRICALLY BRACED FRAMES

dissipative zones. This approach is in accordance with clause 6.2.3(b), in order to guarantee that the dissipative zones form where they are intended to by design. Under this assumption, section 6.2.4 of EC8-1 states that the overstrength factor γ_{ov} may be taken as 1.00 instead of 1.25 in the design verifications. Hence, this condition was considered in the calculations shown hereinafter.

7.1.4 Actions

7.1.4.1 Characteristic values of unit loads

Both permanent (G_k) and live loads (Q_k) considered in this worked example are equal to those adopted in chapter 6, which are reproduced in Table 7.2. The slab weight is inclusive of steel sheeting, infill concrete, finishing floor elements and partition walls. Analogously, the weight of stairwells includes the contribution of the flights, steps and landings.

Table 7.2 – Characteristic values of vertical permanent and live loads

	G_k (kN/m²)	Q_k (kN/m²)
Storey slab	4.20	2.00
Roof slab	3.60	0.50
		1.00 (Snow)
Stairs	1.68	4.00
Claddings	2.00	

7.1.4.2 Seismic action

In this case study both the elastic spectral shape and the reference peak ground acceleration a_{gR} are those described in chapter 6.

A reference peak ground acceleration equal to $a_{gR} = 0.25g$ (where g is the acceleration of gravity), a type C soil, a type 1 spectral shape and importance factor γ_I equal to 1.0 were assumed.

The spectral parameters for soil type C (Table 2.2 or Table 3.2 of EC8-1) are $S = 1.15$, $T_B = 0.20$ s, $T_C = 0.60$ s and $T_D = 2.00$ s. In addition, the lower bound factor β for the design response spectrum (see section 2.3.5) was assumed equal to 0.2, as recommended in section 3.2.2.5 of EC8-1.

As described in section 2.3.5, the design response spectrum is obtained according to equations (2.23), namely by dividing the ordinates of elastic response spectrum by the behaviour factor q. Figure 7.3 depicts both the resulting elastic and the design spectra.

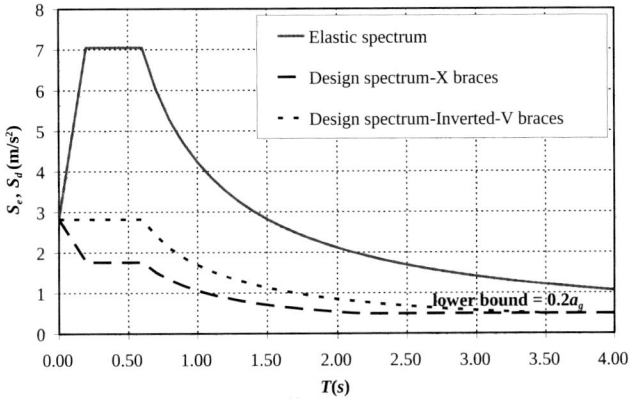

Figure 7.3 – Elastic and design response spectra

The behaviour factor was chosen on the basis of the intended design concept, corresponding to DCH structures (see concept b in sub-chapter 3.1). In addition, the behaviour factor according to clause 6.3.2 (see Figure 3.3 and Table 3.2) is given by:

- $q = 4$ for X-CBFs;
- $q = 2.5$ for inverted-V CBFs.

7.1.4.3 Combination of actions for seismic design situations

The combination of actions for seismic design conditions are equal to those described in sub-section 6.1.4.3.

7.1.4.4 Masses

As shown in sub-section 6.1.4.4, the inertial effects in the seismic design situation were evaluated by taking into account the presence of the masses corresponding to the combination of permanent and variable gravity loads as given by equation (2.25) in section 2.3.6, in accordance with clause 3.2.4 (2)P.

7. Multi-storey Building with Concentrically Braced Frames

Table 7.3 summarizes the seismic weights and the related masses at each floor that incorporate the weight of the pre-designed beams, columns and braces. Seismic weights per unit floor area and the moment of inertia of masses are also reported. Some differences may be noticed between the values reported in Table 7.3 and those summarized in Table 6.4. The slight differences are due to the different dimensions of the steel profiles belonging to seismic resistant systems.

Table 7.3 – Seismic weights and masses

Storey	G_k	Q_k	Seismic Weight		Seismic Mass
(-)	(kN)	(kN)	(kN)	(kN/m²)	(kN s²/m)
VI	3195.63	1326.00	3519.03	4.73	358.72
V	3990.72	1608.00	4196.23	5.64	427.75
IV	4087.66	1608.00	4276.87	5.75	435.97
III	4106.70	1608.00	4283.01	5.76	436.60
II	4187.79	1608.00	4353.15	5.85	443.75
I	4261.26	1608.00	4411.33	5.93	449.68

7.1.5 Pre-design

The building examined in this worked example satisfies the criteria for regularity in plan and in elevation. Figures 7.4 and 7.5 show the cross sections of dissipative (i.e. braces) and non-dissipative (i.e. beams and columns) members for X-CBFs, while Figures 7.6 and 7.7 depict the corresponding cross sections for inverted V-CBFs.

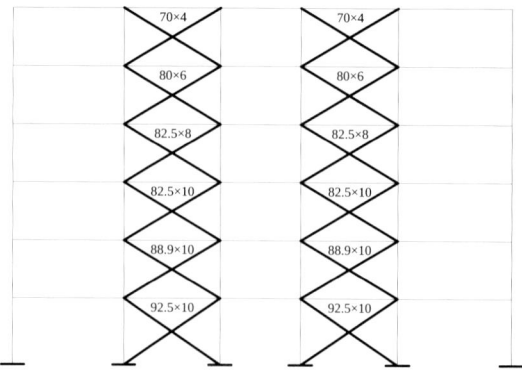

Figure 7.4 – Cross sections of X-braces

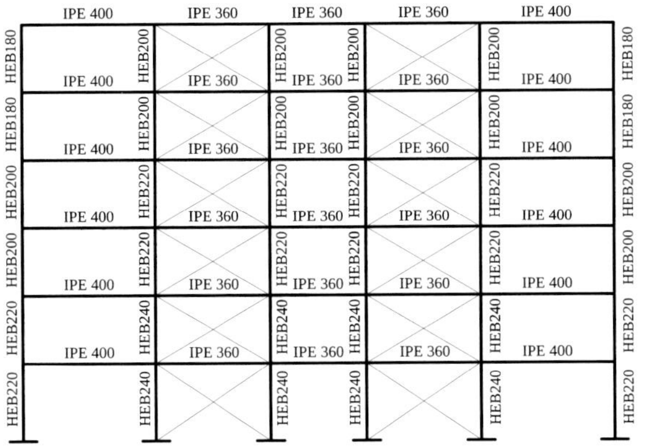

7.1 BUILDING DESCRIPTION AND DESIGN ASSUMPTIONS

Figure 7.5 – Cross sections of beams and columns in X-CBFs

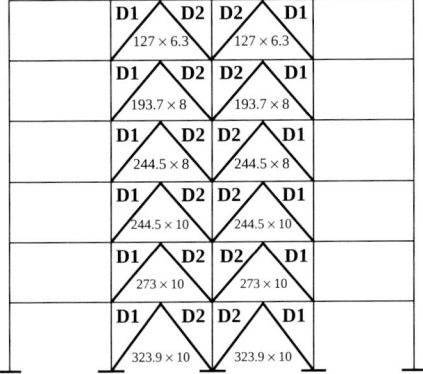

Figure 7.6 – Braces of Inverted V-CBFs

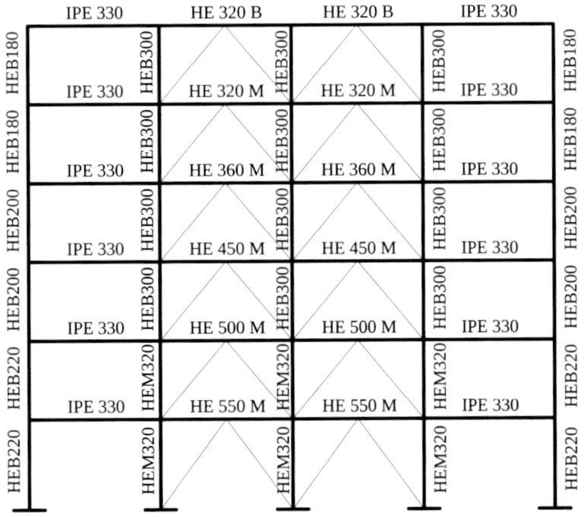

Figure 7.7 – Cross sections of beams and columns in inverted V-CBFs

7. MULTI-STOREY BUILDING WITH CONCENTRICALLY BRACED FRAMES

7.2 STRUCTURAL ANALYSIS AND CALCULATION MODELS

7.2.1 General features

Consistently to the case study described in chapter 6, also the CBF structure satisfies the criteria for regularity both in plan and in elevation. Besides the typological and geometrical features (i.e. the presence of rigid in plan diaphragms, the plan slenderness ratio, the absence of setbacks, etc.), the position of braces is crucial to guarantee plan and elevation regularity. As mentioned in section 7.1.1, the braces were mainly distributed along the perimeter and uniformly along the building height. This arrangement guarantees that the structural eccentricity e_o (which is the nominal distance between the centre of stiffness and the centre of mass) is practically negligible at each level and for both X and Y directions and the torsional radius r is larger than the radius of gyration of the floor mass in plan, as well.

7.2.2 Modelling assumptions

As discussed in sub-chapter 2.6, the structural model must represent the distribution of stiffness and mass so that all significant deformation shapes and inertia forces are properly accounted for under seismic actions (section 4.3.1 of EC8-1).

In this worked example, two separate calculation 2D planar models in the two main plan directions were used, one in the X direction and the other in the Y direction. This approach is allowed by clause 4.3.1(5), since the examined building satisfies the conditions given by clauses 4.2.3.2 and 4.3.3.1(8).

The calculation models assume all pinned connections (brace-to-beam and brace-to-column connections, beam-to-column and column bases), but the columns are considered continuous through each floor beam. These assumptions are frequently accepted and they are conservative in terms of storey displacements and brace axial forces.

The tributary area of masses was evaluated as described in section 2.6.4. Masses are considered as lumped in a selected master joint for each floor, because the floor diaphragms may be taken as rigid in their planes. This assumption implies that the numerical model gives axial forces in the beams equal to zero. Therefore, differently from MRFs, handmade calculations are necessary to evaluate the axial forces in the beams belonging to the braced spans, as described in the following section.

Since the structure is characterized by full symmetry of in-plan distribution of stiffness and masses, in order to ensure a minimum of torsional strength and stiffness and to limit the consequences of unpredicted torsional response, EC8-1 imposes accidental torsional effects (see sub-chapter 2.7). These take into account the possible uncertainties in the distributions of stiffness and mass and/or a possible torsional component of the ground motion about a vertical axis.

EC8 allows a simplified modelling approach if the building is conforming to the criteria for regularity both in plan and in elevation. As two separate 2D planar models are used, the effects of the accidental eccentricity may be estimated by multiplying the effects of the seismic actions (which can be obtained either by the lateral force method or by modal superposition) using the magnification factor δ (clause 4.3.3.2.4(2)) given by equation (2.51). Since in this example the planar 2D perimetric frames are analyzed, the δ factor is equal to 1.6 (being $x/L_e = 0.5$, where x is the in-plan distance between the structural element under consideration and L_e is the in-plan distance between two outermost lateral load resisting systems).

All modelling assumptions previously described are general for both X- and inverted-V CBFs. However, the different seismic behaviour of these two bracing systems needs to make a distinction in terms of design criteria and calculation models. The crucial issue influencing the structural model is the brace schematization. Braces are expected to dissipate energy through cyclic deformation in the post-buckling range, while the adjacent structural elements (beams, columns and connections) have to remain elastic. Obviously, as discussed in chapter 3, the energy dissipation capacity is larger when the brace yields in tension rather than when it buckles in compression. Anyway, the overall seismic performance also depends on the structural scheme, namely the bracing arrangement, thus requiring different modelling assumptions. In case of X-CBFs, EC8-1 stipulates that the energy dissipation capacity of braces in compression is neglected and the lateral forces are assigned to tension braces only. According to clause 6.7.2(2)P, in case of X-CBFs, the structural model shall include the tension braces only (i.e. the generic braced bay is ideally composed by a single brace), unless a non-linear analysis is carried out. In order to make tension alternatively developing in all the braces at any storey, two models must be developed, one with the braces tilted in one direction and another with the braces tilted in the opposite direction (see Figure 3.9 for more details). On the contrary, in frames with V bracings both the tension and compression diagonals should be taken into account.

7.2.3 Numerical models and method of analysis

The calculation models were developed in SAP 2000 on the basis of the assumptions described in section 7.2.2. Schemes of the numerical models used for these calculation examples are shown in Figures. 7.8 and 7.9.

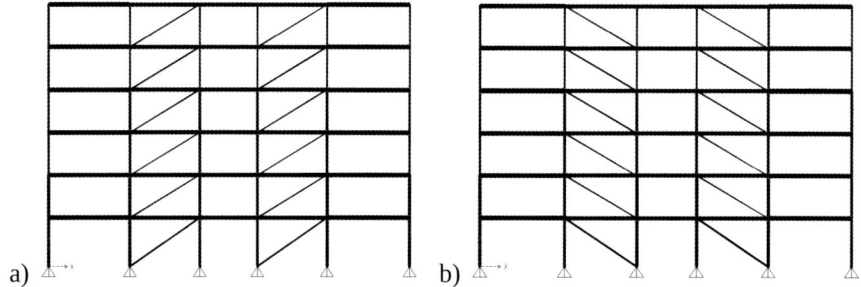

Figure 7.8 – X-CBFs: numerical models of the calculation example with only tension braces tilted in +X direction (a) and in –X direction (b)

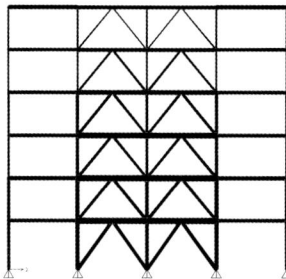

Figure 7.9 – Inverted V-CBFs: numerical models of the calculation example

The effects of seismic actions have been determined by means of a linear-elastic modal response spectrum analysis (clause 4.3.3.1(2)) of the 2D frames schematically shown in Figures. 7.8 and 7.9, using the design spectrum described in sub-section 7.1.4.2. As discussed in section 2.5.3, all vibration modes satisfying that the sum of their effective modal masses gives at least 90 % of the total mass of the structure were accounted for. Therefore, only the

7.2 STRUCTURAL ANALYSIS AND CALCULATION MODELS

first and the second mode of vibration in both the X and Y directions, whose shapes, relevant natural periods T_i and participating mass ratios M_i are shown in Figure 7.10, were considered for the calculation of seismic effects. The SRSS (Square Root of the Sum of the Squares) method is used to combine the modal maxima, since the first and the second modes of vibration in both X and Y direction may be considered as independent (as $T_2 \leq 0.9T_1$, according to section 4.3.3.3.2 of EC8-1).

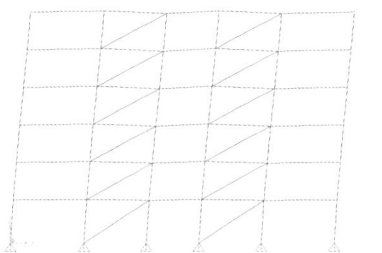
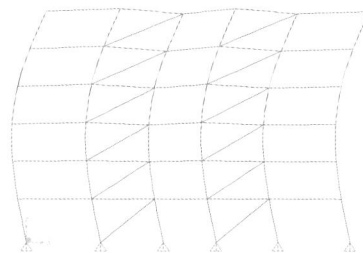

$T_1 = 1.030$ s; $M_1 = 0.780$ $T_2 = 0.366$ s; $M_2 = 0.154$

Dynamic properties in X direction

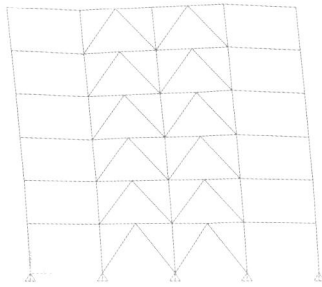
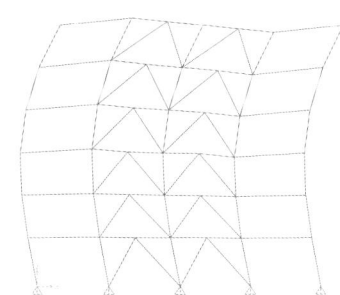

$T_1 = 0.414$ s; $M_1 = 0.761$ $T_2 = 0.160$ s; $M_2 = 0.164$

Dynamic properties in Y direction

Figure 7.10 – Fundamental dynamic properties of the examined structures

The diagrams of internal forces (i.e. bending moments, shear and axial forces) due to both gravity and seismic loads are depicted qualitatively in Figure 7.11 and Figure 7.12 for X-CBFs (N.B. the plots refer the model with braces tilted in +X direction; owing to the structural symmetry, the seismic effects for model with braces tilted in the –X direction are mirrored) and inverted V-CBFs, respectively.

7. MULTI-STOREY BUILDING WITH CONCENTRICALLY BRACED FRAMES

As discussed in chapter 3, the internal forces obtained from the calculation models are used to design the dissipative members (i.e. the bracing members), while the non-dissipative members (i.e. both beams and columns) should be designed for the combination given by gravity load induced effects and seismic induced effects, the latter magnified as given by equation 3.21 (see section 3.6.2).

Figure 7.11 and Figure 7.12 show that the axial forces induced by the seismic actions into the columns are directly provided by the numerical models, while they are missing for the beams due to the presence of rigid diaphragms. Nevertheless, it is possible to calculate the beam axial forces by simple hand calculations, which obviously differ for X-CBFs and inverted V-CBFs. A detailed description about the calculation of beam axial forces is found in sub-sections 7.3.1.2 and 7.3.2.2, respectively.

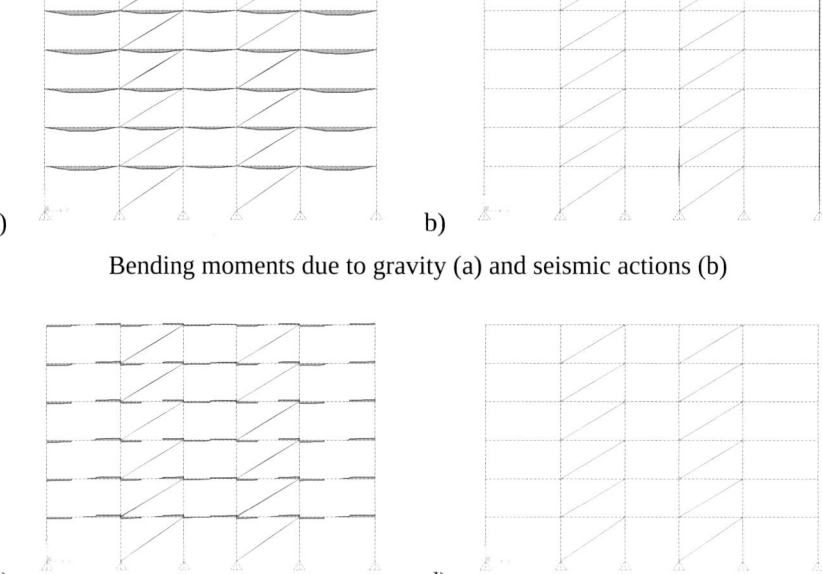

a) b)
Bending moments due to gravity (a) and seismic actions (b)

c) d)
Shear forces due to gravity (c) and seismic actions (d)

Figure 7.11 – Internal forces for X-CBFs
(N.B. the Plots were scaled by the same factor per type of effort)

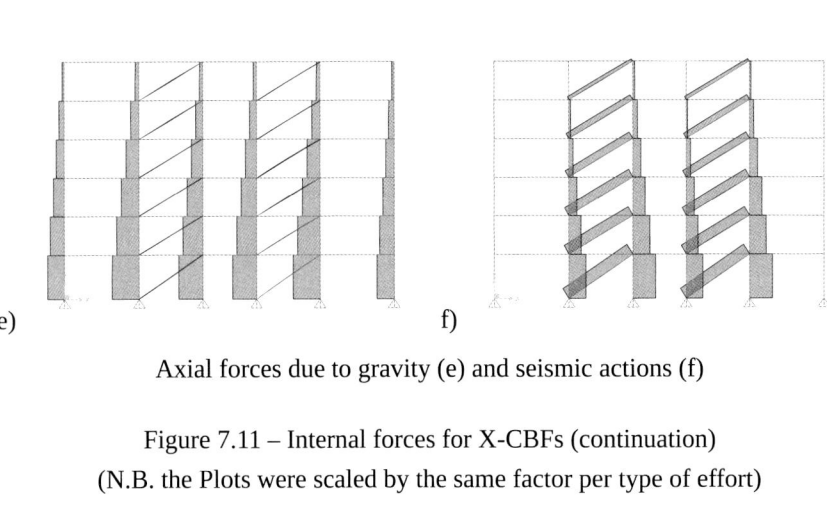

Axial forces due to gravity (e) and seismic actions (f)

Figure 7.11 – Internal forces for X-CBFs (continuation)
(N.B. the Plots were scaled by the same factor per type of effort)

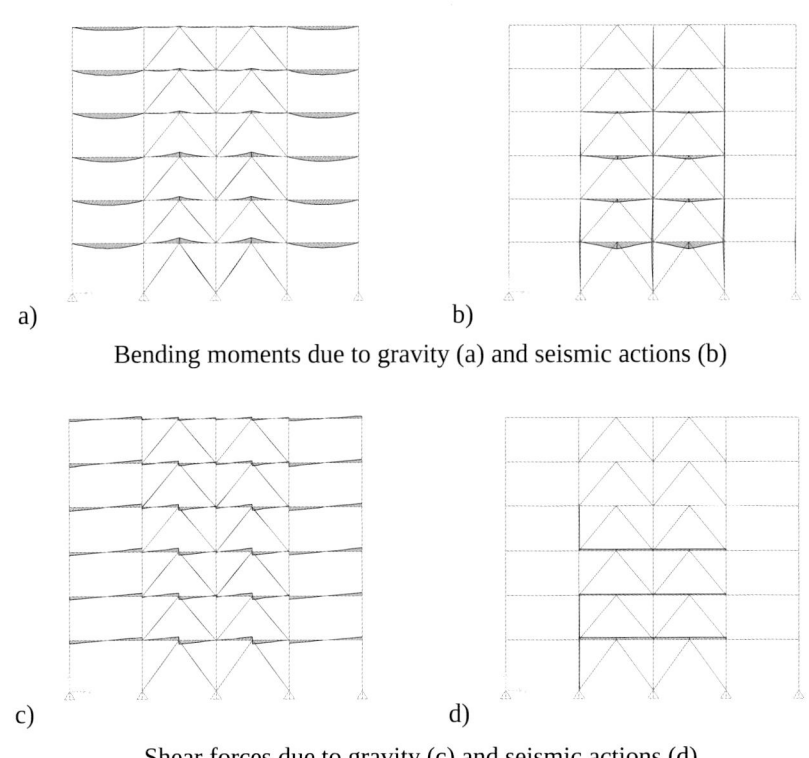

Bending moments due to gravity (a) and seismic actions (b)

Shear forces due to gravity (c) and seismic actions (d)

Figure 7.12 – Internal forces for inverted V-CBFs
(N.B. the Plots were scaled by the same factor per type of effort)

7. MULTI-STOREY BUILDING WITH CONCENTRICALLY BRACED FRAMES

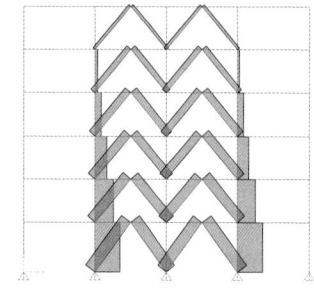

e)   f)

Axial forces due to gravity (e) and seismic actions (f)

Figure 7.12 – Internal forces for inverted V-CBFs (continuation)
(N.B. the Plots were scaled by the same factor per type of effort)

7.2.4 Imperfections for global analysis of frames

The effects induced by constructional imperfections were verified as discussed in chapter 6 according to clause 5.3.2(4) of EN 1993-1-1. Table 7.4 summarizes the calculation of the equivalent forces H_i and the limiting value of the shear force at each storey (i.e. $H_{Ed} = 0.15\ V_{Ed}$, where V_{Ed} is the total design vertical load on the structure at the bottom of the storey) above which it is possible to disregard the effects of global imperfections. The effects of global imperfections should be accounted for in the X direction, while they may be disregarded in the Y direction.

Table 7.4 – Global sway imperfections in the X and Y directions

Storey (-)	V_{Ed} (kN)	$H_{Ed,x}$ (kN)	$H_{Ed,y}$ (kN)	$0.15\ V_{Ed}$ (kN)	$H_{Ed,x}/V_{Ed}$ (-)	$H_{Ed,y}/V_{Ed}$ (-)	$H_i = \phi V_{Ed,i}$ (kN)
VI	879.40	261.87	639.55	131.91	0.30	0.73	2.32
V	1928.03	454.19	1186.41	289.20	0.24	0.62	5.08
IV	2996.81	590.19	1590.77	449.52	0.20	0.53	7.90
III	4067.14	705.02	1904.78	610.07	0.17	0.47	10.72
II	5154.99	798.89	2113.28	773.25	0.15	0.41	13.58
I	6257.38	**868.99**	2251.26	938.61	0.14	0.36	16.49

7.2.5 Frame stability and second order effects

P-Δ effects were specified through the inter-storey drift sensitivity coefficient θ given by equation (2.68), and are summarized in Tables 7.5 and 7.6 for calculation in the X and Y directions, respectively. As the coefficients θ are always smaller than 0.1, second order effects should not be accounted for. This result is not unusual because CBFs generally are less sensitive to second order effects due to their large lateral stiffness when compared to MRFs.

Table 7.5 – Stability coefficients for X-CBFs (i.e. frame in X direction)

Storey (-)	P_{tot} (kN)	V_{tot} (kN)	h (mm)	$d_r = d_e \times q$ (mm)	$\theta = \dfrac{P_{tot} \cdot d_r}{V_{tot} \cdot h}$ (-)	$\alpha = \dfrac{1}{(1-\theta)}$ (-)
VI	879.40	264.19	3500	44	0.04	1.04
V	1928.03	459.27	3500	46	0.06	1.06
IV	2996.81	598.08	3500	45	0.06	1.07
III	4067.14	715.73	3500	42	0.07	1.07
II	5154.99	812.48	3500	39	0.07	1.08
I	6257.38	885.48	4000	39	0.07	1.07

Table 7.6 – Stability coefficients for inverted V-CBFs (i.e. frame in Y direction)

Storey (-)	P_{tot} (kN)	V_{tot} (kN)	h (mm)	$d_r = d_e \times q$ (mm)	$\theta = \dfrac{P_{tot} \cdot d_r}{V_{tot} \cdot h}$ (-)	$\alpha = \dfrac{1}{(1-\theta)}$ (-)
VI	879.40	639.55	3500	14	0.01	1.01
V	1928.03	1186.41	3500	14	0.01	1.01
IV	2996.81	1590.77	3500	13	0.01	1.01
III	4067.14	1904.78	3500	12	0.01	1.01
II	5154.99	2113.28	3500	10	0.01	1.01
I	6257.38	2251.26	4000	10	0.01	1.01

7.3 DESIGN AND VERIFICATION OF STRUCTURAL MEMBERS

7.3.1 Design and verification of X-CBFs

7.3.1.1 Design and verification of braces

Circular hollow sections (see Figure 7.13) in S 235 steel grade are used for the X braces, whose cross sections are class 1, as defined in section 5.6 of EN 1993-1-1 ($d/t \leq 50\varepsilon^2$, where d is diameter, t the relevant thickness and $\varepsilon = (235/f_y)^{0.5}$. Table 7.7 summarizes the calculated d/t ratios.

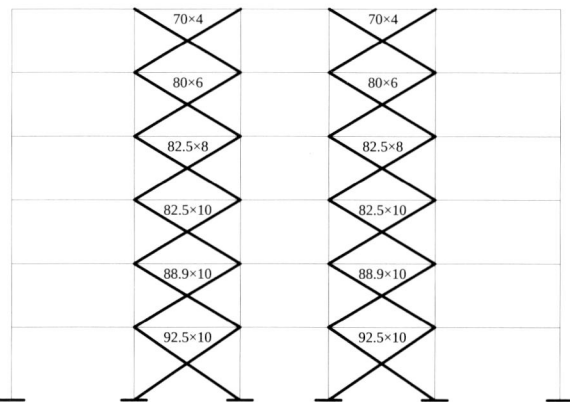

Figure 7.13 – Cross sections of X-braces

Circular hollow sections were used because of the wide range of available profiles that make it easier to satisfy both the slenderness limits ($1.3 < \overline{\lambda} \leq 2.0$) and the requirement of minimizing the variation among the diagonals of the overstrength ratio Ω_i (as defined in section 3.6.1), whose maximum value must not differ from the minimum Ω by more than 25 %. In particular, the first condition mostly influenced the design choices, requiring several iterations to select the adequate cross section for the braces.

The brace slenderness was calculated considering the out-of-plane restraint offered by the tensioned diagonal, as discussed in section 4.4.3, i.e. the out-of-plane buckling length was assumed equal to half of the theoretical working length (i.e. node-to-node distance).

The selected brace cross sections, the relevant brace slenderness (λ and $\overline{\lambda}$), the axial plastic strength ($N_{pl,Rd}$), the axial force due to the seismic load combination (i.e. $N_{Ed} = \alpha \cdot (N_{Ed,I} + N_{Ed,E})$, where α accounts for P-Δ effects, $N_{Ed,I}$, is the force induced by the global imperfections and N_{Ed} is the axial force induced by seismic forces), the overstrength factors Ω_i and their percentage variation along the building height $(\Omega_i - \Omega)/\Omega$ are given in Table 7.8. The minimum overstrength ratio Ω is equal to 1.10 (highlighted in bold in Table 7.8). Hence, according to the capacity design criteria described in section 3.6.2, the seismic induced forces acting in the non-dissipative members should be amplified by $1.1\gamma_{ov}\Omega = 1.10 \times 1.00 \times 1.10 = 1.21$ (N.B. $\gamma_{ov} = 1.00$ because S235 is used for dissipative elements and S355 for all non-dissipative members, according to section 6.2.4 of EC8-1).

Table 7.7 – Cross section properties of X-braces

Storey (-)	Brace cross section $d \times t$ (mm × mm)	d (mm)	t (mm)	d/t (-)	$50\varepsilon^2$ (-)
VI	70 ×4	70	4	17.50	50.00
V	80 ×6	80	6	13.33	50.00
IV	82.5 ×8	82.5	8	10.31	50.00
III	82.5 ×10	82.5	10	8.25	50.00
II	88.9 ×10	88.9	10	8.89	50.00
I	92.5 ×10	92.5	10	9.25	50.00

Table 7.8 – Design checks of X-braces

Storey (-)	Brace (mm × mm)	λ (-)	$\bar{\lambda}$ (-)	$N_{pl,Rd}$ (kN)	$N_{Ed,I}$ (kN)	$N_{Ed,E}$ (kN)	N_{Ed} (kN)	Ω_i (-)	$(\Omega_i - \Omega)/\Omega$ (%)
VI	70 ×4	148.84	1.59	194.90	1.40	151.58	152.98	1.27	15.45
V	80 ×6	132.74	1.41	327.79	4.29	262.91	267.2	1.23	11.82
IV	82.5 ×8	131.85	1.40	440.01	8.86	341.63	350.49	1.26	14.55
III	82.5 ×10	135.49	1.44	535.25	15.11	408.10	423.21	1.26	14.55
II	88.9 ×10	124.50	1.33	582.50	22.94	462.44	485.38	1.20	9.09
I	92.5 ×10	123.61	1.32	609.08	33.60	522.20	555.8	**1.10**	0.00

7.3.1.2 Design and verification of beams

Figure 7.14 shows the cross sections for the beams of the X-CBFs. Since lateral-torsional restraints were assumed, beams were designed to resist

7. Multi-storey Building with Concentrically Braced Frames

both gravity and seismic actions. Besides the ultimate limit state verifications under non-seismic load combinations (Simões da Silva et al, 2016), this section focuses on the verification checks under the seismic load combination.

As previously mentioned, axial forces develop in the beam under seismic action, but the calculation model does not account for those effects, which can be easily obtained by imposing static equilibrium of the forces acting on the beams. Figure 7.15 schematically shows an example of the calculation of the beam axial forces where, in accordance with the overall capacity design requirement expressed by equation (3.21), the quantities $1.1\gamma_{ov}\Omega N_{Ed,i+1}$ and $1.1\gamma_{ov}\Omega N_{Ed,i}$ are the magnified brace axial forces at level "$i+1$" and "i", where $N_{Ed,i+1}$ and $N_{Ed,i+1}$ are obtained from structural analysis (see Figure 7.11) respectively. Consequently, the quantity $1.1\gamma_{ov}\Omega F_{Ed,i}$ is the amplified seismic force at i-th storey transmitted by the shear connectors from the slab to the beams of the braced bays by means of the axially uniform distributed load q_i.

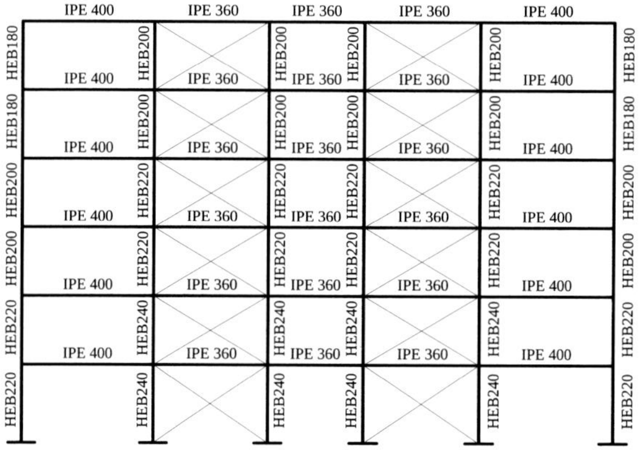

Figure 7.14 – Cross sections of beams and columns in X-CBFs

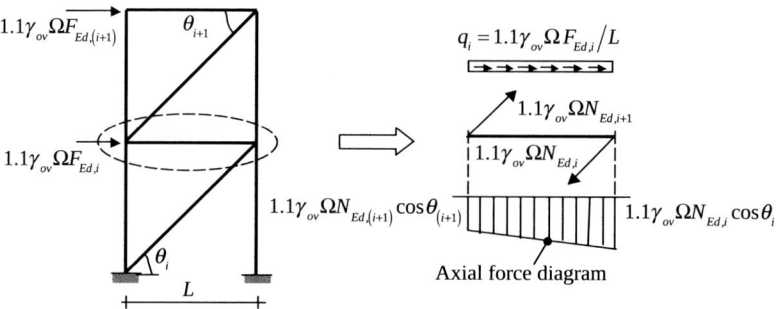

Figure 7.15 – Calculation scheme for seismically induced axial forces into the beam of X-CBFs (Mazzolani et al, 2006)

7.3 Design and Verification of Structural Members

Since the beams were considered as simply supported at both ends, due to the linearity of axial force diagram (see Figure 7.15), the beam end section closer to the lower brace (namely the brace at the i-th storey, as shown in Figure 7.14) and the section at the beam mid-span are the most engaged by axial, shear and combined bending-axial forces. It should be noted that no shear interaction occurs for the beams of this example, because the acting shear forces V_{Ed} are less than $0.5V_{pl,Rd}$, according to section 6.2.10 of EN 1993-1-1.

Table 7.9 reports the beam axial strength checks performed according to EN 1993:1-1 6.2.4 under the seismic load combination with reference to the beam end section connected to the lower brace.

Table 7.9 – Axial strength checks of beams in X-braced bays

Storey	Section	N_{Rd}	$N_{Ed,G}$	$N_{Ed,I}$	$N_{Ed,E}$	N_{Ed}	N_{Rd}/N_{Ed}
(-)	(-)	(kN)	(kN)	(kN)	(kN)	(kN)	(-)
VI	IPE 360		0	1.21	130.94	159.64	16.17
V	IPE 360		0	3.71	227.10	278.50	9.27
IV	IPE 360	2580.85	0	7.65	295.10	364.73	7.08
III	IPE 360		0	13.05	352.52	439.60	5.87
II	IPE 360		0	19.82	399.46	503.16	5.13
I	IPE 360		0	27.96	434.50	553.71	4.66

Table 7.10 reports the beam combined bending-axial strength checks according to section 6.2.9 of EN 1993-1-1, performed at the beam mid-span, where the highest bending moment is acting. In particular, $M_{N,Rd}$ is the plastic moment resistance reduced due to the beam axial force N_{Ed}.

Table 7.10 – Combined bending-axial strength checks of beams in X-braced bays

Storey	$N_{Ed,G}$	$N_{Ed,I}$	$N_{Ed,E}$	N_{Ed}	$M_{Ed,G}$	$M_{Ed,I}$	$M_{Ed,E}$	M_{Ed}	$M_{N,Rd}$	M_{Rd}/M_{Ed}
(-)	(kN)	(kN)	(kN)	(kN)	(kNm)	(kNm)	(kNm)	(kNm)	(kNm)	(-)
VI		0.70	65.47	79.92	64.28			64.28	361.75	5.63
V		2.84	179.02	219.46	86.27			86.27	361.75	4.19
IV	0.00	6.58	261.10	322.51	86.27	0.00	0.00	86.27	361.75	4.19
III		11.99	323.81	403.80	86.27			86.27	361.75	4.19
II		19.03	375.99	473.97	86.27			86.27	361.75	4.19
I		28.27	409.64	523.94	86.27			86.27	361.75	4.19

7. Multi-storey Building with Concentrically Braced Frames

To illustrate the application of the capacity design requirements (see equation (3.21)) for the beams, on the basis of the values reported in Table 7.9, the axial force in the beam end section connected to the lower brace at first storey was computed as follows:

$$N_{Ed} = N_{Ed,G} + \alpha \cdot \left(N_{Ed,I} + 1.10 \cdot \gamma_{ov} \cdot \Omega \cdot N_{Ed,E}\right) =$$
$$= 0 \text{ kN} + 1.00 \cdot \left(27.96 \text{ kN} + 1.10 \cdot 1.0 \cdot 1.10 \cdot 434.50 \text{ kN}\right) = \qquad (7.1)$$
$$= 553.71 \text{ kN}$$

Analogously, the axial force and the bending moment acting at the mid-length section of the beam were computed as follows:

$$M_{Ed} = M_{Ed,G} + \alpha \cdot \left(M_{Ed,I} + 1.10 \cdot \gamma_{ov} \cdot \Omega \cdot M_{Ed,E}\right) =$$
$$= 86.27 \text{ kNm} + 1.00 \cdot \left(0 \text{ kNm} + 1.10 \cdot 1.00 \cdot 1.10 \cdot 0 \text{ kNm}\right) = \qquad (7.2)$$
$$= 86.27 \text{ kNm}$$

$$N_{Ed} = N_{Ed,G} + \alpha \cdot \left(N_{Ed,I} + 1.1 \cdot \gamma_{ov} \cdot \Omega \cdot N_{Ed,E}\right) =$$
$$= 0 \text{ kN} + 1.00 \cdot \left(28.27 \text{ kN} + 1.10 \cdot 1.00 \cdot 1.10 \cdot 409.64 \text{ kN}\right) = \qquad (7.3)$$
$$= 523.94 \text{ kN}$$

Finally, Table 7.11 reports the beam safety checks against shear forces, according to section 6.2.6 of EN 1993-1-1, where A is the gross cross section area, A_v is the shear area, $V_{pl,Rd}$ is the design plastic shear resistance and V_{Ed} is the design value of the shear force.

The verification checks for the ultimate limit state under gravity loads governed the design of the beams for X-CBFs. This explains why constant beam cross sections are adopted along the building height and the large overstrength factors shown in Tables 7.9 and 7.10.

Table 7.11 – Shear strength checks of beams in X-braced bays

Storey	Section	A	A_v	$V_{pl,Rd}$	$V_{Ed,G}$	$V_{Ed,I}$	$V_{Ed,E}$	V_{Ed}	$V_{pl,Rd}/V_{Ed}$
(-)	(-)	(mm²)	(mm²)	(kN)	(kN)	(kN)	(kN)	(kN)	(-)
VI	IPE 360	7270	3510.8	719.57	34.54			34.54	20.83
V	IPE 360	7270	3510.8	719.57	47.50	0.00	0.00	47.50	15.15
IV	IPE 360	7270	3510.8	719.57	47.50			47.50	15.15

7.3 Design and Verification of Structural Members

Table 7.11 – Shear strength checks of beams in X-braced bays (continuation)

Storey (-)	Section (-)	A (mm²)	A_v (mm²)	$V_{pl,Rd}$ (kN)	$V_{Ed,G}$ (kN)	$V_{Ed,I}$ (kN)	$V_{Ed,E}$ (kN)	V_{Ed} (kN)	$V_{pl,Rd}/V_{Ed}$ (-)
III	IPE 360	7270	3510.8	719.57	47.50			47.50	15.15
II	IPE 360	7270	3510.8	719.57	47.50	0.0	0.0	47.50	15.15
I	IPE 360	7270	3510.8	719.57	47.50			47.50	15.15

7.3.1.3 Design and verification of columns

As discussed in section 7.2.3, the forces acting on the columns can be directly obtained from the calculation model. Since the bending moments and the shear forces are negligible (see Figure 7.11), the columns were verified against axial forces only. Since in the structural model of X-CBFs only one diagonal brace per bay was considered, two different numerical models were implemented, one with the braces tilted in the +X direction and another with the braces tilted in the –X direction (Figure 7.8). The axial forces obtained from the elastic model due to seismic forces were amplified by the overstrength factor ($1.1\gamma_{ov}\Omega$), with the value of Ω given in Table 7.8.

Table 7.12 – Axial strength checks for columns in + X direction

column "A" (see Figure 7.1)									
Storey (-)	Section (-)	A (mm²)	χ (-)	$N_{pl,Rd}$ (kN)	$N_{Ed,G}$ (kN)	$N_{Ed,I}$ (kN)	$N_{Ed,E}$ (kN)	N_{Ed} (kN)	$\chi N_{pl,Rd}/N_{Ed}$
VI	HEB 200	7808	0.60	2771.84	104.67	0.00	0.00	104.67	15.82
V	HEB 200	7808	0.60	2771.84	240.71	0.70	76.38	333.83	4.96
IV	HEB 220	9104	0.65	3231.92	376.76	2.86	207.74	630.99	3.33
III	HEB 220	9104	0.65	3231.92	512.35	7.33	374.79	973.18	2.16
II	HEB 240	10600	0.69	3763.00	651.83	14.94	569.38	1355.72	1.92
I	HEB 240	10600	0.62	3763.00	792.17	26.50	787.06	1771.01	1.33

7. MULTI-STOREY BUILDING WITH CONCENTRICALLY BRACED FRAMES

Table 7.12 – Axial strength checks for columns in + X direction (continuation)

\multicolumn{9}{c}{column "B" (see Figure 7.1)}
Storey (-)	Section (-)	A (mm²)	χ (-)	$N_{pl,Rd}$ (kN)	$N_{Ed,G}$ (kN)	$N_{Ed,I}$ (kN)	$N_{Ed,E}$ (kN)	N_{Ed} (kN)	$\dfrac{\chi N_{pl,Rd}}{N_{Ed}}$
VI	HEB 200	7808	0.60	2771.84	91.63	0.70	76.38	184.75	9.88
V	HEB 200	7808	0.60	2771.84	212.77	2.86	207.74	467.00	3.91
IV	HEB 220	9104	0.65	3231.92	335.40	7.33	374.79	796.23	2.89
III	HEB 220	9104	0.65	3231.92	456.94	14.94	569.38	1160.83	1.98
II	HEB 240	10600	0.69	3763.00	581.09	26.50	787.06	1559.94	1.82
I	HEB 240	10600	0.62	3763.00	704.04	45.13	1064.44	2037.14	1.27

Table 7.13 – Axial strength checks for columns in – X direction

\multicolumn{9}{c}{column "A" (see Figure 7.1)}
Storey (-)	Section (-)	A (mm²)	χ (-)	$N_{pl,Rd}$ (kN)	$N_{Ed,G}$ (kN)	$N_{Ed,I}$ (kN)	$N_{Ed,E}$ (kN)	N_{Ed} (kN)	$\dfrac{\chi N_{pl,Rd}}{N_{Ed}}$
VI	HEB 200	7808	0.60	2771.84	104.67	0.70	76.38	197.79	8.37
V	HEB 200	7808	0.60	2771.84	240.71	2.86	207.74	494.94	3.35
IV	HEB 220	9104	0.65	3231.92	376.76	7.33	374.79	837.59	2.51
III	HEB 220	9104	0.65	3231.92	512.35	14.94	569.38	1216.24	1.73
II	HEB 240	10600	0.69	3763.00	651.83	26.50	787.06	1630.67	1.59
I	HEB 240	10600	0.62	3763.00	792.17	45.13	1064.44	2125.27	1.10
\multicolumn{9}{c}{column "B" (see Figure 7.1)}
Storey (-)	Section (-)	A (mm²)	χ (-)	$N_{pl,Rd}$ (kN)	$N_{Ed,G}$ (kN)	$N_{Ed,I}$ (kN)	$N_{Ed,E}$ (kN)	N_{Ed} (kN)	$\dfrac{\chi N_{pl,Rd}}{N_{Ed}}$
VI	HEB 200	7808	0.60	2771.84	91.63	0.00	0.00	91.63	19.92
V	HEB 200	7808	0.60	2771.84	212.77	0.70	76.38	305.89	5.97
IV	HEB 220	9104	0.65	3231.92	335.40	2.86	207.74	589.63	3.90
III	HEB 220	9104	0.65	3231.92	456.94	7.33	374.79	917.77	2.51
II	HEB 240	10600	0.69	3763.00	581.09	14.94	569.38	1284.98	2.20
I	HEB 240	10600	0.62	3763.00	704.04	26.50	787.06	1682.88	1.53

The cross sections adopted for columns are shown in Figure 7.14, while the verification checks are reported for the columns A and B (see Figure 7.1) in Table 7.12 and in Table 7.13, for the braces tilted in the + X and – X directions, respectively. In these tables, χ is the buckling reduction coefficient calculated according to section 6.3.1.2 of EN 1993-1-1.

To illustrate the application of the capacity design requirements (see equation (3.21)) for the columns, on the basis of the values reported in Table 7.9, the axial force at the first level of column A was computed as follows:

$$N_{Ed} = N_{Ed,G} + \alpha \cdot \left(N_{Ed,I} + 1.10 \cdot \gamma_{ov} \cdot \Omega \cdot N_{Ed,E} \right) =$$
$$= 792.17 \text{ kN} + 1.00 \cdot \left(26.50 \text{ kN} + 1.10 \cdot 1.00 \cdot 1.10 \cdot 787.06 \text{ kN} \right) = \quad (7.4)$$
$$= 1771.01 \text{ kN}$$

The designed columns, especially those corresponding to the vertical "B" in the upper storeys, where very small profiles are sufficient to satisfy the strength requirements, are characterized by quite large overstrength factors. The reason relates to the need to guarantee sufficient width of column flange for proper detailing of the beam-to-column connections, brace-to-column connections, and column-to-column splices.

7.3.2 Design and verification of inverted V-CBFs

7.3.2.1 Design and verification of braces

As for the X-CBFs, circular hollow sections and S235 steel grade are used for the braces of the inverted-V CBFs. The adopted brace cross sections belong to class 1, as shown in Table 7.14.

Differently from X-bracing, for chevron braces the slenderness is only limited by $\bar{\lambda} \leq 2.0$. Owing to the symmetry of the structure, the design results are reported only for the diagonal members of one of the two bays of the bracing alignment. Because of the presence of vertical loads and the different axial deformation of the columns, the brace axial force is slightly different for braces "D1" and "D2" (as indicated in Figure 7.16). The design checks for "D1" diagonal members are summarized in Table 7.15 for the braces in tension and in Table 7.16 for those in compression. The symbols appearing in

7. MULTI-STOREY BUILDING WITH CONCENTRICALLY BRACED FRAMES

these tables are defined in section 7.3.1.1. Analogously, Tables 7.17 and 7.18 show the results for "D2" diagonal members.

Since the inverted-V CBFs are not sensitive to global imperfections, these effects are not reported in the calculation tables shown hereinafter for all members.

The minimum overstrength ratio Ω is equal to 2.26, highlighted in bold in Table 7.15. Hence, according to the capacity design criteria described in section 3.6.2, the seismic induced forces acting in the non-dissipative members should be amplified by $1.1\gamma_{ov}\Omega = 1.1 \times 1.00 \times 2.26 = 2.486$ (N.B. $\gamma_{ov} = 1.0$ because S235 is used for dissipative elements and S355 for all non-dissipative members, according to section 6.2.4 of EC8-1).

Table 7.14 – Inverted V-brace cross section properties

Storey (-)	Brace cross section $d \times t$ (mm × mm)	d (mm)	t (mm)	d/t (-)	$50\,\varepsilon^2$ (-)
VI	127 × 6.3	127	6.3	20.16	50.00
V	193.7 × 8	193.7	8	24.21	50.00
IV	244.5 × 8	244.5	8	30.56	50.00
III	244.5 × 10	244.5	10	24.45	50.00
II	273 × 10	273	10	27.30	50.00
I	323.9 × 10	323.9	10	32.39	50.00

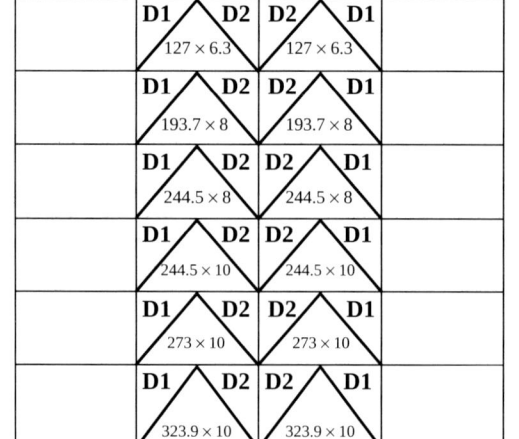

Figure 7.16 – Braces of Inverted V-CBFs

7.3 Design and Verification of Structural Members

Table 7.15 – Design checks of Inverted V-braces (D1) in tension

Storey (-)	Brace (mm × mm)	$N_{pl,Rd}$ (kN)	$N_{Ed,D1}$ (kN)	Ω_i (-)	$(\Omega_i - \Omega)/\Omega$ (%)
VI	127 × 6.3	561.65	245.68	2.29	1.33
V	193.7 × 8	1097.45	457.52	2.40	6.19
IV	244.5 × 8	1395.90	617.25	**2.26**	0.00
III	244.5 × 10	1722.55	748.47	2.30	1.77
II	273 × 10	1941.10	826.79	2.35	3.98
I	323.9 × 10	2317.10	965.46	2.40	6.19

Table 7.16 – Design checks of Inverted V-braces (D1) in compression

Storey (-)	Brace (mm × mm)	λ (-)	$\bar{\lambda}$ (-)	χ (-)	$N_{b,Rd}$ (kN)	$N_{Ed,D1}$ (kN)	$N_{b,Rd}/N_{Ed,D1}$ (-)
VI	127 × 6.3	107.94	1.15	0.46	257.44	245.60	1.05
V	193.7 × 8	70.15	0.75	0.70	762.93	461.96	1.65
IV	244.5 × 8	55.07	0.59	0.79	1107.33	622.87	1.78
III	244.5 × 10	55.53	0.59	0.79	1361.43	756.68	1.80
II	273 × 10	49.51	0.53	0.83	1606.33	843.92	1.90
I	323.9 × 10	45.05	0.48	0.85	1979.19	986.84	2.01

Table 7.17 – Design checks of Inverted V-braces (D2) in tension

Storey (-)	Brace (mm × mm)	$N_{pl,Rd}$ (kN)	$N_{Ed,D2}$ (kN)	Ω_i (-)	$(\Omega_i - \Omega)/\Omega$ (%)
VI	127 × 6.3	561.65	245.68	2.29	1.33 (*)
V	193.7 × 8	1097.45	453.99	2.42	7.08 (*)
IV	244.5 × 8	1395.90	604.94	2.31	2.21 (*)
III	244.5 × 10	1722.55	714.96	2.41	6.64 (*)
II	273 × 10	1941.10	796.83	2.44	7.96 (*)
I	323.9 × 10	2317.10	910.59	2.54	12.39 (*)

(*) the Reader should observe that Ω is the minimum among those calculated for all braces, whose value is given in Table 7.15. Therefore the ratios in this table were obtained using $\Omega = 2.26$.

7. MULTI-STOREY BUILDING WITH CONCENTRICALLY BRACED FRAMES

Table 7.18 – Design checks of Inverted V-braces (D2) in compression

Storey (-)	Brace (mm × mm)	λ (-)	$\bar{\lambda}$ (-)	χ (-)	$N_{b,Rd}$ (kN)	$N_{Ed,D2}$ (kN)	$N_{b,Rd}/N_{Ed,D2}$ (-)
VI	127 × 6.3	107.94	1.15	0.46	257.44	245.68	1.05
V	193.7 × 8	70.15	0.75	0.70	762.93	453.99	1.68
IV	244.5 × 8	55.07	0.59	0.79	1107.33	604.94	1.83
III	244.5 × 10	55.53	0.59	0.79	1361.43	714.96	1.90
II	273 × 10	49.51	0.53	0.83	1606.33	796.83	2.02
I	323.9 × 10	45.05	0.48	0.85	1979.19	910.59	2.17

7.3.2.2 Design and verification of beams

Figure 7.17 shows the cross sections of the beams of the inverted V-CBFs. The basic design assumptions concerning the material properties and the restraints against lateral-torsional buckling made for beams of X-CBFs were also adopted for inverted V-CBFs. However, differently from X-CBFs, EC8-1 requirements significantly influence the design of brace-intercepted beams, as discussed in Section 3.6.2, which should possess adequate strength to resist the potentially significant brace post-buckling force redistribution in combination with appropriate gravity loads.

The design forces acting in the beams of the inverted V-CBFs are different from those obtained by the structural analysis reported in section 7.2.3 (see Figure 7.12). Indeed, according to clause 6.7.4(2), brace-intercepted beams should resist two loading conditions, which correspond to two different ultimate scenarios. In the first scenario, braces are not considered as an intermediate support for the beams, because of the cyclic deterioration of their strength and stiffness. This implies that the beams must be able to carry all non-seismic actions on a span equal to the bay length. In the second scenario, the beams should be designed to carry also the vertical component of the force transmitted by the brace yielded in tension and by the braced in the post-buckling range, as shown in Figure 7.18. This vertical component is calculated assuming that the tension brace transfers a force equal to its yield resistance $N_{pl,Rd}$ and the compression brace transfers a force equal to a percentage of its original buckling strength ($N_{b,Rd}$) (to take into account the strength degradation under cyclic loading). The reduced compression strength is given by $\gamma_{pb} N_{pl,Rd}$ with a value of the factor γ_{pb} specified in the National Annexes. The value recommended by EC8-1 is 0.30 (also assumed in Figure 7.18).

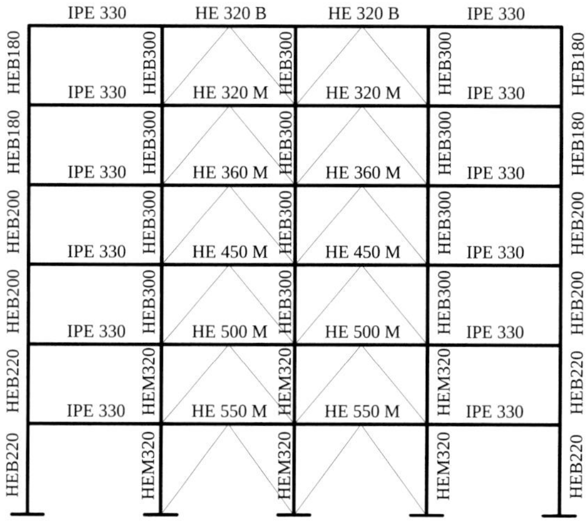

Figure 7.17 – Cross sections of beams and columns in inverted V-CBFs

Vertical component of the force transmitted by the tension and compression braces:

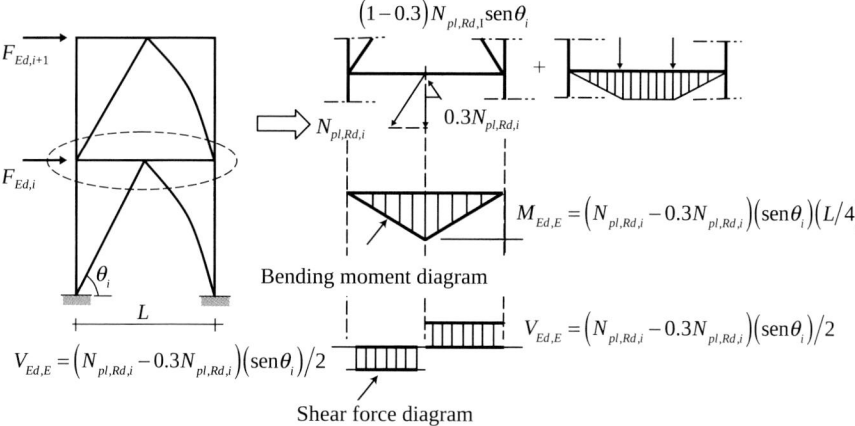

Figure 7.18 – Seismic induced forces into brace-intercepted beams

Lateral forces $F_{Ed,i}$, equilibrating the forces from plastic mechanism:

$$F_{Ed,i} = (1+0.3)\left(N_{pl,Rd,(i+1)} \cos\theta_{(i+1)} - N_{pl,Rd,i} \cos\theta_i\right)$$

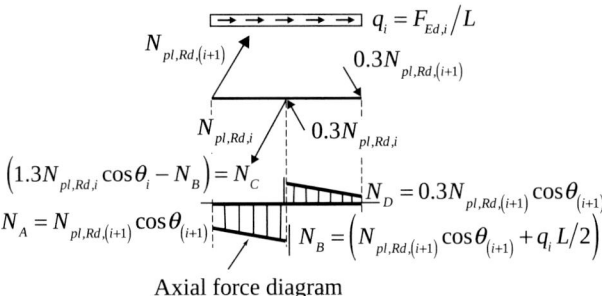

Axial force diagram

Figure 7.18 – Seismic induced forces into brace-intercepted beams (continuation)

Table 7.19 reports the axial forces acting in the critical beam cross sections under seismic conditions. These forces are illustrated in Figure 7.18, where also the calculation procedure is explained (see also section 3.6.2). Table 7.20 reports the beam axial strength checks performed at the beam ends where N_{Ed} is the maximum design axial force that is equal to N_A (see Figure 7.18). Table 7.21 reports the combined bending-axial strength checks, performed at the beam mid-span cross section ($M_{Ed,G}$ is the bending moment generated by the vertical loads, $M_{Ed,E}$ is the bending moment transmitted by the braces, M_{Ed} is the resultant moment, M_{Rd} is the bending resistance eventually reduced to take into account the combined presence of bending, shear and axial forces as indicated by section 6.2.10 of EN 1993-1-1).

Table 7.19 – Seismic induced axial forces into brace-intercepted beams (see Figure 7.18)

Storey	$N_{pl,Rd}$	q_i	N_A	N_B	N_C	N_D
(-)	(kN)	(kN/m)	(kN)	(kN)	(kN)	(kN)
VI	561.65	79.21	0.00	237.63	237.63	0.00
V	1097.45	75.56	365.58	592.27	336.36	109.67
IV	1395.90	42.09	714.33	840.60	340.57	214.30
III	1722.55	46.07	908.59	1046.79	410.78	272.58
II	1941.10	30.82	1121.21	1213.67	428.83	336.36
I	2317.10	27.47	1263.46	1345.88	461.46	379.04

7.3 Design and Verification of Structural Members

Table 7.20 – Axial strength checks of brace-intercepted beams

Storey (-)	Section (-)	A (mm²)	$N_{pl,Rd}$ (kN)	$N_{Ed,G}$ (kN)	$N_{Ed,E}=N_A$(kN) (see Fig. 7.18)	$N_{Ed}=N_{Ed,G}+N_{Ed,E}$ (kN)	$N_{pl,Rd}/N_{Ed}$ (-)
VI	HE 320 B	16130	5726.15	0.00	0.00	0.00	-
V	HE 320 M	31200	11076.00		365.58	365.58	30.30
IV	HE 360 M	31880	11317.40	0.00	714.33	714.33	15.84
III	HE 450 M	33540	11906.70		908.59	908.59	13.10
II	HE 500 M	34430	12222.65		1121.21	1121.21	10.90
I	HE 550 M	35440	12581.20		1263.46	1263.46	9.96

Table 7.21 – Combined bending-axial strength checks of brace-intercepted beams

Storey (-)	Section (-)	N_{Ed} (kN)	$M_{Ed,G}$ (kNm)	$M_{Ed,E}$ (kNm) (see Fig. 7.18)	M_{Ed} (kNm)	M_{Rd} (kNm)	M_{Rd}/M_{Ed} (-)
VI	HE 320 B	475.25	41.90	447.83	489.74	762.90	1.56
V	HE 320 M	928.63	58.13	875.05	933.19	1574.43	1.69
IV	HE 360 M	1181.17	58.35	1113.02	1171.38	1771.10	1.51
III	HE 450 M	1457.57	58.62	1373.48	1432.10	2247.51	1.57
II	HE 500 M	1642.50	59.24	1547.74	1606.98	2518.37	1.57
I	HE 550 M	1807.34	61.28	1946.36	2007.64	2816.22	1.40

Table 7.22 reports the beam shear strength checks according to section 6.2.6 of EN 1993-1-1, where V_{Ed} is the design value of the shear force, obtained by summing up the contributions from the gravitational loads $V_{Ed,G}$ and the seismic forces $V_{Ed,E}$, the latter obtained from the calculation scheme depicted in Figure 7.18.

Table 7.22 – Shear checks of brace-intercepted beams

Storey (-)	Section (-)	A (mm²)	A_v (mm²)	$V_{pl,Rd}$ (kN)	$V_{Ed,G}$ (kN)	$V_{Ed,E}$ (kN) (see Fig. 7.18)	V_{Ed} (kN)	$V_{pl,Rd}/V_{Ed}$ (-)
VI	HE 320 B	16130	5172.75	1060.20	27.93	149.28	177.21	5.98
V	HE 320 M	31200	9450.00	1943.01	38.75	291.69	330.44	5.88
IV	HE 360 M	31880	10240.00	2098.78	38.90	371.01	409.91	5.12
III	HE 450 M	33540	11980.00	2455.41	38.08	457.83	496.90	4.94
II	HE 500 M	34430	12950.00	2654.22	39.49	515.91	555.41	4.78
I	HE 550 M	35440	13960.00	2861.23	40.62	648.79	689.41	4.15

7.3.2.3 Design and verification of columns

Columns were designed to resist the axial forces induced by gravity and seismic actions and combined according to equation (3.21), namely by magnifying the seismic induced effects by the minimum overstrength ratio Ω (given in Table 7.15). As for the previous case, the bending moments and the shear forces obtained from the structural analysis are negligible (see Figure 7.12) and their effects were not accounted for. It should be noted that, differently from brace-intercepted beams, axial forces acting in the columns are directly provided by the numerical model, which includes both the tension and compression braces.

Owing to the structural symmetry about the vertical axis, the verification outcomes are reported for only columns "C" and "D" (see Figure 7.1) in Tables 7.23 and 7.24, respectively.

Table 7.23 – Axial strength checks for the external columns of the braced cantilever

Storey	Section	A	χ	$N_{pl,Rd}$	$N_{Ed,G}$	$N_{Ed,E}$	N_{Ed}	$\chi N_{pl,Rd}/N_{Ed}$
(-)	(-)	(mm²)	(-)	(kN)	(kN)	(kN)	(kN)	(-)
VI	HE 300 B	14910	0.78	5293.05	98.90	0.00	98.90	41.89
V	HE 300 B	14910	0.78	5293.05	239.34	185.17	699.98	5.92
IV	HE 300 B	14910	0.78	5293.05	587.76	525.98	1896.21	2.18
III	HE 300 B	14910	0.78	5293.05	534.34	975.33	2960.63	1.40
II	HE 320 M	31200	0.80	11076.00	697.36	1520.43	4479.66	1.98
I	HE 320 M	31200	0.75	11076.00	850.81	2110.59	6101.21	1.36

Table 7.24 – Axial strength checks for the central column of the braced cantilever

Storey	Section	A	χ	$N_{pl,Rd}$	$N_{Ed,G}$	$N_{Ed,E}$	N_{Ed}	$\chi N_{pl,Rd}/N_{Ed}$
(-)	(-)	(mm²)	(-)	(kN)	(kN)	(kN)	(kN)	(-)
VI	HE 300 B	14910	0.78	5293.05	86.42	0.00	86.42	47.94
V	HE 300B	14910	0.78	5293.05	225.50	0.00	225.50	18.37
IV	HE 300 B	14910	0.78	5293.05	379.73	0.00	379.73	10.91
III	HE 300 B	14910	0.78	5293.05	528.13	0.00	528.13	7.84
II	HE 320 M	31200	0.80	11076.00	702.29	0.00	702.29	12.61
I	HE 320 M	31200	0.75	11076.00	856.93	0.00	856.93	9.69

To illustrate the application of the capacity design requirements (see equation (3.21)) for the columns, on the basis of the values reported in Table 7.23, the axial force at the first level of column C was computed as follows:

$$N_{Ed} = N_{Ed,G} + \alpha \cdot \left(N_{Ed,I} + 1.1 \cdot \gamma_{ov} \cdot \Omega \cdot N_{Ed,E} \right) =$$
$$= 850.81 \text{ kN} + 1.00 \cdot (0 \text{ kN} + 1.10 \cdot 1.00 \cdot 2.26 \cdot 2110.59 \text{ kN}) = \quad (7.5)$$
$$= 6101.211 \text{ kN}$$

Similarly to the case of X-CBFs, the designed columns, especially those corresponding to vertical D (i.e the inner column of the inverted V-CBFs), are characterized by very large overstrength factors at upper levels, particularly where very small profiles are generally sufficient to satisfy the strength requirements. This outcome again results from the need to have adequate width for the column flange for both beam-to-column connections and brace-to-column connections. In addition, column-to-column splices can be properly detailed.

7.4 DAMAGE LIMITATION

As illustrated in section 2.11.2, the damage limitation requirement is satisfied if, under a seismic action having a larger probability of occurrence than the design seismic action, the interstorey drift demand is less than the upper limit expressed by equation (2.66) (see section 4.4.3.2 of EC8-1).

In this worked example, ductile non-structural elements with intestorey drift limit equal to 0.75 % h were considered.

The displacement reduction factor v was assumed equal to the recommended value that is $v = 0.5$. The structural displacements d_s induced by the design seismic actions were computed as follows:

$$vd_s = vq \cdot d_e \quad (7.6)$$

where:
- d_s is the displacement of the structural system induced by the design seismic action,
- q is the behaviour factor and
- d_e is the displacement of the structural system, as determined by a linear elastic analysis under the design seismic forces.

7. MULTI-STOREY BUILDING WITH CONCENTRICALLY BRACED FRAMES

Figures 7.19a,b show the lateral storey displacements for both X-CBFs and inverted V-CBFs, while Tables 7.25 and 7.26 report the relevant drift verification checks that are satisfied at each storey. As the Reader can easily recognize, X-CBFs are more deformable than inverted V-CBFs. This feature strictly depends on two aspects: i) the structural model of X-CBFs accounts for the tension diagonal only, thus halving the lateral stiffness; ii) although the slope of braces in inverted V-CBFs is larger than X-CBFs (i.e. at the same cross section their effective area would be smaller), their cross section are noticeable larger owing to the design criterion based on buckling prevention under design forces, which are also larger due to smaller behaviour factor.

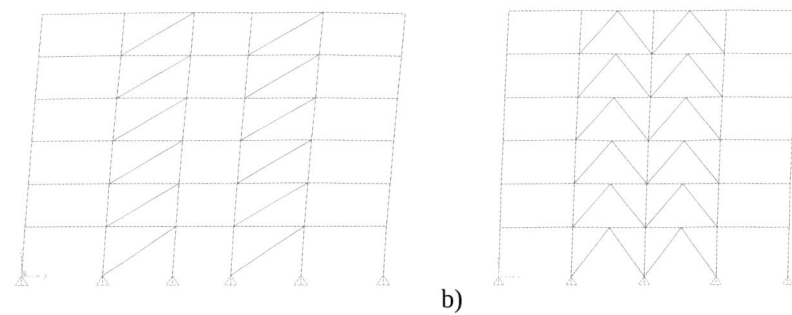

a) b)

Figure 7.19 – Lateral displacement shapes: X-CBFs (a); Inverted V-CBFs (b)

Table 7.25 – Damage limitation check for X-CBFs

Storey	h	$vd_r = v(d_e \times q)$	vd_r/h	Performance Limit
(-)	(mm)	(mm)	(%)	(%)
VI	3500	22.00	0.63	
V	3500	23.00	0.66	
IV	3500	22.40	0.64	0.75
III	3500	21.00	0.60	
II	3500	19.40	0.55	
I	4000	19.40	0.49	

Table 7.26 – Damage limitation check for inverted V-CBFs

Storey	h	$vd_r = v(d_e \times q)$	vd_r/h	Performance Limit
(-)	(mm)	(mm)	(%)	(%)
VI	3500	6.88	0.20	0.75
V	3500	6.75	0.19	
IV	3500	6.50	0.19	
III	3500	5.75	0.16	
II	3500	5.00	0.14	
I	4000	4.75	0.12	

Chapter 8

MULTI-STOREY BUILDING WITH ECCENTRICALLY BRACED FRAMES

8.1 BUILDING DESCRIPTION AND DESIGN ASSUMPTIONS

This chapter illustrates and discusses the design steps and the verifications required by EC8-1 for a multistorey building equipped with Eccentrically Braced Frames (EBFs) in "split-k" configuration and shear links. Firstly, a general description of the geometric features and design assumptions (i.e. materials, loads, seismic action, etc.) is provided. In a second part of the chapter, the assumptions for structural modelling and the methods of analysis adopted to calculate the seismic induced effects are described. Finally, the design and verifications for both ultimate limit state and damage limitation limit state are described and discussed.

This worked example uses the same building architectural definition as in chapter 6, all details being found in section 6.1.1 concerning architectural plan, geometrical properties, diaphragms and gravity load resisting systems. However, in order to allow for an independent reading of this chapter and easy comparison with the other case studies, a similar presentation is used that may include some duplicate information.

8.1.1 Building description

The worked example is a six storey office building with a rectangular plan layout and an area of 31.00 m × 24.00 m. The storey height is equal to 3.50 m with exception of the first floor, which is 4.00 m high.

8. MULTI-STOREY BUILDING WITH ECCENTRICALLY BRACED FRAMES

As discussed in chapter 2, the plan position and the vertical distribution of braces mainly influence the building regularity. However, differently from CBFs, EBFs do not often interfere with the architectural needs, and either windows or doorways can be easily arranged. Hence, the braces were distributed around the perimeter of the building in order to guarantee large torsional rigidity, without limiting architectural functionality.

Figure 8.1 shows the structural layout of the typical floor, illustrating the location of braces, while Figure 8.2 shows the vertical configuration and the cross sections of the structural members of the EBFs in the two main plan directions. It should be noted that link length was varied at each storey in order to fit the relevant theoretical upper bound limit for the shear links. This assumption aims to guarantee a similar dissipative behaviour of the short links for each storey. Indeed, as discussed in section 3.7.1, the link length is the key parameter characterizing the hysteretic performance of EBFs.

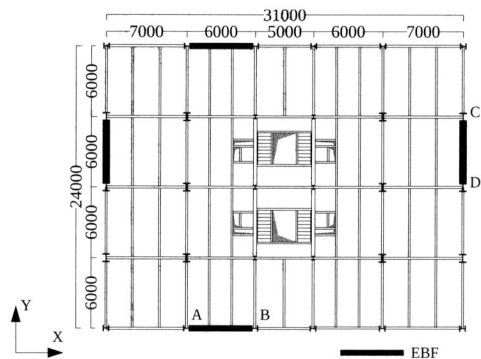

Figure 8.1 – Structural plan of the typical floor

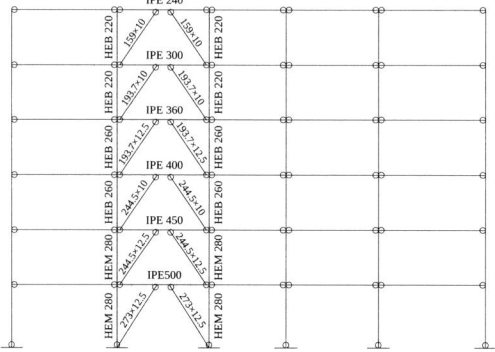

Figure 8.2 – Vertical layout of EBF in X direction

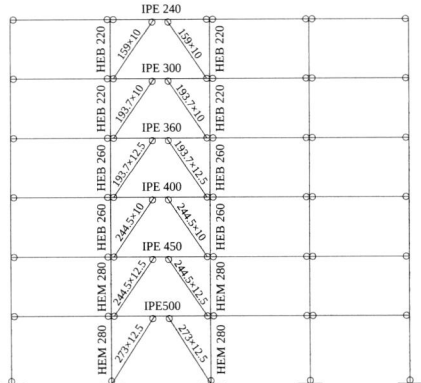

Figure 8.3 – Vertical layout of EBF in Y direction

8.1.2 Normative references

Besides the seismic recommendations, the structural design was carried out according to the following European codes:

- EN 1990 (2002) Eurocode 0: Basis of structural design;
- EN 1991-1-1 (2002) Eurocode 1: Actions on structures - Part 1-1: General actions - Densities, self-weight, imposed loads for buildings;
- EN 1993-1-1 (2005) Eurocode 3: Design of steel structures - Part 1-1: General rules and rules for buildings;
- EN 1993-1-8 (2005) Eurocode 3: Design of steel structures - Part 1-8: Design of joints;
- EN 1994-1-1 (2004) Eurocode 4: Design of composite steel and concrete structures - Part 1.1: General rules and rules for buildings.

For the sake of generality, a specific National Annex is not used. Hence, the calculation examples were carried out using the recommended values of the safety factors.

8.1.3 Materials

Similarly to chapters 6 and 7, the material properties of steel were accounted for according to clause 6.2(3)a. The mechanical properties of the materials used for the structural members are summarized in Table 8.1, where

8. Multi-storey Building with Eccentrically Braced Frames

the partial factors are also indicated. In this worked example, S355 steel grade was used for both dissipative and non-dissipative members.

Table 8.1 – Material properties and partial safety factors

Grade	f_y	f_t	γ_M	γ_{ov}	E
(-)	(N/mm²)	(N/mm²)	(-)	(-)	(N/mm²)
S355	355	510	$\gamma_{M0} = 1.00$ $\gamma_{M1} = 1.00$ $\gamma_{M2} = 1.25$	1.25	210000

8.1.4 Actions

8.1.4.1 Characteristic values of unit loads

Both permanent (G_k) and live loads (Q_k) considered in this worked example are equal to those adopted in chapters 6 and 7, which are reproduced in Table 8.2. The slab weight is inclusive of steel sheeting, infill concrete, finishing floor elements and partition walls. Analogously, the weight of stairwells includes the contribution of the flights, steps and landings.

Table 8.2 – Characteristic values of vertical permanent and live loads

	G_k (kN/m²)	Q_k (kN/m²)
Storey slab	4.20	2.00
Roof slab	3.60	0.50 1.00 (Snow)
Stairs	1.68	4.00
Claddings	2.00	

8.1.4.2 Seismic action

In this worked example both the elastic spectral shape and the reference peak ground acceleration a_{gR} are those described in chapter 6.

8.1 Building Description and Design Assumptions

A reference peak ground acceleration equal to $a_{gR} = 0.25g$ (where g is the acceleration of gravity), a type C soil, a type 1 spectral shape and importance factor γ_I equal to 1.0 were assumed.

The spectral parameters for soil type C (Table 2.2 or Table 3.2 of EC8-1) are $S = 1.15$, $T_B = 0.20$ s , $T_C = 0.60$ s and $T_D = 2.00$ s. In addition, the lower bound factor β for the design response spectrum (see section 2.3.5) was assumed equal to 0.2, as recommended in section 3.2.2.5 of EC8-1.

As described in section 2.3.5, the design response spectrum is obtained according to equations (2.23), namely by dividing the ordinates of elastic response spectrum by the behaviour factor q. Figure 8.4 depicts both the resulting elastic and the design spectra.

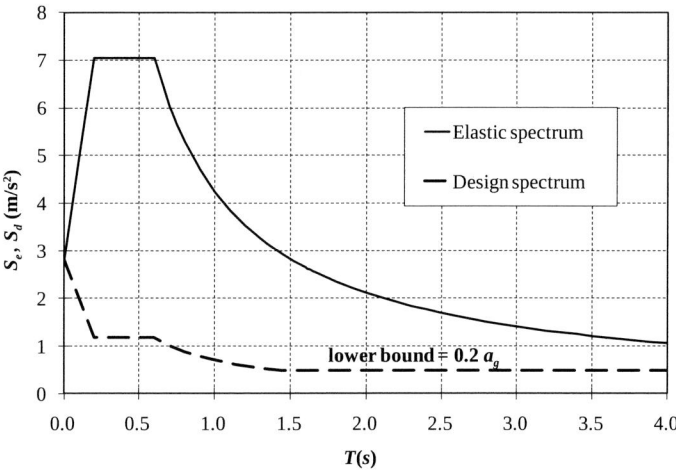

Figure 8.4 – Elastic and design response spectra

The behaviour factor was chosen on the basis of the intended design concept, corresponding to DCH structures (see concept b in sub-chapter 3.1). In addition, the behaviour factor according to clause 6.3.2 (see Figure 3.3 and Table 3.2) is given by:

$$q = \frac{\alpha_u}{\alpha_1} \cdot q_0 = 1.2 \cdot 5 = 6.0 \tag{8.1}$$

where q_0 is the reference value of the behaviour factor for systems regular in elevation (see Table 3.2), while α_u/α_1 is the plastic redistribution parameter (see Figure 3.2d).

8. MULTI-STOREY BUILDING WITH ECCENTRICALLY BRACED FRAMES

8.1.4.3 Combination of actions for seismic design situations

The combination of actions for seismic design conditions are equal to those described in sub-section 6.1.4.3.

8.1.4.4 Masses

As shown in sub-section 6.1.4.4, the inertial effects in the seismic design situation were evaluated by taking into account the presence of the masses corresponding to the combination of permanent and variable gravity loads as given by equation (2.25) in section 2.3.6, in accordance with clause 3.2.4 (2)P.

Table 8.3 summarizes the seismic weights and the related masses at each floor that incorporate the weight of the pre-designed beams, columns and braces. Seismic weights per unit floor area and the moment of inertia of masses are also reported. Some differences may be noticed between the values reported in Table 8.3 and those summarized in Table 6.4 and Table 7.3. The slight differences are due to the different dimensions of the steel profiles belonging to seismic resistant systems.

Table 8.3 – Seismic weights and masses

Storey	G_k	Q_k	Seismic Weight		Seismic Mass
(-)	(kN)	(kN)	(kN)	(kN/m²)	(kN s²/m)
VI	3143.40	1326.00	3466.80	4.66	353.39
V	3871.52	1608.00	4112.72	5.53	419.24
IV	3908.04	1608.00	4149.24	5.58	422.96
III	3914.71	1608.00	4155.91	5.59	423.64
II	3951.44	1608.00	4192.64	5.64	427.38
I	4016.01	1608.00	4257.21	5.72	433.97

8.2   STRUCTURAL ANALYSIS AND CALCULATION MODELS

8.2.1 General features

Consistently to the examples described in chapters 6 and 7, also the EBF structure satisfies the basic principles of conceptual design reported in clause 4.2.1(2). These criteria allow to obtain a regular building both in plan and in elevation, which is a fundamental requirement to achieve a high seismic performance and reliable structural model. The typological and geometrical features (i.e. the presence of rigid in plan diaphragms, the plan slenderness ratio, the absence of setbacks, etc.) and the layout of braces guarantee that the structure is regular both in plan and in elevation because it complies the requirements given in sections 4.2.3.2 and 4.2.3.3 of EC8-1, respectively.

8.2.2 Modelling assumptions

As discussed in sub-chapter 2.6, the structural model must represent the distribution of stiffness and mass so that all significant deformation shapes and inertia forces are properly accounted for under seismic actions (section 4.3.1 of EC8-1).

Similarly to chapter 7, two separate calculation 2D planar models in the two main plan directions were used, one in the X direction and the other in the Y direction. This approach is allowed by clause 4.3.1(5), since the examined building satisfies the conditions given by clauses 4.2.3.2 and 4.3.3.1(8).

The tributary area of masses was evaluated as described in section 2.6.4. Masses are considered as lumped in a selected master joint for each floor, because the floor diaphragms may be taken as rigid in their planes.

Since the structure is characterized by full symmetry of in-plan distribution of stiffness and masses, in order to ensure a minimum of torsional strength and stiffness and to limit the consequences of unpredicted torsional response, EC8-1 imposes accidental torsional effects (see sub-chapter 2.7). These take into account the possible uncertainties in the distributions of stiffness and mass and/or a possible torsional component of the ground motion about a vertical axis.

EC8 allows a simplified modelling approach if the building is conforming to the criteria for regularity both in plan and in elevation. As two separate 2D planar models are used, the effects of the accidental eccentricity may be estimated by multiplying the effects of the seismic actions (which can be obtained either by the lateral force method or by modal superposition) using the magnification factor δ (clause 4.3.3.2.4(2)) given by equation (2.51). Since in this example the planar 2D perimetric frames are analyzed, the δ factor is equal to 1.6 (being $x/L_e = 0.5$, where x is the in-plan distance between the structural element under consideration and L_e is the in-plan distance between two outermost lateral load resisting systems).

All modelling assumptions previously described were also adopted for the CBFs described in chapter 7. However, some specific remarks are useful to clarify the differences with respect to CBFs. Indeed, EC8-1 stipulates that the seismic performance of EBFs is dominated by the links, which are the zones where inelastic action is concentrated, differently from the case of CBFs where the braces are the dissipative members. This implies that the modelling approximations used for the braces in CBFs should not be used for EBFs, because in these frames diagonal braces are part of non-dissipative zones and should be designed to remain stable under seismic conditions.

The calculation models developed for the EBFs assume pinned connections (brace-to-link, brace-to-column, beam-to-column connections and column bases), but the links are considered continuous along the beam and the columns were assumed continuous through each floor, as shown in Figures 8.2 and 8.3. These assumptions are frequently accepted and they are conservative in terms of storey displacements and brace axial forces.

8.2.3 Numerical models and method of analysis

The calculation models were developed in SAP 2000 on the basis of the assumptions described in section 8.2.2. Schemes of the numerical models used for these calculation examples are shown in Figure 8.5.

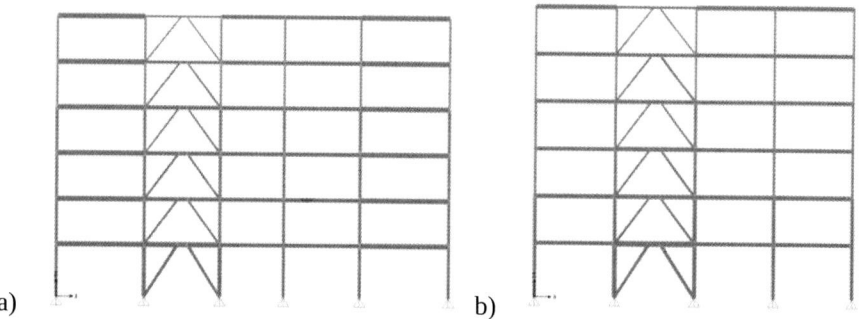

Figure 8.5 – EBFs calculation models in X direction (a) and in Y direction (b)

The effects of seismic actions have been determined by means of a linear-elastic modal response spectrum analysis (clause 4.3.3.1(2)) of the 2D frames

8.2 Structural Analysis and Calculation Models

schematically shown in Figure 8.5, using the design spectrum described in sub-section 8.1.4.2. As discussed in section 2.5.3, all vibration modes satisfying that the sum of their effective modal masses gives at least 90 % of the total mass of the structure were accounted for. Therefore, only the first and the second mode of vibration in both the X and Y directions, whose shapes, relevant natural periods T_i and participating mass ratios M_i are shown in Figure 8.6, were considered for the calculation of seismic effects. The SRSS (Square Root of the Sum of the Squares) method is used to combine the modal maxima, since the first and the second modes of vibration in both X and Y direction may be considered as independent (as $T_2 \leq 0.9 T_1$, according to section 4.3.3.3.2 of EC8-1).

The models practically have the same dynamic properties (i.e. fundamental periods and participanting mass ratios). Indeed, the structural members of the EBFs are the same in both directions. The slight differences are due to the presence of gravity resisting members that differ between the two directions.

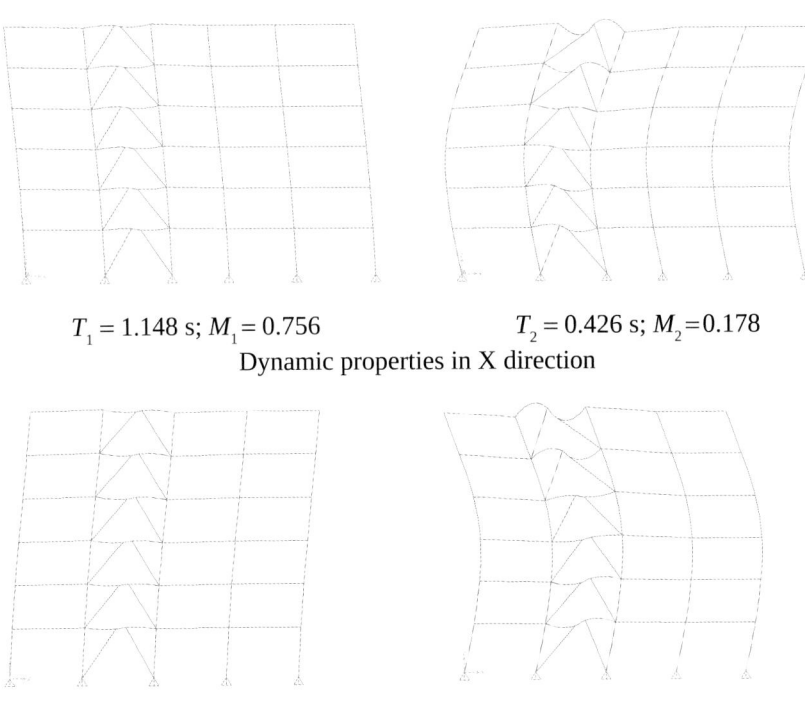

$T_1 = 1.148$ s; $M_1 = 0.756$ $T_2 = 0.426$ s; $M_2 = 0.178$
Dynamic properties in X direction

$T_1 = 1.148$ s; $M_1 = 0.776$ $T_2 = 0.426$ s; $M_2 = 0.178$
Dynamic properties in Y direction

Figure 8.6 – Fundamental dynamic properties of the examined structures

8. MULTI-STOREY BUILDING WITH ECCENTRICALLY BRACED FRAMES

The diagrams of internal forces (i.e. bending moments, shear and axial forces) due to both gravity and seismic loads are depicted qualitatively in Figure 8.7 and Figure 8.8 for EBFs in X and Y directions, respectively.

As discussed in chapter 3, the internal forces obtained from the calculation models are used to design the dissipative members (i.e. the bracing members), while the non-dissipative members (i.e. both beams and columns) should be designed for the combination given by gravity load induced effects and seismic induced effects, the latter magnified as given by equation (3.40) (see section 3.7.2).

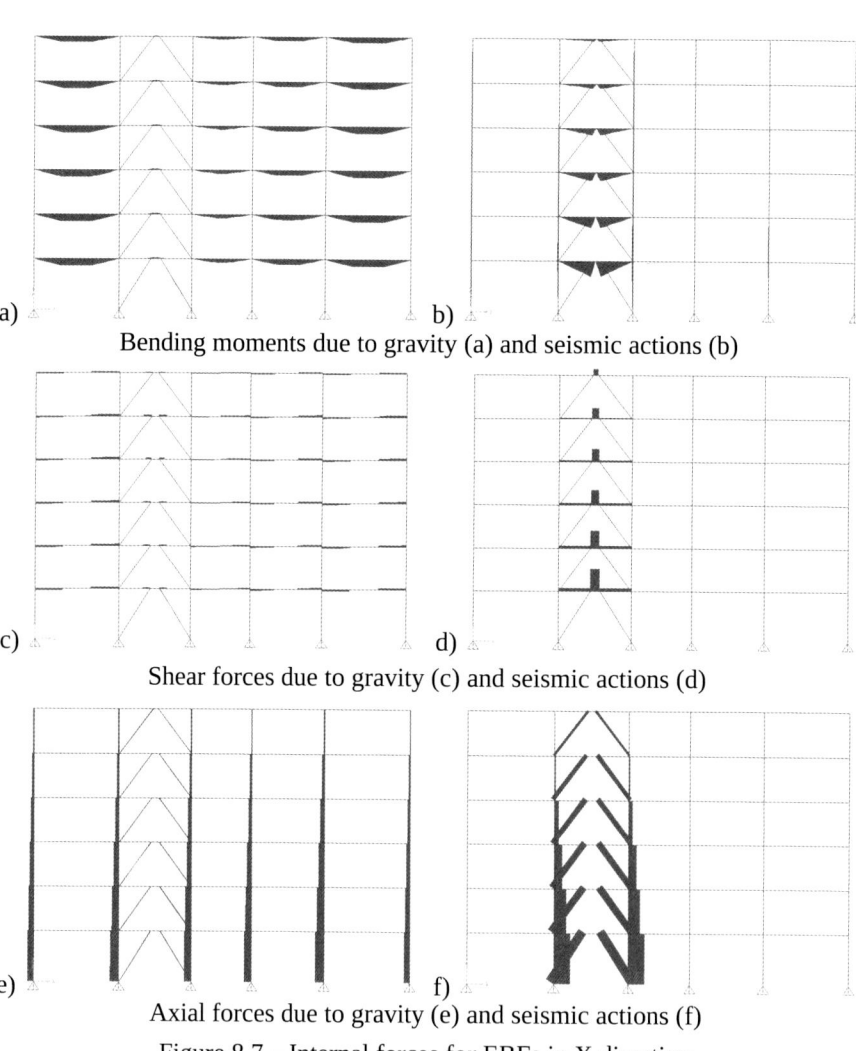

Bending moments due to gravity (a) and seismic actions (b)

Shear forces due to gravity (c) and seismic actions (d)

Axial forces due to gravity (e) and seismic actions (f)

Figure 8.7 – Internal forces for EBFs in X direction
(N.B. the Plots were scaled by the same factor per type of effort)

8.2 STRUCTURAL ANALYSIS AND CALCULATION MODELS

Figures 8.7 and 8.8 show that both columns and braces are mostly subject to axial forces, while bending and shear forces are negligible. The beam segments outside the link are mainly subject to bending moments, but the peak value occurs at the intersection with link, implying that if the link strength is verified the beam segments are verified as well. Additional details about the verification checks for the beams are reported in section 8.3.2.

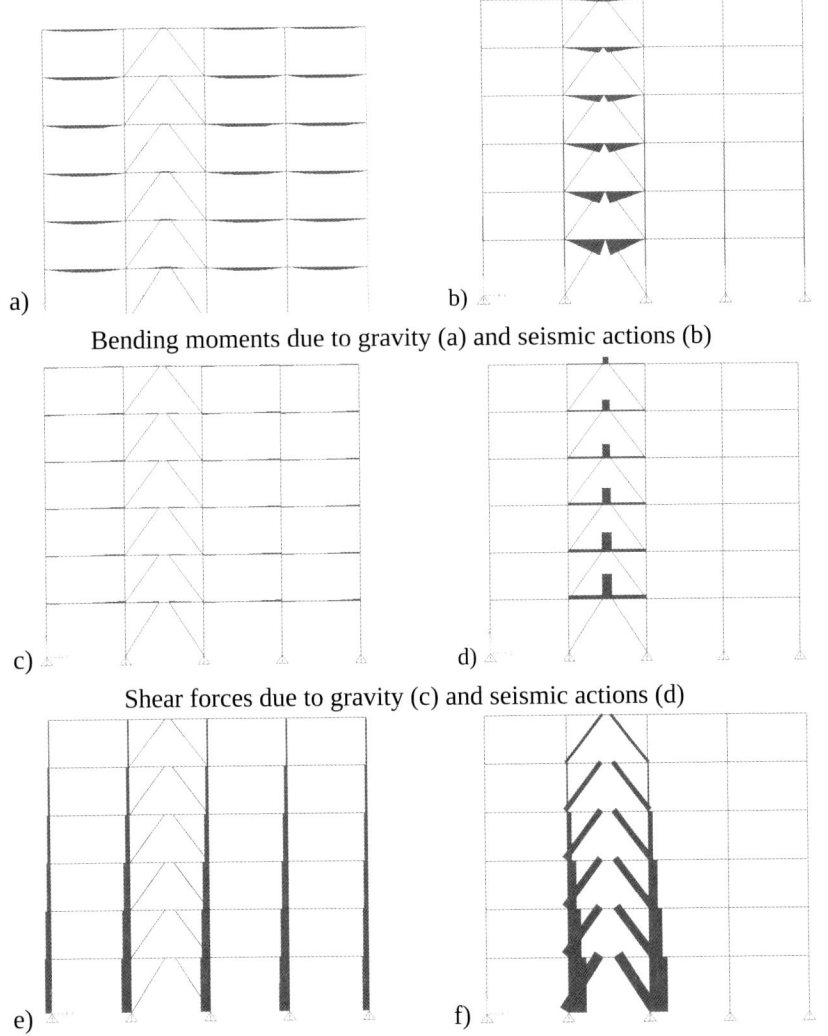

Bending moments due to gravity (a) and seismic actions (b)

Shear forces due to gravity (c) and seismic actions (d)

Axial forces due to gravity (e) and seismic actions (f)

Figure 8.8 – Internal forces for EBFs in Y direction

(N.B. the Plots were scaled by the same factor per type of effort)

8. MULTI-STOREY BUILDING WITH ECCENTRICALLY BRACED FRAMES

8.2.4 Imperfections for global analysis of frames

The effects induced by constructional imperfections were verified as discussed in chapter 6 according to clause 5.3.2(4) of EN 1993-1-1. Table 8.4 summarizes the calculation of the equivalent forces H_i and the limiting value of the shear force at each storey (i.e. $H_{Ed} = 0.15\ V_{Ed}$, where V_{Ed} is the total design vertical load on the structure at the bottom of the storey) above which it is possible to disregard the effects of global imperfections. The effects of global imperfections should be accounted for both the X and Y directions.

Table 8.4 – Global sway imperfections in both X and Y direction

Storey (-)	V_{Ed} (kN)	$H_{Ed,x}$ (kN)	$H_{Ed,y}$ (kN)	$0.15\ V_{Ed}$ (kN)	$H_{Ed,x}/V_{Ed}$ (-)	$H_{Ed,y}/V_{Ed}$ (-)	$H_i = \phi\ V_{Ed,i}$ (kN)
VI	1732.72	321.24	320.91	259.91	0.19	0.19	4.57
V	3788.25	543.81	544.46	568.24	0.14	0.14	9.98
IV	5862.03	686.66	686.90	879.30	0.12	0.12	15.45
III	7939.13	806.86	807.39	1190.87	0.10	0.10	20.92
II	10034.58	928.01	928.42	1505.19	0.09	0.09	26.44
I	12161.79	1018.97	1018.16	1824.27	0.08	0.08	32.05

8.2.5 Frame stability and second order effects

P-Δ effects were specified through the inter-storey drift sensitivity coefficient θ given by equations (2.68), and are summarized in Tables 8.5 and 8.6 for calculation in the X and Y directions, respectively.

Table 8.5 – Stability coefficients for EBF in X direction

Storey (-)	P_{tot} (kN)	V_{tot} (kN)	h (mm)	$d_r = d_e \times q$ (mm)	$\theta = \dfrac{P_{tot} \cdot d_r}{V_{tot} \cdot h}$ (-)	$\alpha = \dfrac{1}{(1-\theta)}$ (-)
VI	1732.72	325.81	3500	49	0.07	1.08
V	3788.25	553.79	3500	55	0.11	1.12
IV	5862.03	702.10	3500	53	0.13	1.14
III	7939.13	827.78	3500	47	0.13	**1.15**
II	10034.58	954.45	3500	40	0.12	1.14
I	12161.79	1051.01	4000	38	0.11	1.12

Table 8.6 – Stability coefficients for EBF in Y direction

Storey (-)	P_{tot} (kN)	V_{tot} (kN)	h (mm)	$d_r = d_e \times q$ (mm)	$\theta = \dfrac{P_{tot} \cdot d_r}{V_{tot} \cdot h}$ (-)	$\alpha = \dfrac{1}{(1-\theta)}$ (-)
VI	1732.72	325.48	3500	49	0.07	1.08
V	3788.25	554.44	3500	55	0.11	1.12
IV	5862.03	702.34	3500	53	0.13	1.14
III	7939.13	828.31	3500	47	0.13	**1.15**
II	10034.58	954.86	3500	40	0.12	1.13
I	12161.79	1050.20	4000	38	0.11	1.13

The coefficient θ ranges within $0.1 \div 0.2$ for the lower storeys (namely from first to fourth), and the maxima are indicated in bold. Hence, to take into account second order effects, the seismic effects were magnified using the maximum Merchant-Rankine multiplier α, which is the maximum amongst the different α calculated for each storey separately per direction according to equations (2.62). In this case, α is equal to 1.15 for both the X and Y directions (see Tables 8.5 and 8.6).

8.3 DESIGN AND VERIFICATION OF STRUCTURAL MEMBERS

8.3.1 Design and verification of shear links

The design of links consists in selecting the appropriate profiles to match the seismic demand and the relevant lengths per link to guarantee the expected dissipative behaviour.

IPE profiles made of S355 steel grade were used for the beams containing the shear links, whose shapes are depicted in Figures 8.2 and 8.3.

Since short links were adopted, the length "e" of each link was selected to round down the theoretical limit value "e_s", where "e_s" should be calculated with equations (3.25) (see section 3.7.1). Obviously, as the link cross sections at each level are different, the limit length "e_s" differs for each link. For example, the design length for the shear link at the first storey (i.e. IPE 500) is obtained as follows:

$$e \simeq e_s = 1.6 \cdot \frac{M_{p,link}}{V_{p,link}} = 1.6 \cdot \frac{549.82}{1011.84} = 0.869 \text{ m} \rightarrow e = 0.860 \text{ m} \quad (8.2)$$

where $M_{p,link}$ and $V_{p,link}$ are the plastic bending moment and the plastic shear strength, calculated according to clause 6.8.2(3) (see equations (3.22) and (3.23) in section 3.7.1)

The cross section, the upper bound shear length and the chosen length of shear links are reported in Table 8.7. The assumed link lengths are the same for the frames in the X and Y directions. It is worth noting that in some cases length "e" slightly differs from "e_s" in order to harmonize the geometry from a technological and constructional point of view.

Table 8.7 – Link sections, upper bound shear length and selected length per storey (X and Y direction)

Storey (-)	Section (-)	$M_{p,link}$ (kNm)	$V_{p,link}$ (kN)	e_s (mm)	e (mm)	e/e_s
VI	IPE 240	96.10	292.53	525.65	520	0.99
V	IPE 300	164.84	420.99	626.47	620	0.99
IV	IPE 360	266.19	569.46	747.90	740	0.99
III	IPE 400	333.41	681.26	783.05	780	1.00
II	IPE 450	428.77	838.85	817.82	810	0.99
I	IPE 500	549.82	1011.84	869.42	860	0.99

Besides the ultimate limit state verifications under non-seismic load combinations (Simões da Silva et al, 2016), the link shear strengths were checked under the seismic load combination. It is worth noting that all beams containing links were assumed restrained against lateral-torsional buckling as required by clause 6.8.2(14). The link strength verifications are summarized in Tables 8.8 and 8.9 for the links belonging to the frames in the X and Y directions, respectively. These tables list the design shear forces $V_{Ed,i}$, obtained directly from the output of the elastic calculation model, and the link plastic shear strength $V_{p,link,i}$, derived by applying the equation (3.23) (see section 3.7.1). In addition, these tables report the minimum shear overstrength ratio Ω, the overstrength ratio Ω_i per link and their percentage variation $(\Omega_i-\Omega)/\Omega$ along the height of the building, which should not be larger than 25 %. The value of Ω is equal to 1.68 in the X direction and 1.67 in the Y direction.

To illustrate the calculation procedure of Ω_i, the link overstrength at the first storey (X direction) was obtained as follows:

8.3 Design and Verification of Structural Members

$$\Omega_1 = 1.5 \cdot \frac{V_{p,link,1}}{V_{Ed,1}} = 1.5 \cdot \frac{1011.84}{866.98} = 1.75 \qquad (8.3)$$

where

$$V_{p,link,1} = (d - t_f) \cdot t_w \cdot f_y / 3^{0.5} = (500 - 16.0) \cdot 10.2 \cdot 355 \cdot 0.58 = 1011843 \text{ N} = $$
$$= 1011.84 \text{ kN},$$

(see equation (3.23) in section 3.7.1) and

$$V_{Ed,1} = \left(V_{Ed,G,1} + \alpha \cdot \left(V_{Ed,I,1} + V_{Ed,E,1}\right)\right) = 2.92 + 1.15 \cdot (72.52 + 679.31) = $$
$$= 866.98 \text{ kN}.$$

Table 8.8 – Strength verifications of links of frames in X direction

Storey (-)	$V_{Ed,G,i}$ (kN)	$V_{Ed,I,i}$ (kN)	$V_{Ed,E,i}$ (kN)	$V_{Ed,i}$ (kN)	$V_{p,link,i}$ (kN)	$\Omega_i = 1.5\, V_{p,link,i}/V_{Ed,i}$ (-)	Ω (-)	$\frac{\Omega_i - \Omega}{\Omega}$ (%)
VI	0.73	2.75	187.39	219.25	292.53	2.00		19.19
V	1.75	8.49	317.22	376.08	420.99	1.68		0.00
IV	2.24	17.59	400.55	482.80	569.46	1.77	1.68	5.37
III	2.38	29.56	470.67	577.28	681.26	1.77		5.42
II	2.66	45.51	541.34	677.11	838.85	1.86		10.67
I	2.92	72.52	679.31	866.98	1011.84	1.75		4.26

Table 8.9 – Strength verifications of links of frames in Y direction

Storey (-)	$V_{Ed,G,i}$ (kN)	$V_{Ed,I,i}$ (kN)	$V_{Ed,E,i}$ (kN)	$V_{Ed,i}$ (kN)	$V_{p,link,i}$ (kN)	$\Omega_i = 1.5\, V_{p,link,i}/V_{Ed,i}$ (-)	Ω (-)	$\frac{\Omega_i - \Omega}{\Omega}$ (%)
VI	1.97	2.74	187.20	220.24	292.53	1.99		19.33
V	3.50	8.49	317.60	378.23	420.99	1.67		0.00
IV	4.23	17.58	400.69	484.89	569.46	1.76	1.67	5.51
III	4.50	29.58	470.98	579.73	681.26	1.76		5.58
II	4.71	45.48	541.58	679.34	838.85	1.85		10.94
I	5.06	72.56	678.77	868.47	1011.84	1.75		4.68

8.3.2 Design and verification of beam segments outside the link

Owing to both the structural configuration and the applied vertical loads, the maximum flexural and shear forces act at both link ends. This implies that the beam verification is implicitly satisfied once the strength verifications of all links is satisfied.

Clause 6.8.2(15) mandates to verify the shear buckling resistance of the beam web panels outside of the links according to section 5 of EN 1993-1-5. The aim of this check is to avoid shear buckling under the design shear forces transmitted by the links. Since the webs of the designed beams are characterized by ratios d_w/t (where d_w is the web depth and t is the web thickness) smaller than $72\varepsilon/\eta$ (where $\varepsilon = (235/f_y)^{0.5}$ and $\eta = 1.20$), the beams should be verified for plastic shear strength, buckling not being relevant. This implies that the strength verifications are directly satisfied, because the shear forces outside the links are smaller than those acting in the link.

8.3.3 Design and verification of braces

Figures 8.2 and 8.3 show the cross sections adopted for the braces at each level belonging to the frames in both the X and Y directions. Hot-rolled circular hollows profiles made of S355 steel grade were assumed. The brace cross sections were also selected to be class 1 according to section 5.6 of EN 1993-1-1 and to provide similar lateral stiffness in the two main directions of the building, which implies similar modal properties (fundamental modes of vibrations and relevant periods) in both main directions of the building.

The brace design axial forces were obtained from the calculation models shown in section 8.2.3 (see Figures 8.7 and 8.8) and combined according to the capacity design criterion given by equation (3.40).

Tables 8.10 and 8.11 summarize the strength verifications of the braces in the X and Y directions, respectively. In particular, since in this case the compression forces are more burdensome for the braces, the braces are verified under axial compression. Therefore, these tables report the mechanical parameters of the braces as the non-dimensioanl slenderness $\overline{\lambda}$, the buckling reduction factor χ, the buckling strength of the braces $N_{b,Rd}$ and the design axial force N_{Ed} calculated according to equation (3.40).

For examplification, the design axial force for the brace (273×12.5) in the first storey in the X direction was calculated as follows:

$$N_{Ed} = N_{Ed,G} + \alpha \cdot \left(N_{Ed,I} + 1.1 \cdot \gamma_{ov} \cdot \Omega \cdot N_{Ed,E} \right) = $$
$$= 54.84 + 1.15 \cdot \left(100.62 + 1.1 \cdot 1.25 \cdot 1.68 \cdot 942.53 \right) = 2656.42 \text{ kN} \quad (8.4)$$

Hence, as the buckling resistance is given by $N_{b,Rd} = \chi \cdot N_{pl,Rd} = 0.86 \cdot 3621$ kN $= 3105.70$ kN, the strength verification is satisfied:

$$N_{b,Rd} = 3105.70 \text{ kN} > N_{Ed} = 2656.42 \text{ kN} \quad (8.5)$$

Table 8.10 – Verification of braces in compression (X direction)

Storey (-)	Section (-)	$\bar{\lambda}$ (-)	χ (-)	$N_{b,Rd}$ (kN)	$N_{Ed,G}$ (kN)	$N_{Ed,I}$ (kN)	$N_{Ed,E}$ (kN)	N_{Ed} (kN)	$N_{b,Rd}/N_{Ed}$ (-)
VI	159×10	1.11	0.59	978.86	36.91	3.82	260.57	732.14	1.34
V	193.7×10	0.89	0.74	1513.67	48.48	11.94	446.19	1244.37	1.22
IV	193.7×12.5	0.90	0.73	1854.56	51.77	25.09	571.52	1593.37	1.16
III	244.5×10	0.69	0.85	2225.30	52.32	42.39	674.86	1885.42	1.18
II	244.5×12.5	0.70	0.85	2744.29	52.08	65.50	779.36	2185.58	1.26
I	273×12.5	0.68	0.86	3105.70	54.84	100.62	942.53	2656.42	1.17

Table 8.11 – Verification of braces in compression (Y direction)

Storey (-)	Section (-)	$\bar{\lambda}$ (-)	χ (-)	$N_{b,Rd}$ (kN)	$N_{Ed,G}$ (kN)	$N_{Ed,I}$ (kN)	$N_{Ed,E}$ (kN)	N_{Ed} (kN)	$N_{b,Rd}/N_{Ed}$ (-)
VI	159×10	1.11	0.59	978.86	19.92	3.81	260.30	621.29	1.58
V	193.7×10	0.89	0.74	1513.67	29.91	11.94	446.73	1067.40	1.42
IV	193.7×12.5	0.90	0.73	1854.56	31.82	25.08	571.72	1369.38	1.36
III	244.5×10	0.69	0.85	2225.30	34.26	42.41	675.30	1626.94	1.37
II	244.5×12.5	0.70	0.85	2744.29	34.56	65.45	779.42	1889.31	1.45
I	273×12.5	0.68	0.86	3105.70	38.06	100.68	941.78	2300.76	1.35

8.3.4 Design and verification of columns

As for the braces, the forces acting on the columns were directly obtained from the calculation model. Since the bending moments and the

8. MULTI-STOREY BUILDING WITH ECCENTRICALLY BRACED FRAMES

shear forces are negligible (see Figures 8.7 and 8.8), the columns were only verified against axial forces, which were combined according to the capacity design criterion given by equation (3.40).

Hot-rolled wide flange HE profiles made of S355 steel grade were used and the relevant cross sections are shown in Figures 8.2 and 8.3, while the verification checks are reported for the columns "A" and "B" (see Figure 8.1) of the EBF in the X direction in Table 8.12 and for the columns "C" and "D" (see Figure 8.1) of the EBF in the Y direction in Table 8.13, respectively. In these tables, χ is the buckling reduction coefficient calculated according to section 6.3.1.2 of EN 1993-1-1. Table 8.13 shows that owing to the symmetry of the applied gravity loads and the structural scheme the same axial forces act on column "C" and "D".

Table 8.12 – Axial strength checks for columns of EBF in X direction

\multicolumn{9}{c}{column "A" (see Figure 8.1)}
Storey (-)	Section (-)	A (mm²)	χ (-)	$N_{pl,Rd}$ (kN)	$N_{Ed,G}$ (kN)	$N_{Ed,I}$ (kN)	$N_{Ed,E}$ (kN)	N_{Ed} (kN)	$\chi N_{pl,Rd}/N_{Ed}$ (-)
VI	HE 220 B	9100	0.65	3230.50	76.45	0.26	17.78	123.89	16.94
V	HE 220 B	9100	0.65	3230.50	207.19	1.77	151.87	611.94	3.43
IV	HE 260 B	11840	0.73	4203.20	345.89	8.77	449.80	1548.18	1.97
III	HE 260 B	11840	0.73	4203.20	487.58	24.41	825.79	2703.19	1.13
II	HE 280M	24020	0.77	8527.10	633.83	51.28	1249.73	4001.21	1.65
I	HE 280M	24020	0.72	8527.10	778.28	91.77	1710.06	5407.61	1.14
\multicolumn{9}{c}{column "B" (see Figure 8.1)}
Storey (-)	Section (-)	A (mm²)	χ (-)	$N_{pl,Rd}$ (kN)	$N_{Ed,G}$ (kN)	$N_{Ed,I}$ (kN)	$N_{Ed,E}$ (kN)	N_{Ed} (kN)	$\chi N_{pl,Rd}/N_{Ed}$ (-)
VI	HE 220 B	9100	0.65	3230.50	60.72	0.26	17.78	108.16	19.41
V	HE 220 B	9100	0.65	3230.50	170.67	1.77	151.87	575.42	3.65
IV	HE 260 B	11840	0.73	4203.20	288.49	8.77	449.80	1490.78	2.05
III	HE 260 B	11840	0.73	4203.20	409.52	24.41	825.79	2625.13	1.16
II	HE 280M	24020	0.77	8527.10	534.62	51.28	1249.73	3902.00	1.69
I	HE 280M	24020	0.72	8527.10	658.80	91.77	1710.06	5288.13	1.16

Table 8.13 – Axial strength checks for columns of EBF in Y direction

column "C" (see Figure 8.1)									
Storey (-)	Section (-)	A (mm²)	χ (-)	$N_{pl,Rd}$ (kN)	$N_{Ed,G}$ (kN)	$N_{Ed,I}$ (kN)	$N_{Ed,E}$ (kN)	N_{Ed} (kN)	$\chi N_{pl,Rd}/N_{Ed}$ (-)
VI	HE 220 B	9100	0.65	3230.50	94.88	0.26	17.76	141.99	14.78
V	HE 220 B	9100	0.65	3230.50	232.01	1.76	151.63	633.80	3.31
IV	HE 260 B	11840	0.73	4203.20	376.91	8.76	449.97	1572.74	1.94
III	HE 260 B	11840	0.73	4203.20	522.00	24.39	826.11	2725.76	1.12
II	HE 280M	24020	0.77	8527.10	673.02	51.28	1250.17	4022.39	1.64
I	HE 280M	24020	0.72	8527.10	822.21	91.72	1710.63	5426.77	1.13
column "D" (see Figure 8.1)									
Storey (-)	Section (-)	A (mm²)	χ (-)	$N_{pl,Rd}$ (kN)	$N_{Ed,G}$ (kN)	$N_{Ed,I}$ (kN)	$N_{Ed,E}$ (kN)	N_{Ed} (kN)	$\chi N_{pl,Rd}/N_{Ed}$ (-)
VI	HE 220 B	9100	0.65	3230.50	94.88	0.26	17.76	141.99	14.78
V	HE 220 B	9100	0.65	3230.50	232.01	1.76	151.63	633.80	3.31
IV	HE 260 B	11840	0.73	4203.20	376.91	8.76	449.97	1572.74	1.94
III	HE 260 B	11840	0.73	4203.20	522.00	24.39	826.11	2725.76	1.12
II	HE 280M	24020	0.77	8527.10	673.02	51.28	1250.17	4022.39	1.64
I	HE 280M	24020	0.72	8527.10	822.21	91.72	1710.63	5426.77	1.13

To illustrate the application of the capacity design requirements (see equation (3.40)) for the columns, on the basis of the values reported in Table 8.12, the axial force into the column "A" at first level was computed as follows:

$$N_{Ed} = N_{Ed,G} + \alpha \cdot \left(N_{Ed,I} + 1.1 \cdot \gamma_{ov} \cdot \Omega \cdot N_{Ed,E}\right) =$$
$$= 778.28 \text{ kN} + 1.15 \cdot \left(91.77 \text{ kN} + 1.1 \cdot 1.25 \cdot 1.68 \cdot 1710.06 \text{ kN}\right) = \quad (8.6)$$
$$= 5407.61 \text{ kN}$$

Hence, the buckling strength verification is given by:

$$N_{b,Rd} = 6138.34 \text{ kN} > N_{Ed} = 4917.48 \text{ kN} \quad (8.7)$$

8. MULTI-STOREY BUILDING WITH ECCENTRICALLY BRACED FRAMES

The columns designed for upper storeys are characterized by quite large overstrength factors. As shown also in chapter 7, this design choice depends on the need to guarantee sufficient width of column flange for proper detailing of the beam-to-column connections, brace-to-column connections, and column-to-column splices.

8.4 DAMAGE LIMITATION

As illustrated in section 2.11.2, the damage limitation requirement is satisfied if, under a seismic action having a larger probability of occurrence than the design seismic action, interstorey drift demand is less than the upper limit expressed by equation (2.66) (see section 4.4.3.2 of EC8-1).

In this worked example, differently from those shown chapters 6 and 7, non-structural elements fixed in a way so as not to interfere with structural deformations were assumed. Under this assumption, the EC8-1 recommended value for the interstorey drift is equal to 1 % of h (see clause 4.4.3.2(c)).

The displacement reduction factor v was assumed equal to the recommended value that is $v = 0.5$. The structural displacements d_s induced by the design seismic actions were computed as follows:

$$vd_s = vq \cdot d_e \qquad (8.8)$$

where d_s is the displacement of the structural system induced by the design seismic action, q is the behaviour factor and d_e is the displacement of the structural system, as determined by a linear elastic analysis under the design seismic forces.

Figures 8.9a,b show the lateral storey displacements for EBFs in X and Y directions, while Tables 8.14 and 8.15 report the relevant drift verification checks that are satisfied at each storey. As the structural properties of EBFs are the same in both directions, the displacement demand is almost the same for both frames.

8.4 DAMAGE LIMITATION

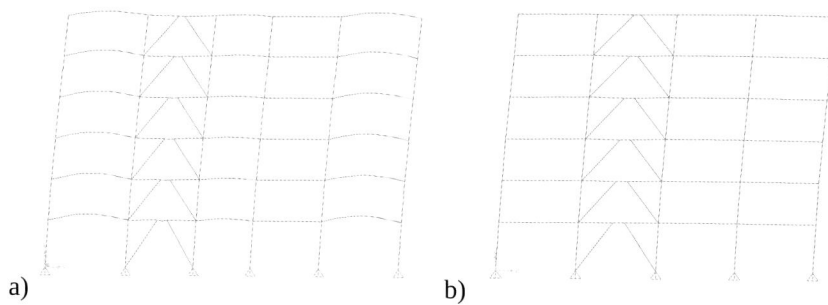

Figure 8.9 – Lateral displacement shapes: X direction (a); Y direction (b)

Table 8.14 – Damage limitation check for EBF in X direction

Storey (-)	h (mm)	$vd_r = v(d_e \times q)$ (mm)	vd_r/h (%)	Performance Limit (%)
VI	3500	24.60	0.70	
V	3500	27.60	0.79	
IV	3500	26.40	0.75	1
III	3500	23.70	0.68	
II	3500	19.80	0.57	
I	4000	19.20	0.48	

Table 8.15 – Damage limitation check for EBF in Y direction

Storey (-)	h (mm)	$vd_r = v(d_e \times q)$ (mm)	vd_r/h (%)	Performance Limit (%)
VI	3500	24.60	0.70	
V	3500	27.60	0.79	
IV	3500	26.40	0.75	1
III	3500	23.70	0.68	
II	3500	19.80	0.57	
I	4000	19.20	0.48	

Chapter 9

CASE STUDIES

9.1 INTRODUCTION

This chapter presents three examples of real buildings erected in high seismicity regions. The first two case studies intend to illustrate conceptual design aspects and specific design details that are inevitably conditioned by the seismic action and correspond to recent buildings that were designed according to EC8-1. In particular, the quantification of actions is given with greater detail in order to show the relative importance of the various types of actions. The third case study, designed and erected before EC8-1 existed, illustrates the pioneering application of seismic isolation, based on sound engineering principles and a deep understanding of the behaviour of structures under seismic conditions.

All case studies show that in real life engineering, there is no such thing as a straightforward building project whereby all objectives and conditions are clearly established from the beginning. In fact, in two case studies, the design concept was changed in the middle of design (and in one case even in the middle of construction) because of the adoption of upgraded seismic standards and seismic zonation, therefore imposing difficult restraints for the development of the structural designs.

The first building is the tallest building in Romania (TCI - Tower Centre International), a 26 storey office building located in Bucharest, designed by a team led by one of the authors (Dan Dubina). TCI was awarded in 2007 with the European Steel Design Award by the European Convention for Constructional Steelwork with the following statement by the International Jury: *"Tallest building in Romania, this 26 stories tower rises to 106 meters. Steel was chosen to withstand the high seismic*

9. Case Studies

risk in the area and the structure was designed to fulfill the new Eurocode seismic code. A "dual-steel" configuration was adopted using different grades for the steel members. Architectural and aesthetic demands are related to interior open space, bracings position and configuration". It is not intended to reproduce the detailed design process of the building but rather to present and discuss specific options for seismic design in tall buildings and to illustrate some relevant structural details, following a general description of the building.

The second example is an industrial building also located in a high seismicity region but designed according to the non-dissipative concept (concept a, see section 3.1) because of the low mass of the building and the specific structural typology. An important proportion of steel buildings that are built in the world is represented by low rise buildings. Low rise buildings are defined as those single storey and other structures that can be erected using mobile cranes and accessed using mobile elevating platforms. Generally, they do not require columns to be spliced. Possible applications of single storey steel buildings include industrial halls, warehouses, commercial spaces, exhibition spaces or sport halls. A general description of the building is presented, followed by some design considerations and relevant detailing options.

Finally, the third example is the Fire Station of Naples, in Italy, a centre composed of eight buildings built in the 1980's and 1990's that represent a catalogue of seismic resistant structural solutions. In particular, the main aspects related to the design of the main building (hereinafter referred as "building A") of the Fire Brigade of Naples are described and discussed. It is characterized by a suspended seismically isolated steel structure designed according to the ECCS recommendations for steel structures in seismic zones, developed by the ECCS TC13 committee chaired by one of the authors (Federico Mazzolani), ECCS (1988), which were incorporated as the "Steel Section" of the first edition of EC8-1, and constituted the framework of chapter 6 of the current version of EC8-1. Therefore, the design of this structure represents a forerunner example of the application of the EC8 design philosophy. It is worth to highlight that building A was the first seismically isolated building in Europe, after few experiences in New Zealand and USA, but before Japan. Owing to its innovative features, in 1986, this building, designed by a team led by one of the authors (Federico Mazzolani), received the prestigious ECCS European Steel Design Award, with the following statement of the International Jury: *"Due to its seismic safety, this building is the perfect demonstration of the possibility of steelwork. At the same time it is rich of expressive values and demonstrates that the fire prevention authorities can overcome the criticisms which sometimes could be done against steel constructions".*

9.2 THE BUCHAREST TOWER CENTRE INTERNATIONAL

9.2.1 General description

The **Tower Centre International, TCI** is a 26-storey steel office building located in Bucharest (Romania), close to the city centre, in a very densely built-up area. The building has three basement levels, 26 floors and a total height of 106.4 m. Figure 9.1 depicts a structural render of the building and external and internal photos of the finished building.

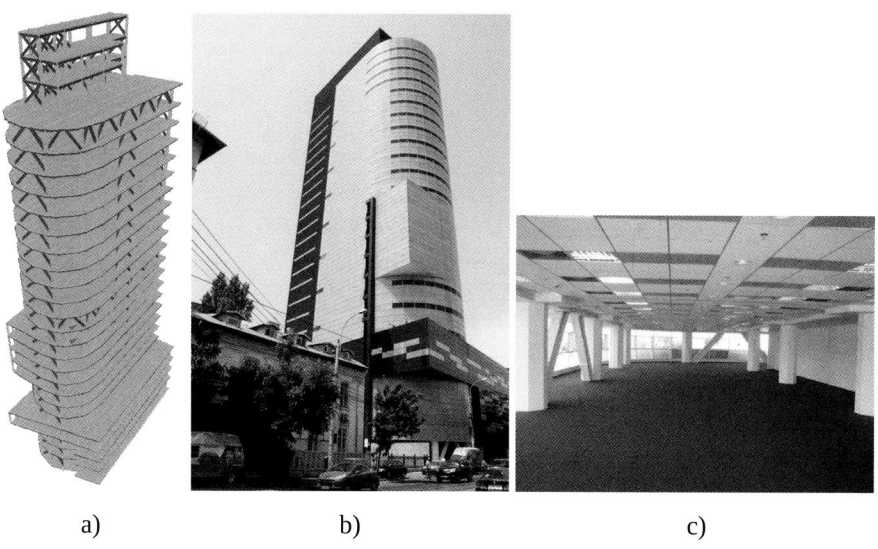

a) b) c)

Figure 9.1 – Structural rendering (a), external view of the building (b) and interior of the building (c)

The storey's height is 4.0 m, except in the first storey that is 5.4 m, and the 2nd to 4th storeys, that are 4.2 m. Between the 7th and 11th storeys, the floors cantilever out by about 4.5 m on one side. The building measures 25.5 m × 41.5 m in plan and has a total construction gross area of approx. 24 000 m². The lowest basement level is a plant floor; the other two levels are for parking. The ground floor and the first two floors are reserved for banking use. Offices occupy the remaining floors (Figure

9. CASE STUDIES

9.1c). The building frame system consists of braced and unbraced steel frames (dual structural configuration, see section 3.3). Figure 9.2 shows typical transverse and longitudinal frames, typical floor plans and the foundation system. As it can be observed in Figure 9.2c, two outrigger girders with inverted V bracings (one at the top and another at the mid-height of the part upper the setback) were adopted to limit the lateral drifts under both seismic and wind actions. Steel decking acting as lost formwork was used for the in situ concrete floors. A curtain wall system was used for three sides of the façade while corrugated metal wall cladding was used for the taller core. The foundation system comprises mat foundations and concrete piles that were driven to 28 m below the surface. The top level of the slab foundation was -12.0 m, and there were three levels below ground level. "Top-down" technology was used to construct the foundation.

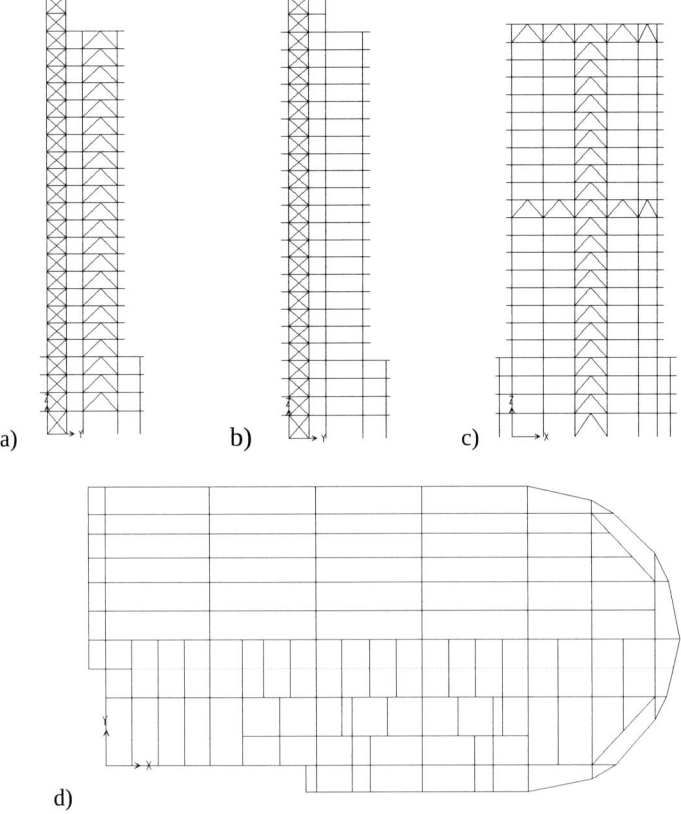

Figure 9.2 – Structural system: a) end transversal frame; b) current transversal frame; c) end longitudinal frame; d) current floor; e) infrastructure and soil layers

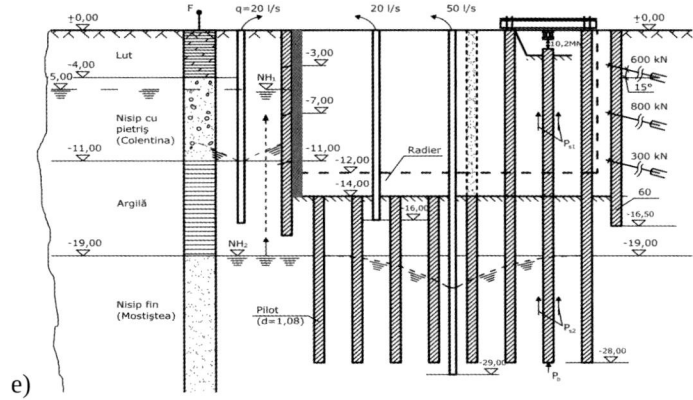

Figure 9.2 – Structural system: a) end transversal frame; b) current transversal frame; c) end longitudinal frame; d) current floor; e) infrastructure and soil layers (continuation)

The cross sections of cruciform columns vary along the height of the building from 2 ×HEM 800 at the base to 2 ×HEA 800 at the top for square sections, and from HEB 1000 ×HEM 500 to HEB 1000 ×HEB 500 for rectangular sections (see Figure 9.3). The cruciform cross section columns, made of hot rolled profiles, were partially encased in reinforced concrete to increase their strength and stiffness (Figure 9.3a). The X cross section columns have similar characteristics in both directions, large torsional stiffness and allow for moment connections in both directions executed by direct bolting to the beams. Partially encasing of the columns (see Figure 9.3a) also increases their fire resistance (details about the fire design may be found in Zaharia and Dubina, 2014). As such, fire resistant paint was only applied to the columns of the lower part, below the fifth floor. For the rest of the columns, the partial encasement supplied the necessary fire resistance.

Beams were made of hot-rolled I-sections (Figure 9.3b). Shear studs were welded on the main beams and the secondary beams but composite action was considered only in the secondary beams in order to apply the weak beam strong column design criteria referring only to the bare steel plastic moment. In fact, composite action was negligible in the primary beams due to the limited number of studs. that played the role of preventing the out-of-plane deflection of the top flange in compression. Three types of profiles were used, i.e. IPE sections for the moment resisting frame spans, and wide flange HEA and HEB sections for the braced spans and the coupling beams (see Figure 9.6), respectively.

Braces were made from hot-rolled I-sections, using wide flange profiles, such as HEA and HEB.

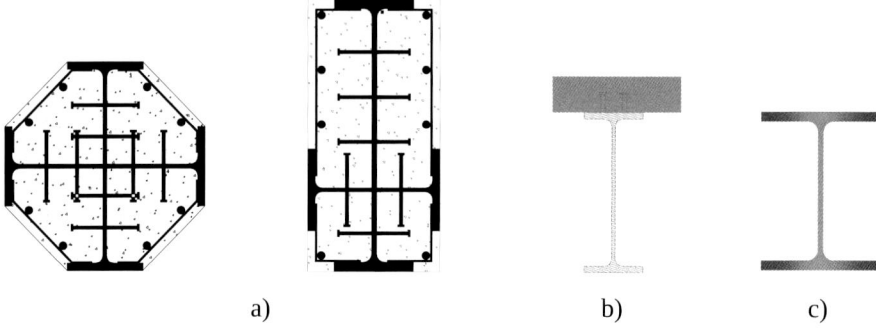

a)   b)   c)

Figure 9.3 – Cross section of the members: a) columns; b) beams; c) braces

All site joints, including beam-to-column joints, column splices and brace joints, were bolted (Figure 9.4). European practice recommends shop welding and site bolting for joining the steel elements. In case of large thicknesses like the column flanges, bolted connections were considered more appropriate than welded ones owing to the difficulties in execution and quality control of onsite welding.

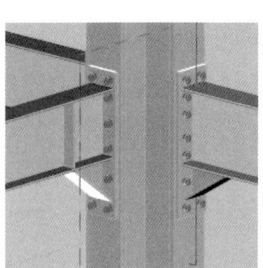 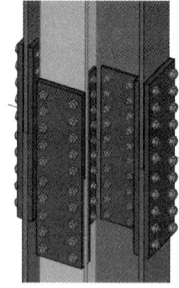 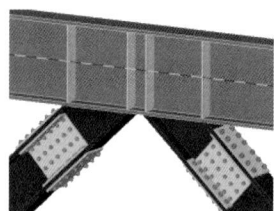

Figure 9.4 – Typical connections

A dual-steel concept was applied in the design of the structure, e.g. mild carbon steel was used for main ductility supplier components, while for members expected to behave predominantly elastic during seismic events, higher grades were used (Dubina, 2010). Grade S355 J2 steel was used for the frame members, while grade S235 J2 was adopted for the dissipative X and V braces.

9.2.2 Design considerations

9.2.2.1 Introduction

Owing to the strong seismic hazard affecting Bucharest, the structure must comply with special conditions in order to perform well during ground motions. The local site conditions and the specific features of the seismic source (i.e. deep source with several acceleration peaks) aggravate the seismic risk in the Bucharest region, particularly for medium-rise frame buildings. The city has been hit by several strong earthquakes during the last 40 years. Therefore, one important concern is the safety of tall buildings under strong ground motions. The Romanian seismic design code, P100-1/2006 (MTCT, 2006) was issued before the structure was completed, to address concerns regarding the seismic risk in Romania. This code was basically built on the EC8-1 platform, but with some particularities and differences, such as the shape of amplification spectra or the explicit allowance of partial strength beam-to-column joints. Other standards were also in the process of revision and alignment to Eurocodes, such as the standard for the design the steel structures, evaluation of loads on structures (wind, snow) and therefore their possible implications in design were also considered.

Multistorey buildings in seismic regions should be designed and constructed to meet both the no-collapse and the damage limitation requirements under specific seismic hazards. In order to avoid explicit inelastic structural analysis in design, the capacity of the structure to dissipate energy is taken into account by performing an elastic analysis based on the design response spectrum, which is reduced from the elastic spectrum by means of the behaviour factor q. For complex structures or structures that combine different systems (i.e. dual systems) or different steel grades (e.g. mild carbon steel and high-strength steel), there is an uncertainty on the reliability of q factors provided in EC8-1. Moreover, in order to achieve a favourable plastic mechanism (global mechanism), specific structural members are designed to dissipate seismic energy (dissipative elements), whereas others are designed to remain predominantly elastic (i.e. non-dissipative elements). To get as close as possible to the global plastic mechanism configuration, it is necessary to control – through design – the history of the appearance of plastic hinges in the dissipative members. For this purpose, a good balance between the strength,

9. Case Studies

stiffness and ductility of the members and connections in the structure has to be ensured. In real structures, this requirement is in many cases difficult or impossible to accomplish because the system of lateral resistance may be designed for conditions other then seismic ones, e.g. wind loads. Even in areas with strong seismic activity where high values of q are used, for multistorey buildings of more than 25-30 storeys, wind forces control the design of the system for resisting lateral forces. Of course, seismic detailing is required and should be carried out carefully. In such cases, the seismic performance should be checked by means of non-linear static or dynamic analysis and, if necessary, the element sizes corrected.

Owing to its irregular shape, the building was tested in a wind tunnel. Both rigid and aero-elastic models were constructed and tested in order to evaluate the distribution of pressure coefficients on the building envelope and the dynamic behaviour of the building (Dubina *et al*, 2009a).

The efficiency of the structural systems for multistory buildings can be quantified in terms of fulfilling the requirements for architectural details, building services, vertical transportation, and fire safety and also to its ability to resist gravity and lateral loads. The gravity load system should weight less, because this leads to savings in cost, and allows a faster and easier installation. The lateral load system should have adequate stiffness against sidesway when subjected to horizontal loads from winds or earthquakes. In general, for tall buildings, the design of the lateral load system is governed by wind loads, even if a seismic design philosophy (structural system, local detailing, etc.) has to be considered. For multistorey buildings between 25 and 30 storeys, the effects of wind and earthquakes come very close. As the TCI structure was on the border between earthquake- and wind-resistant structures, both actions contributed to the final choice of the sections (Figure 9.5). This last statement challenged the design team because some wind design findings contrasted with the seismic results. For example, in order to assure an adequate lateral stiffness (i.e. to limit the lateral drift) against wind load, heavy bracing was necessary, much heavier than would be necessary for seismic design (in terms of both stiffness and strength). The solution adopted by the design team was to retain the bracing cross sections, thus assuring an adequate lateral stiffness but reducing the brace yield strength f_y from 355 N/mm² to 235 N/mm² in order to decrease the level of design overstrength in the bracing members.

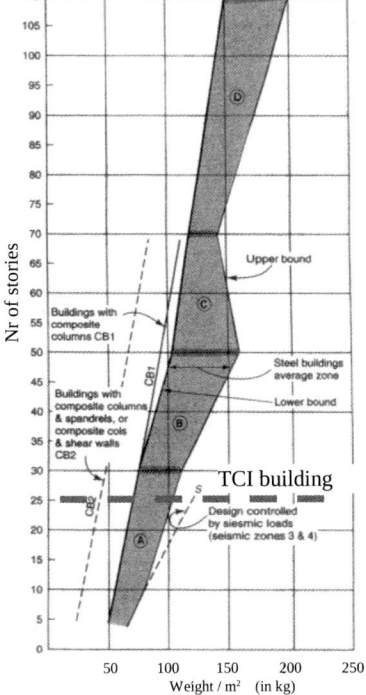

Figure 9.5 – Wind load vs. seismic load (Taranath, 2005)

9.2.2.2 Pre-design alternatives

The original design project was developed in 1997 by a different design office, and afterwards the construction works began. Some 95 % of the infrastructure was completed between 1997 and 1998, with some minor works remaining at level –3.20 m. Afterwards the construction works were halted. In 2006, after a pause of eight years, the works for the structure started, but based on a different project. This new project was developed based on the existing infrastructure but adapted to new design code requirements, which came into force during the period 2004-2006, including the new seismic code P100/1-2006. The upgrade of the structure to the new code requirements posed a real challenge to the design team, due to the constraints imposed by the existing infrastructure. Thus, three structural configurations were studied by varying the position and type of braces (see Figure 9.6). In all cases, the tubular column sections already installed in the basement, which were made from welded plates, were substituted above ground by cruciform or X-shaped cross sections.

9. CASE STUDIES

- Configuration 1 (C1): this configuration consisted of steel plate shear walls (SPSW) in the transverse direction and inverted V braces in the longitudinal direction. Columns were 1000 × 500 mm and 800 × 800 mm, respectively;
- Configuration 2 (C2): this configuration is similar to C1, with the difference that inverted V braces from the round side of the building were removed, while the SPSWs from the 2nd and 5th axis were extended. Additionally, belt trusses were used at mid-height and top of the structure. The same sections for the columns as for C1 were used;
- Configuration 3 (C3): this configuration used inverted V braces in the transverse and longitudinal directions and belt trusses at mid-height and top of the structure. Girders from the round side (marked with bold continuous lines) were strengthened compared with the two previous configurations, in order to increase the stiffness of the structure and to reduce its tendency to twist.

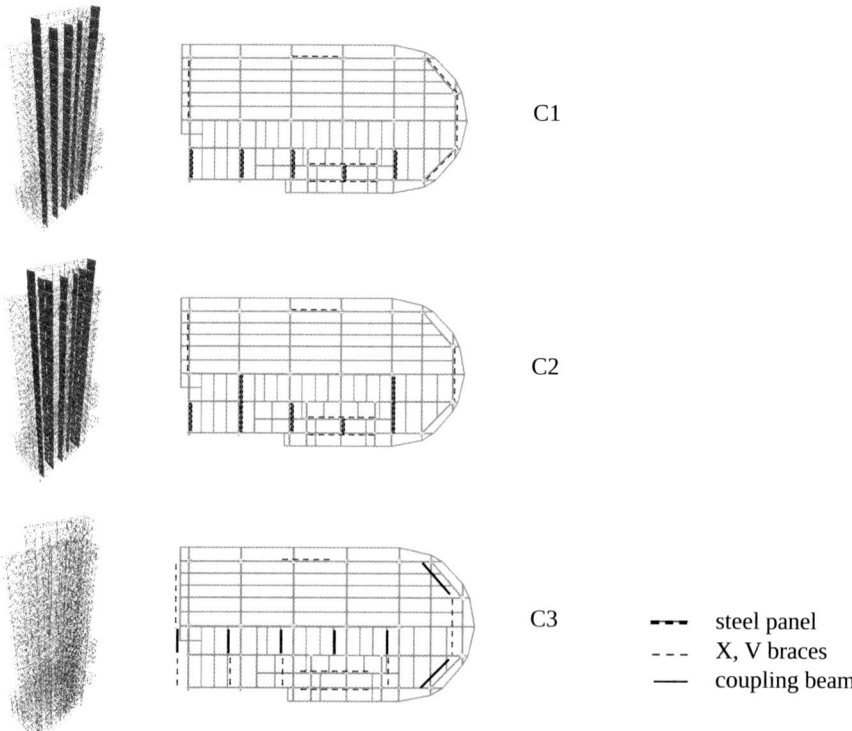

- - -   steel panel
- - -   X, V braces
——   coupling beam

Figure 9.6 – The preliminary analysis considering three structural systems

Table 9.1 – Comparison between the configurations in the preliminary analysis

Structural configuration	1st Period (s)	Base shear force (kN)		Top displacement (m)	
		transversal	longitudinal	transversal	longitudinal
Configuration 1	2.83	15436	11810	0.345	0.263
Configuration 2	2.69	16861	14453	0.256	0.233
Configuration 3	2.86	14805	12413	0.244	0.218

Table 9.1 shows the period of vibration, base shear force and top displacement for the three systems. Configuration C3 exhibits a good behaviour, the base shear force from the seismic combination has the lowest value in the transverse direction and in the longitudinal direction it is significantly lower than for C2. The top displacement both for the transverse and longitudinal directions has the lowest values for C3. From the three systems, C3 was selected. Combined with the architectural demands, another reason for this decision was that the configuration allowed for more clear space while assuring a favorable global behaviour of the structure. The two configurations with SPSWs were considered inappropriate also because the installation of the panels could have posed serious problems in the narrow space available on site.

9.2.2.3 Actions and load combinations

National standards as well as the Eurocodes were considered in the evaluation of the actions and the combination of actions. The following loads were considered in the design:

Dead load
Dead load (G) includes the self-weight of the structure, permanent fittings and equipment. For the common areas, like offices and conference rooms, the gravity load is given by the specific weights of the layers and permanent fittings:

120 mm concrete slab	3.00
Corrugated steel sheet	0.10
Ceiling	0.25
Raised floor (170 mm)	1.80

Sprinklers	0.30
Partitions	1.00
Mechanical and miscellaneous (beams and girders)	0.40
Total dead load	6.85 kN/m²

Imposed load

Imposed loads on floors and roofs (Q) are due to occupancy and depend on the category of use for different areas of the building. For the determination of the imposed loads, floor and roof areas in buildings should be sub-divided into categories according to their use. According to EN 1991-1-1 (CEN, 2002b), office areas are classified as category B of use. The value of uniformly distributed load ranges from 2.0 to 3.0 kN/m². For this specific application, the imposed load was 3 kN/m².

Snow load

On the roofs, imposed loads and snow loads should not be applied together simultaneously. The value of load due to snow for persistent design situation was determined as follows (EN 1991-1-3 (CEN, 2003)):

$$s = \mu_i C_e C_t s_k$$

where:

- μ_i is the snow load shape coefficient; for flat roofs, $\mu_i = 0.8$;
- s_k is the characteristic value of snow load on the ground; for the specific location of the building, $s_k = 2$ kN/m²;
- C_e is the exposure coefficient; considering the site topography and the height of the roof, $C_e = 0.8$;
- C_t is the thermal coefficient; $C_t = 1.0$.

The characteristic value for snow loading on the roof (S) amounts to 1.28 kN/m². Therefore, the imposed load on roof was considered instead of snow.

Wind load

Wind loading on the walls (W) was calculated with a mean wind speed $V_{b,0} = 26$ m/s and an urban area (EN 1991-1-4 (CEN,2005f)). Wind actions were considered both in the transverse (WX) and longitudinal (WY) directions, see Figure 9.7.

9.2 THE BUCHAREST TOWER CENTRE INTERNATIONAL

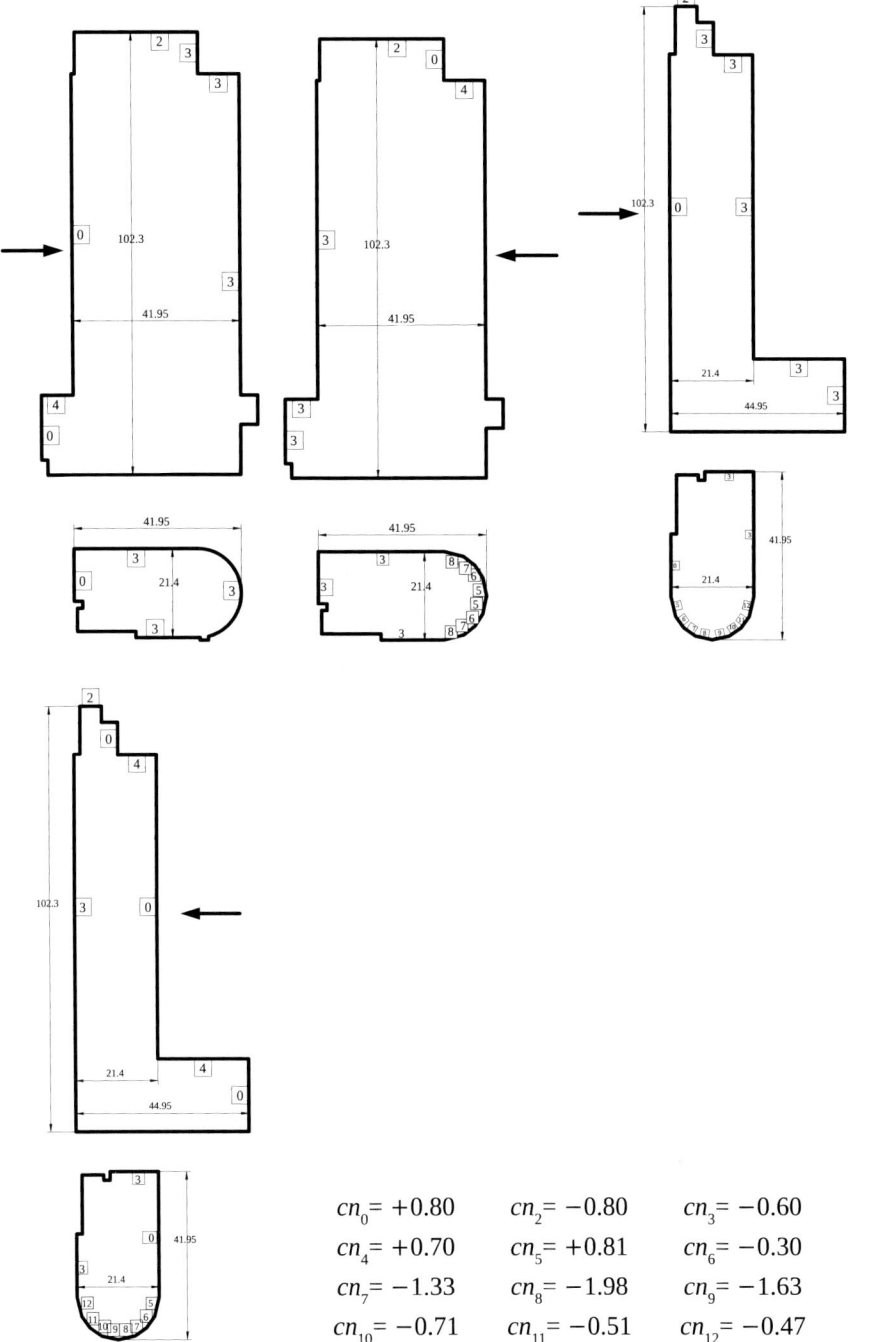

$cn_0 = +0.80 \quad cn_2 = -0.80 \quad cn_3 = -0.60$
$cn_4 = +0.70 \quad cn_5 = +0.81 \quad cn_6 = -0.30$
$cn_7 = -1.33 \quad cn_8 = -1.98 \quad cn_9 = -1.63$
$cn_{10} = -0.71 \quad cn_{11} = -0.51 \quad cn_{12} = -0.47$

Figure 9.7 – Pressure coefficients for different wind directions on the building

9. CASE STUDIES

The external pressure coefficients are a function of the location on the building and building shape. Positive coefficients represent a positive (inward-acting) pressure, and negative coefficients represent negative (outward-acting or suction) pressure. The building shape affects the value of the pressure coefficients and the loads applied to the building surfaces. Figure 9.7 shows the pressure coefficients for each direction of the wind and location on the building surface. Note that some values are interpolated from different cases given in the code.

Tall buildings may experience high winds and therefore precise calculations are required. Moreover, extreme local pressures may be experienced on their walls or glass facades. In case of the TCI building, owing to its irregular shape, there were no precise relations in the code to evaluate the wind pressure (pressure coefficients), and therefore a boundary layer wind tunnel test was carried out with a rigid model. The wind tunnel tests were performed at the "Laboratory of Technical University for Construction in Bucharest – UTCB" (Sandu *et al*, 2004), in the preliminary phase of structural design. The scale of the model was 1:100 (Figure 9.8).

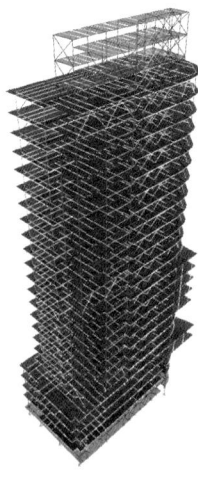

Figure 9.8 – Rigid wind model, length scale 1:100

9.2 THE BUCHAREST TOWER CENTRE INTERNATIONAL

The profile of mean wind velocity is described by Davenport's power law (Figure 9.9):

$$U(z) = U_r \left(\frac{z}{z_r} \right)^\alpha \qquad (9.1)$$

where $U(z)$ is the mean wind speed at height z, U_r is the known wind speed at a reference height z_r, and α is an empirically derived coefficient.

Eight different wind directions were considered, at 45° intervals between 0° and 360° (Table 9.2).

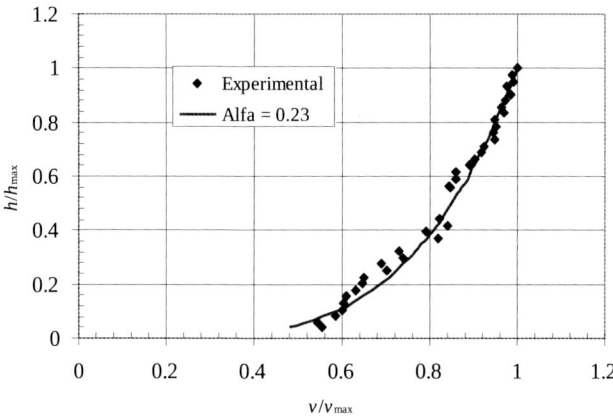

Figure 9.9 – Mean wind profile, experimental vs. theoretical

Table 9.2 – Wind direction on the scaled model

Wind direction	N	NE	E	SE	S	SW	W	NW
Wind – model incident angle θ	230°	185°	140°	95°	50°	5°	320°	275°

The wind tunnel test confirmed the values of the pressure coefficients given by code provisions (Figure 9.10). Attention should be paid to the transition between round and flat areas, where pressures tend to intensify significantly (Figure 9.11).

9. Case Studies

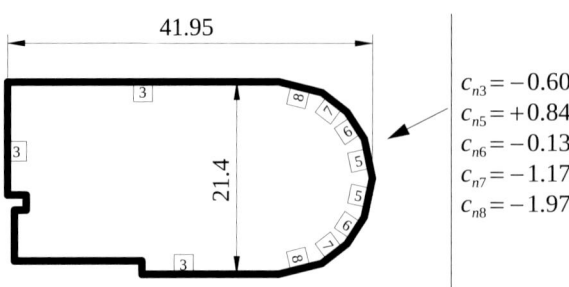

Figure 9.10 – Design pressure coefficients, E-W

Resonant dynamic response in along-wind or cross-wind conditions is a feature of the overall structural loads experienced by these structures. Therefore, in order to obtain a more accurate estimation of the dynamic behaviour of the building under wind loads, aeroelastic model tests in the boundary layer wind tunnel were carried out. These tests provide the overall mean and dynamic loads, displacements, rotations, and accelerations. In order to achieve a correct evaluation of the wind effects, it is necessary to have good similarity between the behaviour of the model and the full-scale structure. This may be achieved by means of a group of variables. This group of variables should be numerically equal for the model (wind tunnel) and the real situation. In the case of the TCI building, a set of eight variables was considered:

U = mean wind speed at some reference position
ρ = density of air
μ = dynamic viscosity of air
l_s = characteristic length of the structure
E_s = Young's modulus
δ = logarithmic decrement of structural damping
g = acceleration due to gravity

With the similarity conditions selected, the scaled model was tested in the turbulent boundary-layer flow. In order to obtain the appropriate turbulence intensity and roughness, five obstacles and discrete surface roughness treatments were installed (Figure 9.12). The results reveal accelerations and displacements within acceptable levels.

9.2 THE BUCHAREST TOWER CENTRE INTERNATIONAL

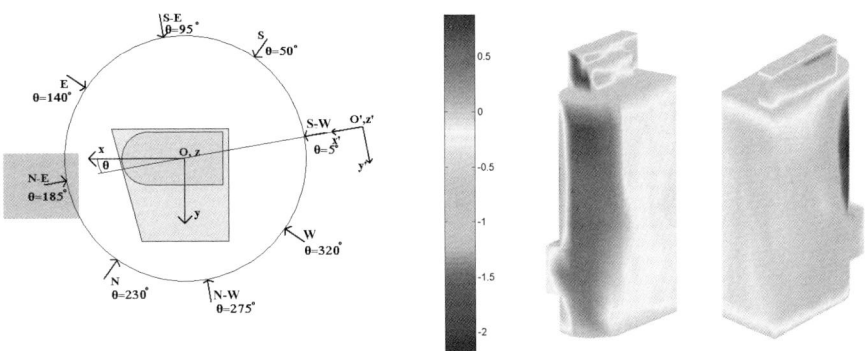

Figure 9.11 – Distribution of the pressure coefficients on the envelope of the rigid model scaled to 1:100, NE direction of the wind

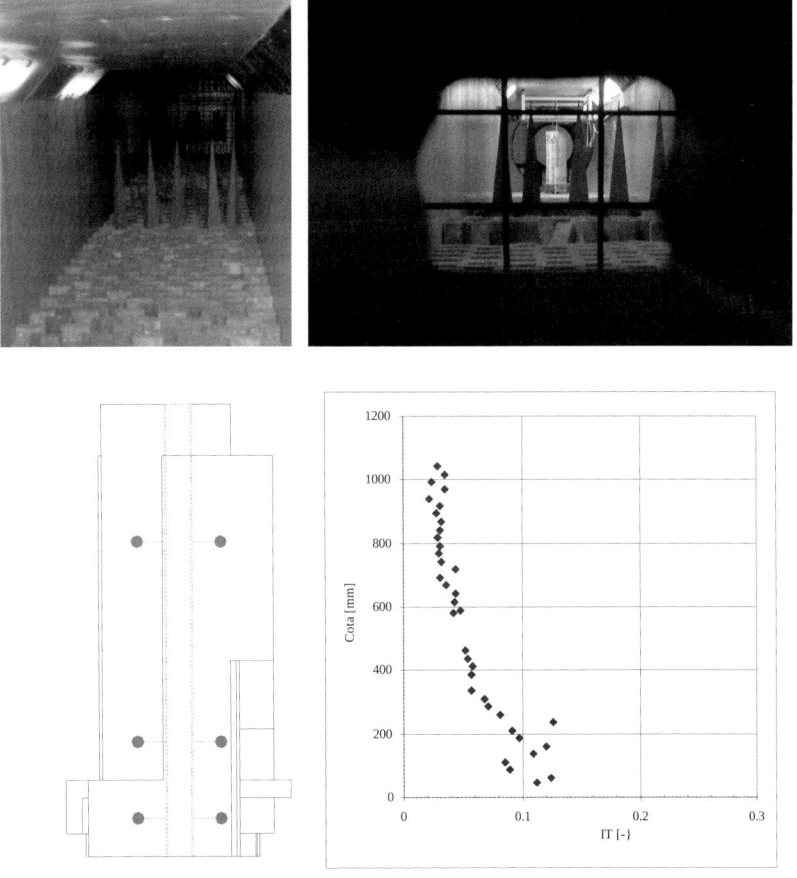

Figure 9.12 – Turbulence and roughness treatments

9. Case Studies

<u>Seismic action</u>

The seismic action was calculated according to the previous Romanian seismic design code, P100/1-2006. The main parameters of the design seismic action are:

- $a_{g,R} = 0.24\ g$ (Bucharest)
- $q = 4$ (dual frame structure)
- $T_B = 0.16$ s; $T_C = 1.6$ s; $T_D = 2.0$ s
- $\beta_0 = 2.75$
- $\gamma_1 = 1.2$ (class III)

where:

- a_g is the design ground acceleration;
- q is the reduction factor;
- T_B is the lower limit of the period of the constant spectral acceleration branch;
- T_C is the upper limit of the period of the constant spectral acceleration branch;
- T_D is the value defining the beginning of the constant displacement response range;
- γ_1 is the importance class factor.

For the horizontal components of the seismic action, the normalised response spectrum according to the Romanian seismic design code, P100/1-2006 is defined by the following expressions:

$T \leq T_B$ $\beta(T) = 1 + \dfrac{(\beta_0 - 1)}{T_B} T$

$T_B < T \leq T_C$ $\beta(T) = \beta_0$

$T_C < T \leq T_D$ $\beta(T) = \beta_0 \dfrac{T_C}{T}$

$T > T_D$ $\beta(T) = \beta_0 \dfrac{T_C T_D}{T^2}$

The resulting normalized response spectrum for Bucharest is illustrated in Figure 9.13.

9.2 THE BUCHAREST TOWER CENTRE INTERNATIONAL

The resulting normalized response spectrum for Bucharest is illustrated in Figure 9.13.

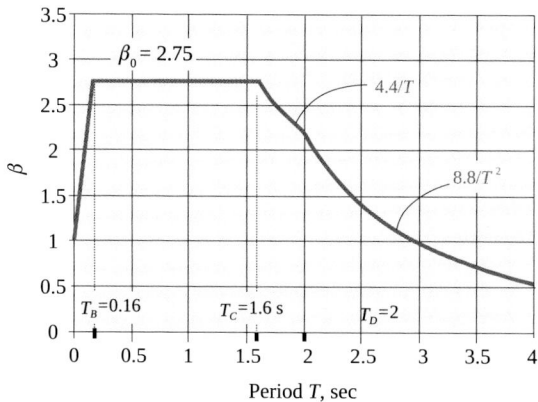

Figure 9.13 – Normalised response spectrum for Bucharest

For the horizontal components of the seismic action, the design spectrum, $S_d(T)$, was defined by the following expressions:

$$0 < T \leq T_B \qquad S_d(T) = a_g \left[1 + \frac{\frac{\beta_0}{q} - 1}{T_B} T \right]$$

$$T > T_B \qquad S_d(T) = a_g \frac{\beta(T)}{q}$$

The load combinations are defined as follows:

ULS combinations

→ Fundamental combinations (persistent or transient design situations)
→ Seismic design situations

According to the capacity design principles, brittle failure or other types of undesirable failure mechanisms shall be prevented, assuming that under the seismic induced effects A_E, plastic hinges are formed in dissipative members and primary seismic columns satisfy the following design requirements:

9. CASE STUDIES

E_ULS = $1.0G + 0.4Q + A_E$ (dissipative members - beams)
E_ULS_O = $1.0G + 0.4Q + 1.1 \times \gamma_{ov} \times \Omega_{MRF(CBF)} \times A_E$
(non-dissipative members - columns, excepting the base of the column, where the dissipative combination is used)

SLS combinations:

→ Non-seismic combinations
→ Seismic combination
E_SLS = $1.0G + 0.4Q + q \cdot v \cdot A_E$

9.2.2.4 Global analysis

The analysis of the structure was performed on a 3D structural model (Figure 9.14) developed using ETABS (Dubina *et al*, 2009b).

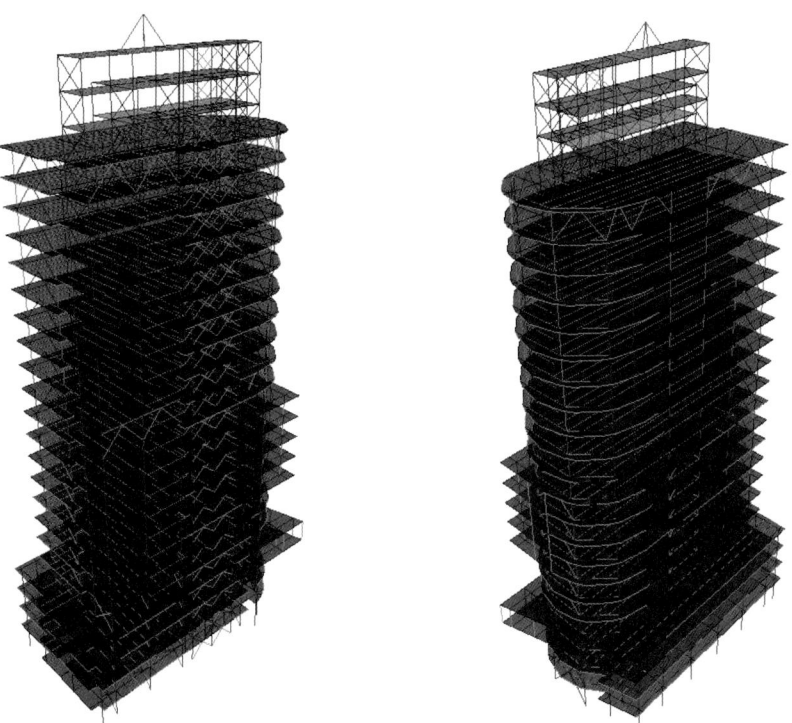

Figure 9.14 – 3D model, south view (left) and north view (right)

Static analysis for the gravity loadings and a modal response spectrum analysis for the earthquake action were carried out. The diaphragm effect of the concrete slab was considered in the analysis. The internal forces and displacements were determined using an elastic analysis. Due to the reduced sensitivity of the structure to 2nd order effects ($\alpha_{cr} > 10$), a first order elastic analysis was performed accounting for the global imperfections. A total number of 26 modes of vibration were considered in the modal analysis in order to excite at least 90 % of the total mass of the structure (i.e. the sum of modal participating masses ≥ 90 % of the total mass) (Table 9.3).

Table 9.3 – Periods of vibration and modal participating mass ratio

Mode of vibration	$T_i(s)$	ΣM_X, % (transversal)	ΣM_Y, % (longitudinal)
1	2.865	2	58
2	2.684	64	61
3	1.762	65	61
4	0.819	72	69
5	0.796	79	77
6	0.648	79	77
7	0.634	79	77
8	0.633	79	77
9	0.627	79	77
10	0.551	79	77
11	0.543	79	77
12	0.418	85	78
13	0.401	86	84
14	0.377	86	84
15	0.360	86	84
16	0.296	86	85
17	0.283	86	85
18	0.275	88	86
19	0.260	88	87
20	0.258	88	87
21	0.186	90	88
22	0.178	91	89
23	0.176	91	90
24	0.087	91	93
25	0.085	92	93
26	0.080	94	93

9. Case Studies

The first two modes of vibration were translational. The first mode was in the longitudinal direction with a period $T_1 = 2.86$ s, while the second mode was in transverse direction and had a period $T_2 = 2.68$ s. The third mode was torsional and had a period $T_3 = 1.76$ s. (Figure 9.15).

In order to account for uncertainties in the location of the masses and in the spatial variation of the seismic motion, the calculated centre of mass at each floor i was considered as being displaced from its nominal location in each direction by an accidental eccentricity (see section 2.7.1 in chapter 2):

$$e_{ai} = \pm 0.05 \times L_i$$

where:

- e_{ai} is the accidental eccentricity of storey mass i from its nominal location, applied in the same direction at all floors;
- L_i is the floor-dimension perpendicular to the direction of the seismic action.

Figure 9.16 shows the elastic response spectrum for the site with the indication of the periods of the first three modes of vibration.

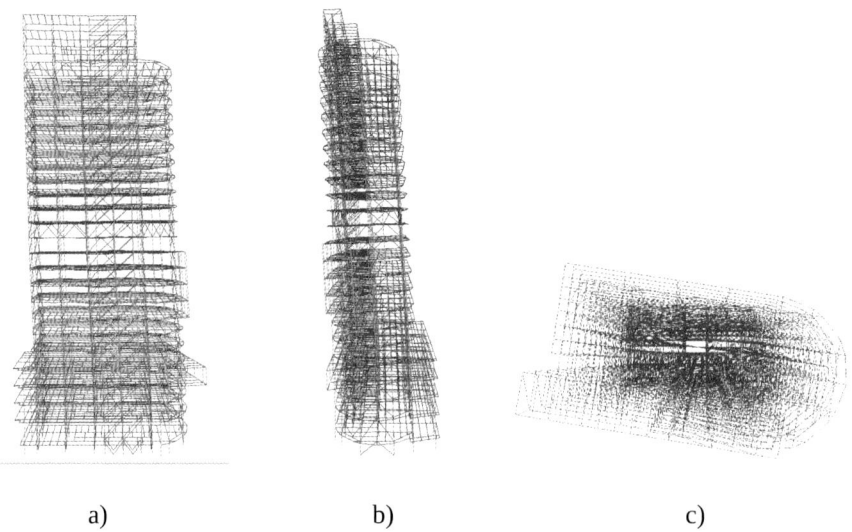

a)   b)   c)

Figure 9.15 – First three modes of vibration: a) mode 1, $T = 2.86$ s; mode 2, $T = 2.68$ s; c) mode 3, $T = 1.76$ s

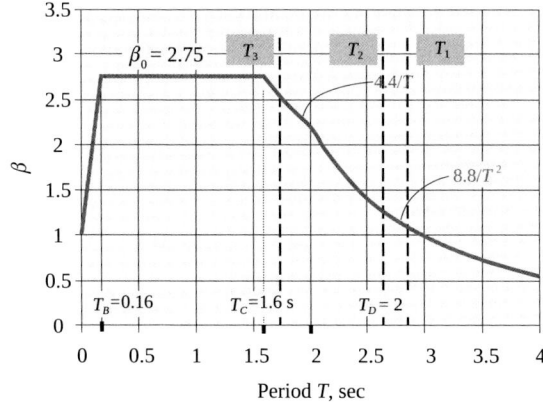

Figure 9.16 – Normalised elastic response spectrum of the site and periods of first three vibration modes

If the fundamental mode is larger than the control period (corner period), T_c, the problem of higher mode effects should be accounted for. In such cases, the masses of some floors may move in opposite directions or the torsion effect can be amplified. Because the actual pattern of lateral load including possible higher-mode effects might be missed by a static non-linear analysis (push-over analysis), a time-history analysis is preferable, as it directly incorporates the higher-mode effects in the results. The results of this analysis are presented in sub- section 9.2.2.7.

9.2.2.5 Structural design

The seismicity of the Bucharest area was one of the main important aspects in design. The objective was to obtain a building with a fundamental period large enough to reduce the base shear force but at the same time to keep the lateral displacements under wind load within allowable limits (i.e. H/500). Perimeter belt trusses mounted at mid-height and at the top of the building have a beneficial effect because they reduce the lateral displacement at the top under wind and seismic actions. They also reduce the torsional effects of the structure and improve the seismic behaviour.

As the seismic design concept was dissipative, the goal of the design was to locate the plastic hinges in the beams, near their connections to the columns (for moment-resisting frames, MRF), or in the braces (for centrically braced frames, CBF) while the columns needed to remain elastic.

9. Case Studies

The design verifications were carried out according to EC8-1 and the other relevant Eurocodes (Eurocode 3 and Eurocode 4). They are not given in detail here because their application is straightforward, as it was explained in detail in the worked examples of chapters 6 and 7.

9.2.2.6 Damage limitation

The "damage limitation requirements" is generally associated to the serviceability limit state and is concerned with ensuring that deflections are not excessive under normal conditions of use. In some cases, for example tall buildings or long span floors, it may also be necessary to ensure that the structure is not subject to excessive vibrations to avoid discomfort to the occupants. Both deflections and vibrations are associated with the stiffness rather than strength of the structure. For steel structures, adequate stiffness is generally ensured by calculating deflections and ensuring that these are less than stipulated limits. For multi-storey buildings, the verification at the serviceability limit state is expressed by limiting the interstorey drifts for lateral loads coming from wind and earthquake but also by limiting the floor member deflections. Table 9.4 summarizes the acceptance criteria for damage limitation for the various actions.

Table 9.4 – Acceptance criteria

	Wind drift	Earthquake drift	Floor mid-span deflections
Total building lateral displacement	$H/500$		
Maximum inter-storey displacement	$h/300$	$H_{floor}/125$	
Allowable inter-storey drift for SLS		$0.008h$	
Deflection of main girder			$L/350$
Deflection of other beams			$L/250$

a) Wind drift:
   - Total building lateral displacement $H/500$
   - maximum inter-story displacement $h/300$

9.2 THE BUCHAREST TOWER CENTRE INTERNATIONAL

where:
- H total height of the building;
- h story height.

For a current story height of 4.0 m and a total height of 106.4 m, the two limitations are as follows:

$\Delta = 237.0$ mm $= H_{total}/500$
$\delta = 13.3$ mm $= H/300$

b) Earthquake drift:
   - Maximum inter-story displacement $H_{floor}/125$

The interstorey drift calculation for SLS is given by:

$$d_s = d_e \times v \times q \leq d_{ra} = 0.008h$$

where:
- d_{ra} is the allowable interstorey drift;
- d_s is the horizontal displacement under the seismic action of a point of the structure;
- q behaviour factor depending on the structural type;
- d_e horizontal displacement of the same point determined by elastic analysis;
- v reduction factor equal to 0.4 taking into account the lower return period for buildings of Class III importance factor;
- h storey height.

c) Floor members mid-span deflection:
   - Main girders deflection $L/350$
   Other beams deflection $L/250$

For wind action, the maximum top displacements (level 106.4 m) and the maximum interstorey drift are:

$\Delta_X = 78$ mm ≤ 237.0 mm $= H_{total}/450$
$\Delta_Y = 237$ mm $= H_{total}/450$

9. Case Studies

$\delta_X = 0.123\ \% \leq 0.33\ \% = 1/300$
$\delta_Y = 0.273\ \% \leq 0.33\ \% = 1/300$

Figure 9.17 shows the total lateral displacement under wind load, in the transverse and longitudinal directions. The top displacement in the transverse direction is close to the allowable value. However, the discomfort to the occupants is not affected, because the vibrations are below the acceptable limit, which for non-residential spaces can be considered $0.024g$ - in case of this project, the maximum acceleration induced by wind reached $0.0106g$.

It is noted that for this structure, the wind criteria dictated the structural stiffness and, consequently, the sections for the braces.

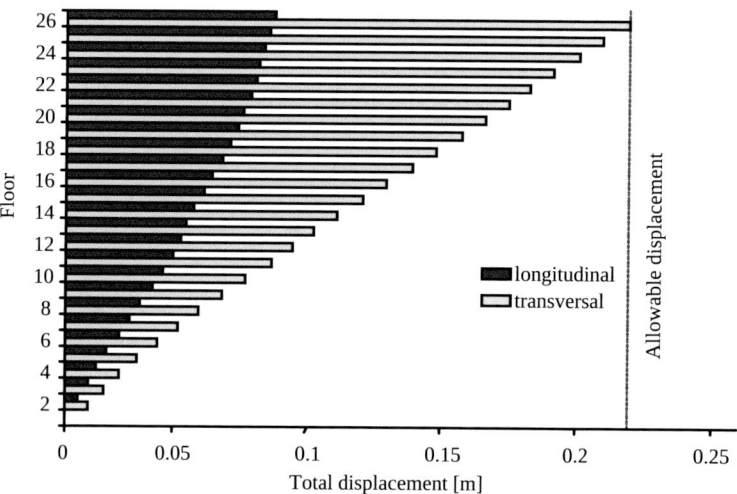

Figure 9.17 – Total lateral displacement under wind load, transverse and longitudinal directions

For the seismic design situation, the "damage limitation requirement" is expressed by limiting the interstorey drifts which for SLS combination is given by:

$$d_r \times \upsilon \times q \leq d_{ra}$$

q – behaviour factor = 4
d_r – interstorey drift under design seismic loads

v – reduction factor, $v = 0.4$ for importance class III
d_{ra} – allowable interstorey drift $= 0.008h$ (ductile non-structural elements)
$\delta_X = 0.52\% \leq 0.8\%$
$\delta_Y = 0.55\% \leq 0.8\%$.

9.2.2.7 Performance-based seismic evaluation

The application of the seismic provisions from EC8-1 and P100-1 are intended to satisfy two performance levels, i.e. damage limitation and no-collapse, respectively. These two limit states specify the minimum stiffness and strength that the structure should provide. Although these prescriptive criteria are intended to result in buildings capable of providing certain levels of performance, the real performance of the building is not assessed. Performance based design is a design philosophy in which the design criteria are expressed in terms of performance objectives, like lateral deflections, interstorey drifts or element ductility, when the structure is subjected to different levels of seismic hazard. To reduce high costs, due to loss of use and repair of heavily damaged structures, different performance objectives need to be taken into account. The results of the application of such an approach are very much influenced by the method of analysis, modeling parameters and definition of acceptance criteria (chapter 2).

Firstly, it is required to estimate the behaviour of the structure, the location of plastic hinges and the amount of plastic deformation that develops in the structure. For structures designed according to the dissipative concept, the post-elastic performance is very important and therefore, nonlinear analysis methods are recommended. Nonlinear static (pushover) analysis provides the location and sequence of plastic hinges in the structure, and can be used to estimate the displacement demand when combined with capacity-demand spectrum. However, the selection of the lateral force pattern and the use of equivalent linear viscous damping in the demand spectrum to represent inelastic hysteretic energy dissipation are only two of the problems that can influence the results. Also, the effects of the higher modes on the response of the structure cannot be accounted for. Nonlinear time-history analysis produces a more reliable prediction of the response of the structure during an earthquake. However, some problems still exist when a time-history is used, for example the selection and scaling of the ground motions. Nevertheless, nonlinear time-history is considered the most accurate approach for evaluating the performance of a structure.

9. CASE STUDIES

Secondly, the definition of the modeling parameters for nonlinear analysis and acceptance criteria corresponding to each performance level may also influence the results. Thus, a reliable prediction of the response is very much dependent on the modeling of the elements and connections, both for static or dynamic analyses. The acceptance criteria can be defined based on global or local response at local level (component-based) or at global level (overall structure-based).

For the seismic evaluation of the TCI building, a performance-based procedure was employed. Three performance levels (limit states) were considered: serviceability limit state (SLS), ultimate limit state (ULS) and collapse prevention limit state (CPLS). The intensity of earthquake action at ULS is equal to the design one (intensity factor $\lambda = 1.0$). The ground motion intensity at SLS was reduced to $\lambda = 0.5$ (similar to $v = 0.5$ in EC8-1), whereas for CPLS the intensity was increased to $\lambda = 1.5$ (P100-1 (MTCT (2006)).

Beams and columns were modeled with fibre hinge beam-column elements, with plastic hinges located at both ends. In order to take into account the buckling of the compression diagonal, the post-buckling resistance of the brace in compression was set equal to $0.3 N_{b,Rd}$ (Figure 9.18), where $N_{b,Rd}$ is the buckling resistance in compression. A strain-hardening ratio of 0.03 was used for all the analyses. P-Delta effects were taken into account by modelling the geometrical non-linearities.

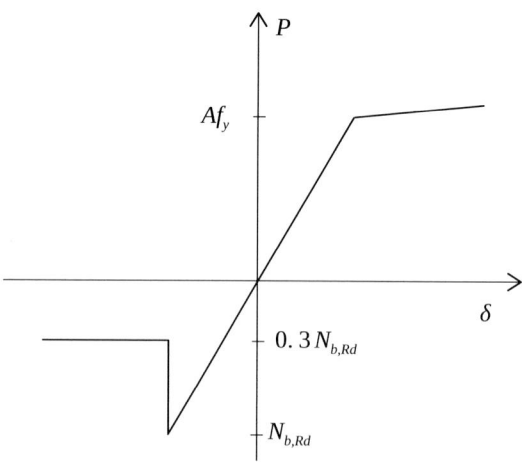

Figure 9.18 – Response of bracing members

9.2 THE BUCHAREST TOWER CENTRE INTERNATIONAL

Non-linear dynamic analyses were carried out in order to assess the structural performance. A set of seven ground motions was used in the analysis (see Figure 9.19). The spectral characteristics of the ground motions were modified by scaling the Fourier amplitudes to match the target spectrum. This results in a group of semi-artificial records representative of the seismic source affecting the building site and soft soil conditions of Bucharest.

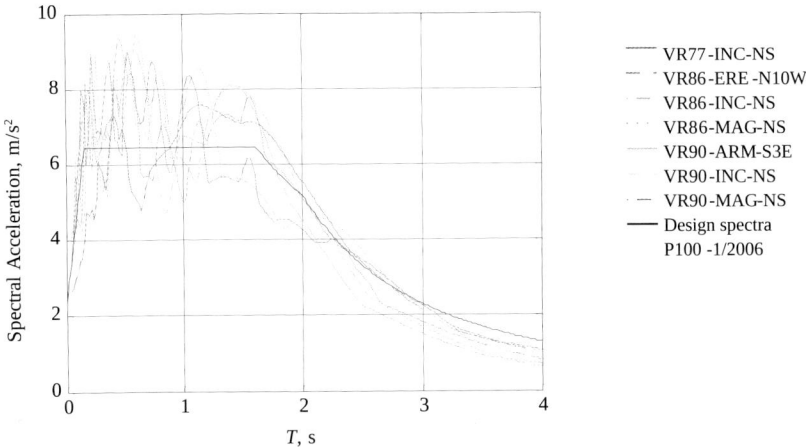

Figure 9.19 – Elastic acceleration response spectra of the semi artificial accelerograms and design spectra (P100-1/2006, $a_g = 0.24g$, $T_C = 1.6s$)

Based on FEMA 356 (2000) Table 5-6, the following acceptance criteria were adopted:
- For the braces in compression the plastic deformations at SLS, ULS and CPLS should be less than $0.25\Delta c$, $5\Delta c$ and $7\Delta c$, respectively, where Δc is the axial deformation at the anticipated buckling load;
- For the braces in tension the plastic deformations at SLS, ULS and CPLS should be less than $0.25\Delta t$, $7\Delta t$ and $9\Delta t$, respectively, where Δt is the axial deformation at the anticipated tensile yielding load;
- For the beams in flexure the plastic rotations at ULS and CPLS should be less than $6\theta y$ and $8\theta y$, respectively, where θy is the yield rotation;
- For the columns in flexure the plastic rotations at ULS and CPLS should be less than $5\theta y$ and $6.5\theta y$, respectively, where θy is the yield rotation.

The results obtained from non-linear time history analyses show that in the post-elastic range the frames progress towards a "full plastic mechanism"

9. Case Studies

configuration, as was considered in the design strategy and effectively realized by the detailing of the structure's steel/composite members and connections (Figure 9.20, Table 9.5). Figure 9.20 shows the evolution of damage pattern at each limit state. As it can be seen, the plastic damage is consistent with the expected performance. Indeed, at SLS ($\lambda = 0.5$) there is minor damage due to the buckling of some braces and the achievement of the yield rotation in some beams. At ULS ($\lambda = 1.0$) there is a uniform plastic engagement of the flexural plastic hinges of the beam as well as the most of braces are yielded in tension. At CPLS ($\lambda = 1.5$) the structure exhibit more sever plastic demand as respect to ULS, without brittle failure modes. In the dual configuration structure, the energy dissipation capacity of the MRFs brings important benefits to the overall energy dissipated by the structure. Plastic deformations in the braces and the beams indicate early damage to the structure at SLS ($a_g = 0.12g$), whereas plastic deformation in non-dissipative elements (columns) are completely avoided. With the ground motion scaled to the design acceleration ($a_g = 0.24g$), the maximum plastic rotation in the beams of an MRF is 0.01 rad and this rotation demand increases to 0.015 rad for $a_g = 0.36g$.

Figure 9.21 shows the distribution of the maximum inter-storey drifts in the transverse direction for SLS ($a_g = 0.12g$), ULS ($a_g = 0.24g$) and CPLS ($a_g = 0.36g$).

For frequent earthquakes, associated with SLS, the maximum interstorey drift ratio is less than 0.005, which was the limit adopted in the design. Again, the effectiveness of the perimeter truss belt for reducing the top lateral displacement is worth mentioning.

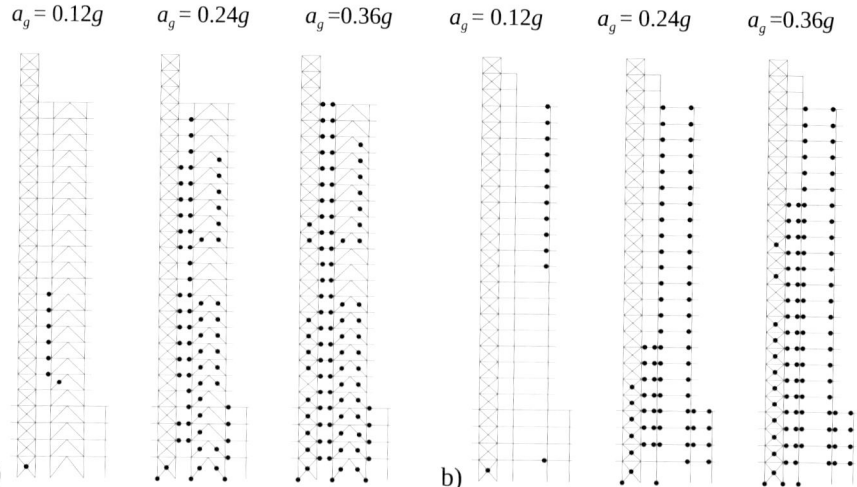

Figure 9.20 – Plastic hinges: a) side transversal frame; b) current transversal frame

Table 9.5 – Plastic rotation in beams and columns (in rad) and plastic deformation in braces (in %) at SLS, ULS and CPLS, average values

	Braces, [%]	Beams, [rad]	Columns, [rad]
SLS	0.002	0.002	-
ULS	0.006	0.01	0.002
CPLS	0.009	0.015	0.0035

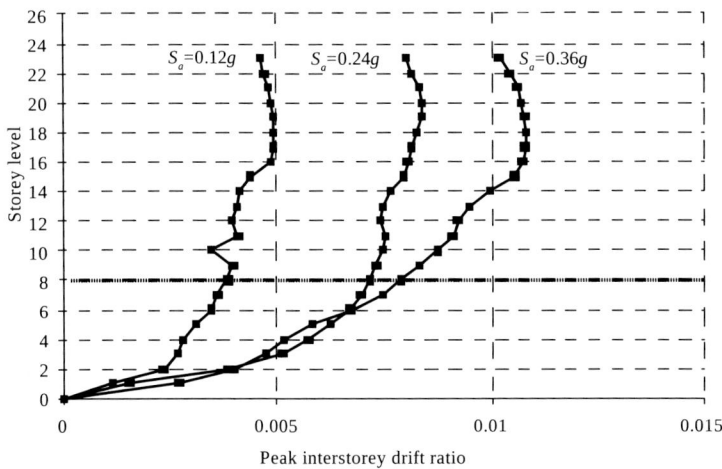

Figure 9.21 – Peak interstorey drift ratio vs. storey level for transversal direction, average of records

9.2.3 Detailing

Connection details are very important in order to assure the required earthquake resistance. Connection details that cannot exhibit good ductility and toughness have major consequences on the structural performance (Jaspart and Weynand, 2016). In this case, considering that very deep beams were designed, haunched extended end-plate joints were used in order to fulfill the overstrength requirement and to guarantee the formation of plastic hinges in the beams, near the end of the haunch.

Figure 9.22 shows a typical beam-to-column joint used in the TCI structure for the moment resisting frames. Shear studs were welded to the column webs near the beam-to-column joints to ensure composite action in the column and the transfer of shear forces from the column web to concrete encasement.

9. CASE STUDIES

Figure 9.22 – Detail of a beam-to-column joint in the MRF of TC1 structure

For X braces and inverted V braces (Figure 9.23), the connection overstrength was achieved by using a higher steel grade (S355) in the gusset plates and cover plates attached to the beams or columns, while S235 was used in the braces.

Figure 9.23 – Some details of the brace connections

9.2.4 Construction

9.2.4.1 Introduction

The original design project was developed in 1997 and the construction works began. About 95 % of the infrastructure was completed between 1997 and 1998, with some minor works remaining at level −3.20 m. Afterwards, the construction works were halted. In 2006, after a pause of eight years, the works for the structure started, but based on a different project.

Before restarting the works, the existing infrastructure was verified and strengthened to suit the new loading conditions, also taking into account the level of degradation due to the prevailing environmental conditions during the period of suspension of the construction. Settlements of nearby buildings (located more than 10 m from the excavation) were continuously monitored during construction. Measurements indicated values of less than 6 mm when the building was completed.

In this re-design, for reasons of easy assembly and good quality control, bolted connections were preferred to on-site welding, with the exception of the welded connection between the existing RHS columns and the new X-shaped columns. The columns were continuous over three storeys and, as a result, the number of splice connections was reduced. As the splices can cause some deviations, the reduction also had a beneficial effect on the imperfections due to the assembly.

Construction of the steel structure (see Figure 9.24) was completed in November 2006. The building was officially opened in 2007.

Figure 9.24 – Views of the building in successive phases of the construction

9.2.4.2 Control of execution

The complexity of the structure required an extensive quality control in the fabrication and construction. Thus, a complete 3D CAD-CAM

9. Case Studies

computer model was produced and then interactively adjusted during fabrication by taking into account the deviations of the structure from the original position. This also enabled interoperability between the design model and the cutting, drilling and welding operating programs used by the steel fabricator.

The beams and braces were made of hot rolled profiles. The columns were made of cruciform shaped profiles (Austrian Cross) by joining two H steel sections. In order to ensure adequate behaviour against brittle fracture, steel quality was J2. The through-thickness properties of steel should prevent lamellar tearing in the steel assemblies.

The susceptibility of the material for lamellar tearing was determined by measuring the through-thickness ductility quality according to EN 10164 (CEN, 2004f), which is expressed in terms of quality classes identified by Z-values (see Table 9.6). According to EN 1993-1-10 (CEN, 2005e), lamellar tearing can be neglected in a detail if:

$$Z_{Ed} \leq Z_{Rd} \tag{9.2}$$

where:

Z_{Ed} is the required design Z-value resulting from the magnitude of strains from restrained metal shrinkage under the weld beads, obtained as given in Table 3.2 of EN 1993-1-10. The required design value Z_{Ed} is determined as follows:

$$Z_{Ed} = Z_a + Z_b + Z_c + Z_d + Z_e \tag{9.3}$$

Being Z_a the value depending on effective weld depth (a_{eff}) relevant for straining from metal shrinkage; Z_b is related to the shape and the position of welds (e.g. in T- and cruciform- and corner connections); Z_c is related to the effect of material thickness s as restraint to shrinkage; Z_d accounts for the restraint of shrinkage after welding by other portions of the structure; Z_d accounts for the influence of preheating.

Z_{Rd} is the available design Z-value for the material, and can be determined according to EN 10164 (Z15, Z25 and Z35 classes are established).

9.2 THE BUCHAREST TOWER CENTRE INTERNATIONAL

Table 9.6 – Choice of quality class according to EN 10164

Target value of Z_{Ed} obtained according to EN 1993-1-10	Required value of Z_{Rd} obtained according to EN 10164
$Z_{Ed} \leq 10$	-
$10 < Z_{Ed} \leq 20$	Z15
$20 < Z_{Ed} \leq 30$	Z25
$Z_{Ed} > 30$	Z35

The evaluation of the required design Z-value is presented for two cases. The first case refers to the end plate of the bolted beam-to-column joints. The second refers to the cap plate used to connect the columns from the pre-existing infrastructure to the columns of the superstructure.

<u>End plate of the bolted beam-to-column joint</u>
Figure 9.25 shows a sketch of the adopted bolted beam-to-column connection. The force in the tension flange may induce lamellar tearing and therefore, it is necessary to calculate the required design Z-value.

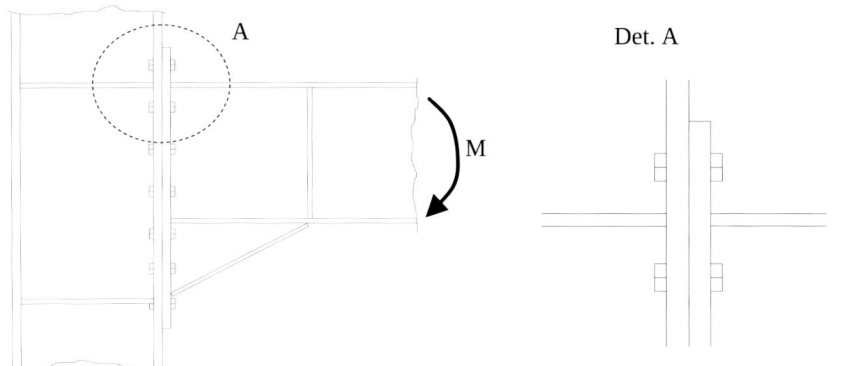

Figure 9.25 – Strengthened connection with plastic hinge located in the beam (left) and correction of the design bending moment (right)

The following values of partial Z factors are obtained:

$Z_a = 6$ (10 mm $< a_{eff} \leq$ 20 mm);

9. CASE STUDIES

$Z_b = 3$ (partial and full penetration welds with appropriate sequence to reduce shrinkage effects);

$Z_c = 8$ (30 mm $< a_{eff} \leq$ 40 mm);

$Z_d = 0$ (low restraint);

$Z_e = 0$ (without preheating).

Hence, the required design value, Z_{Ed}, is then obtained as follows:

$Z_{Ed} = Z_a + Z_b + Z_c + Z_d + Z_e \quad \rightarrow Z_{Ed} = 17 \rightarrow$ Required value of Z_{Rd} is Z15 (see Table 9.6).

Welded splice connection of the columns using cap plate

Through-thickness property of steel is of high importance for the welded splice connections between the existing columns in the infrastructure and the columns of the superstructure. The columns of the existing infrastructure (Figure 9.26) comprised steel hollow sections (1) encased in concrete (3) and connected through reinforced concrete beams. The cruciform cross sections made of hot-rolled profiles (2) were partially encased in reinforced concrete to increase strength, rigidity and fire resistance. The connections between the infrastructure and superstructure columns were welded, including a thick cap plate (4) at the top of the hollow section column.

The following values of partial Z factors are obtained (REF):

$Z_a = 9$ (20 mm $< a_{eff} \leq$ 30 mm);

$Z_b = 3$ (partial and full penetration welds with appropriate sequence to reduce shrinkage effects);

$Z_c = 8$ (20 mm $< a_{eff} \leq$ 30 mm);

$Z_d = 0$ (high restraint);

$Z_e = 0$ (without preheating).

The required design value, Z_{Ed}, is then obtained as follows:

$Z_{Ed} = Z_a + Z_b + Z_c + Z_d + Z_e$ → $Z_{Ed} = 23$ → Required value of Z_{Rd} is Z25
(see Table 9.6).

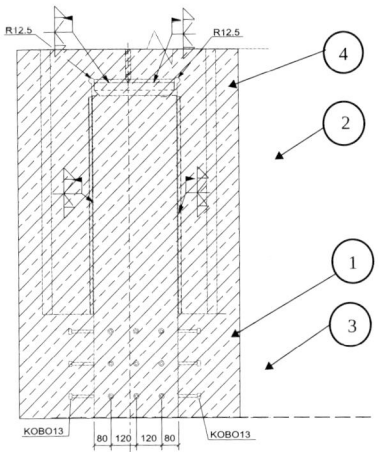

1. Steel hollow section column
2. Cruciform cross sections column
3. Concrete part
4. Thick cap plate

Detail of welds between the thick cap plate and the cruciform column

Figure 9.26 – Connection between columns of infrastructure and superstructure

After joint completion, ultrasonic tests (UT) of the welding revealed discontinuities and intrusions in the cap plates, adjacent to the welds (Figure 9.27), caused mainly by the poor ductility of the plate in the through-thickness direction. The internal flaws were located in the mid-thickness of the plates. In order to minimize the risk of lamellar tearing (some columns can be subjected to tension), the cap plates were removed and replaced with new plates. This intervention caused large gaps between the columns of the existing infrastructure and the new ones. Taking into account the tensile stress in the cap plate for the columns loaded in tension and the stress concentration in the welding, the welding and the cap plate material were strictly controlled to avoid lamellar tearing and microcracking of welds. For each connection, detailed welding procedures were drawn up and verified based on both destructive and non-destructive testing.

9. Case Studies

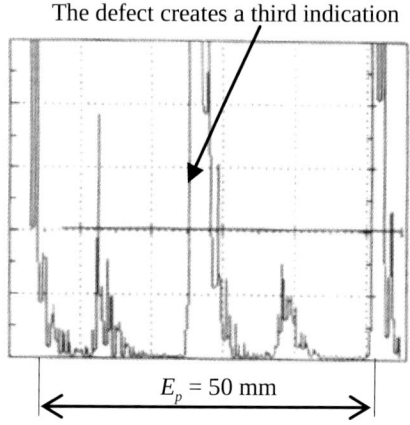

Figure 9.27 – Ultrasonic testing report: principle (left); the defect in the thick plate during the inspection on site (right)

9.2.4.3 Erection

The building superstructure was completed in 10 months. The structural steel erection sequence and the zteel parts delivery were organised considering there was a very limited space for storage and handling. The first pieces hoisted were the columns, then the beams and the braces (Figure 9.28a). A multiple lift rigging process was used to speed the erection (Figure 9.28b).

Figure 9.28 – Erection of the structure: a) columns are continuous over 3 stories and are hoisted first; b) multiple lift rigging

Shear studs were welded on the main beams and the secondary beams but the composite action was considered for secondary beams only. For the main girders, the number of shear studs is not sufficient to have a composite action with the slab and they are located to restrain the top beam flange from out of plane deflection, namely to prevent lateral-torsional buckling. Steel decking acting as lost formwork was used for the in situ concrete floors. Floor slabs were poured after the structural steel frame was in the appropriate position and the bolts were tightened. Plumbing up the structural frame was crucial because of the very low allowable tolerances. Theodolites as well as Global Position System (GPS) technology were used to verify the alignment of columns and the level of the floors. Normally, the surveying on the erection of high-rise buildings was done by geodetic electro-optical total stations. However, the precision of the procedure depends on the reference points serving as fixed points for the total station. Points that are close to ground are suitable because they are not subject to influences. However, increasing construction heights, possibly aggravated by densely built-up surroundings, which was the case for TCI building, give rise to difficulties in the use of ground-level fixed points. As a result, beyond a certain threshold height, it becomes impossible to use ground-level reference points (Van Cranenbroeck, 2005).

A Global Positioning System (GPS) can determine with millimeter precision the position of a point from the structure using spatial trilateration (Van Cranenbroeck, 2005). This process implies that, at a certain point in time, the distance to at least four satellites is measured and their position is determined in the chosen reference system. These elements must be determined with maximum precision, considering also a series of errors and their corresponding corrections that can be specific or general, depending on the case. Subsequent calculus determines the final coordinates which represent the spatial position of the point, determined for the specific reference system. A relevant example for the use, on a large scale, of the GPS technology for the positioning of the specific points of a building, is the Burj Dubai Tower (800 meters high and finalized in the year 2008). In that case, the land survey, as well as the tracing of the constructive elements, was done exclusively using GPS detection.

This method comes with some advantages over the classical methods (which consists of angular and length measurements, using the theodolite or the total station):

- Unitary determinations for both planimetric and altimetric measurements are possible;
- Lack of visibility between the reference points and the object is not a problem;
- Possible to perform irrespective of the weather or the moment of the day;
- The coordinates are determined as a result of multiple measurements, thus there is an additional control;
- Point determination precisions result directly from the compensation process;
- Real time monitoring can be granted by the use of appropriate software.

A general approximation of the planimetric precision of the positioning varies between the values of: +/- 1...3 mm, and for the altimetric precision values of: +/- 3...10 mm are achieved.

In order to reach the desired precision, it is necessary to use a high-performance GPS receiver. These receivers are often referred to as geodetic devices. The system delivers ellipsoidal coordinates (ellipsoidal latitude, ellipsoidal longitude and ellipsoidal altitude) on the WGS84 ellipsoid (World Geodetic System, 1984). In order to obtain the planar coordinates, determined in the building's reference system, a series of alterations should be made. In order to attain these transformations, GPS receivers need to be positioned in each point of the tracing network. The use of static positioning method is recommended; each point should be stationed for a period of at least 2 hours and the measurements frequency should be 15 seconds. These transformations take place during the first series of measurements. For this session, planar coordinates in the building's reference system were determined for two new points. These points, i.e. "800" point and "900" point were found in the building's influence area (Figure 9.29a). The last point works as a permanent station point. Afterwards, specific points for the steel structure were determined by placing two receivers on "800" and "900" points while the third one was placed in every point that needs to be measured. The mobile receiver was successively placed in every specific point of the metallic structure (Figure 9.29b). The final results (X, Y, Z coordinates) were used to determine the deviations from the theoretical position (inclinations, rotations, translations).

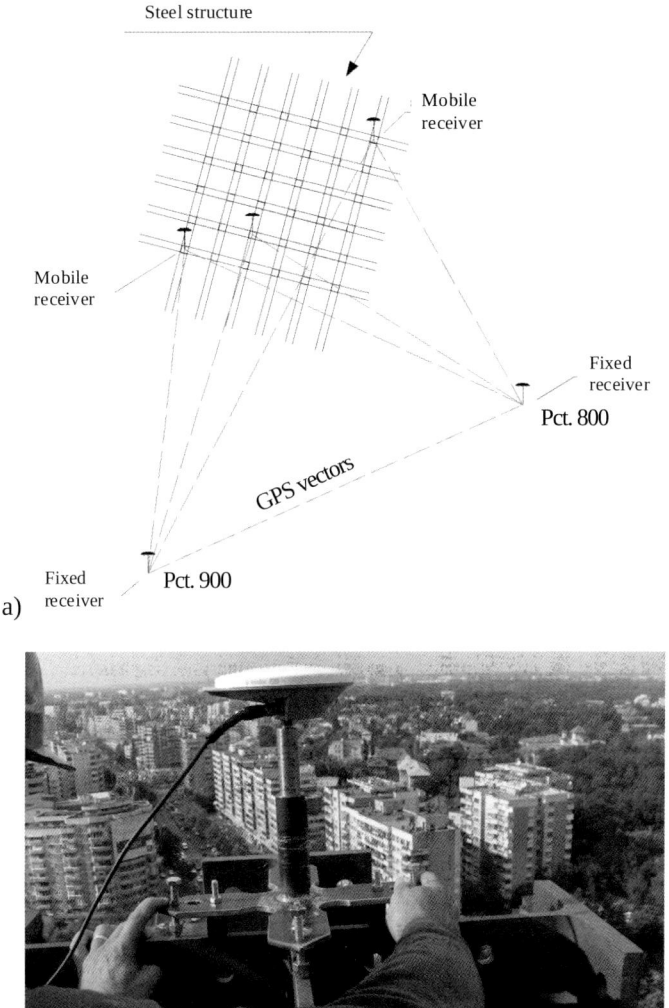

Figure 9.29 – GPS system to control the position of members: a) reference points serving as fixed points for the total station; b) positioning the antenna of the receiver in a point to be surveyed

When the steel structure was completed, the total deviation from the vertical was about 20 mm and was much less than the 70 mm allowable deviation (Table 9.7).

9. CASE STUDIES

Table 9.7 – Deviations from the theoretical position

Level	Allowable total translation	Total measured deviation	Allowable interstory translation	Maximum measured interstory deviation
+106.4 m	71.28 mm	18 mm	8 mm	7 mm
+102.3 m	69.76 mm	16 mm	8 mm	5 mm
+98.3 m	68.67 mm	14 mm	8 mm	3 mm

9.3 SINGLE STOREY INDUSTRIAL WAREHOUSE IN BUCHAREST

9.3.1 General description

This example presents the design of a single storey steel building, located in the capital city of Romania, Bucharest. The single storey building (Production Unit) is part of a larger complex that includes a production area and a storage area, respectively (Figure 9.30).

The main dimensions of the buildings are summarized in Table 9.8.

Table 9.8 – Main dimensions of the buildings

	Production Unit (Building A)	Warehouse Unit (Building B)
Building width (m)	78 (15 + 24 + 24 + 15)	72 (24 + 24 + 24)
Building length (m)	180	120
Bay length (m)	6	6
No of bays	30	20
Building height to caves (m)	8.2	11.8
Roof pitch	2 %	2 %

9.3 SINGLE STOREY INDUSTRIAL WAREHOUSE IN BUCHAREST

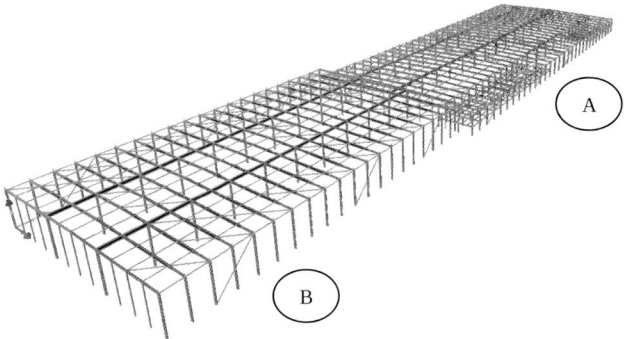

Figure 9.30 – General view with the two buildings

Production unit

The primary framing system consists of portal frames with parallel flange columns and partially tapered rafters. The columns have fixed bases and rafters are rigidly connected to the columns. The building also includes a mezzanine that is partially supported by the main framing and by additional columns. In order to increase the clear space at the interior, every second transverse rafter is supported by longitudinal beams. The transverse stability of the building is assured by the rigidity of the main frames. The end transverse walls have no vertical braces. The longitudinal stability of the building is assured by bracings located in the walls and roof in several bays. The mezzanine is also stabilized by a longitudinal bracing system. Due to its length, an expansion joint is used at the building's mid-length and this separates the structure into two independent structures (A1 and A2). Figures 9.31 to 9.34 show a general view of the structure, as well as the transverse and longitudinal frames.

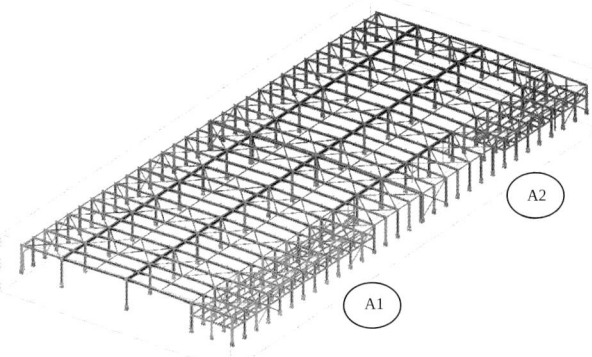

Figure 9.31 – General view with the production unit

9. Case Studies

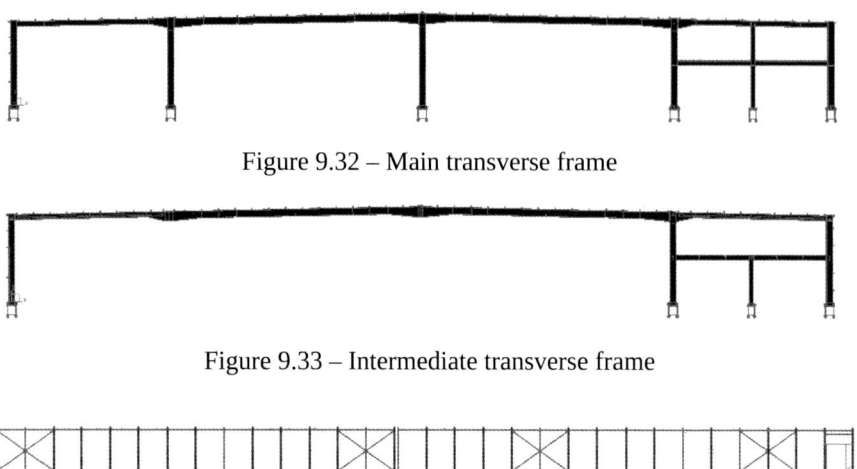

Figure 9.32 – Main transverse frame

Figure 9.33 – Intermediate transverse frame

Figure 9.34 – Longitudinal frame

Warehouse unit

The primary framing system consists of portal frames with parallel flange columns and partially tapered rafters (Figure 9.35). The columns have fixed bases and rafters are rigidly connected to the columns.

The transverse stability of the building is assured by the rigidity of the main frames (Figures 9.36 and 9.37). The end walls have no vertical braces. The longitudinal stability of the building (Figure 9.38) is assured by bracings located in the roof and walls in several bays. In order to increase the clear space at the interior, every second transversal rafter was supported by longitudinal beams.

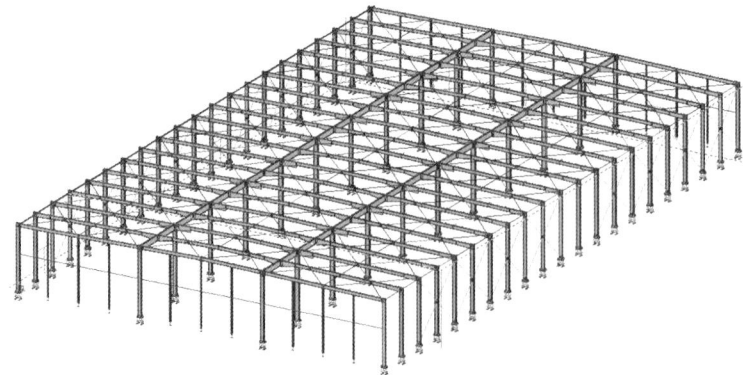

Figure 9.35 – General view with the warehouse unit

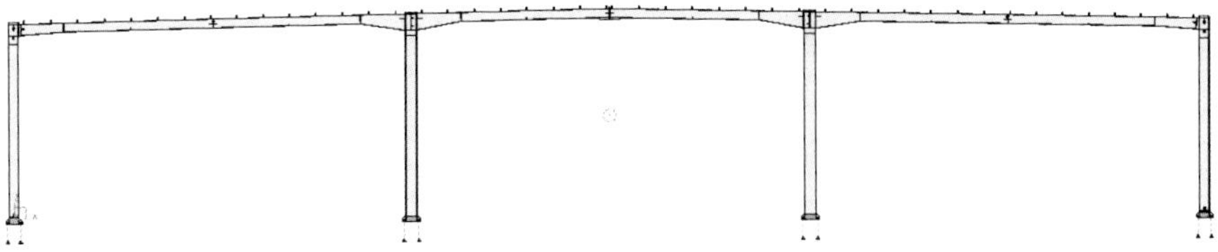

Figure 9.36 – Main transverse frame

Figure 9.37 – Intermediate transverse frame

Figure 9.38 – Marginal longitudinal frame

9.3.2 Design considerations

9.3.2.1 Design concept

Single storey industrial buildings are generally designed to withstand gravity loads and wind forces. However, many European countries are located in seismic regions and therefore the response to earthquakes needs also to be evaluated.

Seismic resistant steel structures can be designed according to concept a), low dissipative structural behaviour (see sub-chapter 3.1), whereby no account is taken of any hysteretic energy dissipation but the q factor may be larger than 1.0, to account for overstrength. For a steel structure, the upper value of the q factor may be taken as 1.5. Alternatively, a steel structure can be designed according to concept b), dissipative structural behaviour (see sub-chapter 3.1), whereby the behaviour factor q is larger than 2 and depends on the typology of the primary resisting system (see sub-chapter 3.3.).

As discussed in sub-chapter 3.1, designing single-storey large span buildings as dissipative usually result in largely uneconomic structures because of the overstrength requirements for the non-dissipative members that lead to very expensive detailing. Therefore, for low-rise buildings with a reduced weight, it is important to take a decision as early in the design process as possible about the advantages of using the dissipative concept. Single storey portal frames are generally characterized by reduced inertial masses (weight of the building) and therefore the inertial forces that are generated are also low. This can suggest that, for such structural systems, the ductility demand is reduced and therefore the benefit from using a high structural ductility class (and large reduction factors q) is not justified. Such structures, or at least the elements of the portal frames (beams, columns), are generally designed from other conditions, like variable or climatic loads, e.g. snow.

Fulop and Dubina (2007) developed a methodology that indicates the optimal selection of the seismic design concept for single-span portal frames with pinned base connection. Based on simplified calculations, design charts were proposed for the evaluation of the relative importance, for the ULS design, of horizontal and vertical loads. The charts can be used, in the pre-design stage, to evaluate the importance of earthquake loads, compared to other load cases (e.g. self-weight, snow) and can guide the designer in his/her decision to propose a low-dissipative (Concept a) or dissipative (Concept b) portal frame. Such choice has important consequences on the compactness of the cross sections, on the acceptable slenderness and required detailing; and ultimately on the cost of the frame. If dissipative concept has to be used in the design, the required reduction factor (q) is easily evaluated from the charts. Hence, it is possible to calibrate the design process (i.e. choose cross sections and details) so as to achieve the required/target value of the q factor.

In this case, for the production building A, despite being located in a high seismicity region, concept a) was chosen for design. A q factor of 1.5 was chosen and the structure was designed in accordance with the low-dissipative structural behaviour concept (DCL). This value of q factor accounts for overstrength and does not require the fulfillment of the requirements for dissipative structures, i.e. compact cross section for the elements or parts where inelastic deformations are expected, overstrength requirements for non-dissipative members, i.e. columns, overstrength requirements for connections located near dissipative zones and requirements for concentric bracings. Therefore, the structure was designed considering the design rules from EN 1993-1-1 for members and EN 1993-1-8 for connections. It should be noted that class 3 cross sections were adopted for the main structural elements. In addition, non-dissipative members (i.e. columns) and connections were designed to the maximum internal forces from the seismic and non-seismic combinations, without considering the overstrength.

9.3.2.2 Actions and load combinations

Permanent loads (G)

The permanent load was evaluated separately for the roof and for the walls. The permanent loads on the floors of the mezzanine structure are also presented. The self-weight of main structural system (beams, columns, braces) was taken into account automatically by the design software.

Roof

Purlins	0.03 kN/m²
Sheeting + accessories	0.12 kN/m²
Insulation	0.05 kN/m²
Services	0.15 kN/m²
Total	0.35 kN/m²

Walls

Rails	0.03 kN/m²
Sheeting + accessories	0.12 kN/m²
Insulation	0.05 kN/m²
Total	0.20 kN/m²

Mezzanine floor

Reinforced concrete slab	4.5 kN/m²
Services	0.3 kN/m²
Ceiling	0.2 kN/m²
Total	5.0 kN/m²

Live load (L)

The imposed load includes the weight of the structure's occupants and contents and is equal to 3 kN/m².

Snow load (S) (according to EN 1991-1-3)

Due to the difference in height of 3.6 meters between building A and building B, the snow may drift and accumulate so that different snow load arrangements need to be considered.

9. CASE STUDIES

Snow load on the roof for the persistent design situations, s:

$$s = \mu_i C_e C_t s_k = 1.60 \text{ kN/m}^2$$

where:

- characteristic snow load value on the ground, s_k, for Bucharest city: $s_k = 2.0$ kN/m²

- exposure coefficient, C_e (Table 5.1): $C_e = 1.0$ (normal topography)

- thermal coefficient, C_t (section 5.2 (8)): $C_t = 1.0$ (usual insulation)

- snow load shape coefficient: $\mu_i = \mu_1$; $\mu_1 = 0.8$ (undrifted snow load, according to Table 5.2), see Figure 9.39

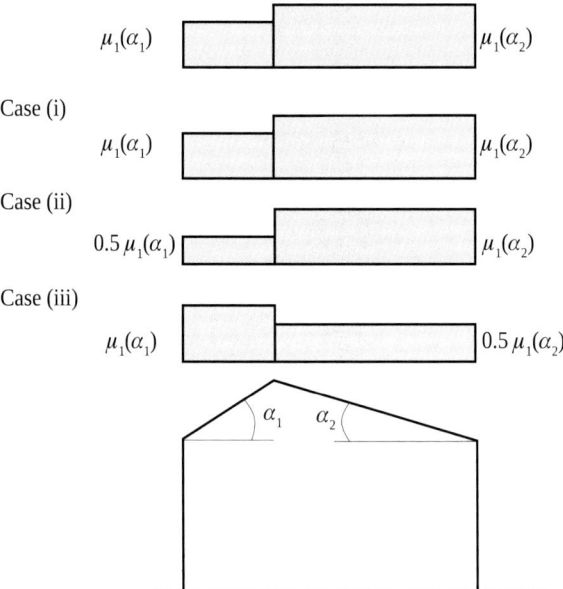

Figure 9.39 – Snow load shape coefficients – pitched roofs, for undrifted load arrangement

$\mu_i = \mu_2$; $\mu_2 = \mu_s + \mu_w$ (roofs abutting to taller construction work, according to section 5.3.6), see Figure 9.40.

- snow load shape coefficient due to sliding of snow from the upper roof: $\mu_s = 0$ for $\alpha = 0° < 15°$

- snow load shape coefficient due to wind:

$$\mu_w = (b_1 + b_2)/2h < \gamma h/s_{k,o};$$

$b_1 = 120$ m; $b_2 = 181$ m; $h = 3.6$ m;

γ is the weight density of snow, which for this calculation may be taken as 2 kN/m³

→ $(b_1 + b_2)/2h = (120 + 181)/7.2 = 41.6$

$\gamma h/s_{k,o} = 2 \times 3.6/2 = 3.6$

→ $\mu_w = 3.6$

The drift length is taken as $l_s = 2h$ (limited to minimum 5 m and maximum 15 m) ⇒ $l_s = 7.2$ m.

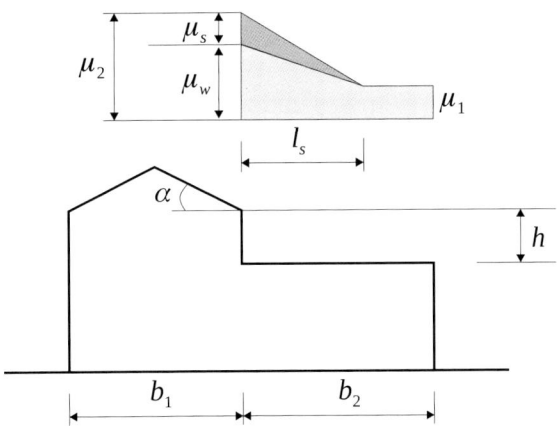

Figure 9.40 – Snow load shape coefficients - roofs abutting to taller construction, for drifted load arrangement

<u>Wind load (W) (according to EN 1991-1-4)</u>
Wind has been considered both in the transversel (WX) and longitudinal (WY) directions.

9. Case Studies

Wind pressure at height z above the ground, on external or internal surfaces of the structure:

$$w(z) = q_p(z_e) \cdot c_{pe} \tag{9.4}$$

where:
- $q_p(z_e)$ is the peak velocity pressure;
- z_e is the reference height for the external pressure;
- c_{pe} is the pressure coefficient for the external pressure.

For the specific location of the building, the peak velocity $q_p(z)$ at height z, can be evaluated as follows:

$$q_p(z) = c_e(z) \cdot q_b \tag{9.5}$$

where:
- $c_e(z)$ is the exposure factor;
- q_b is the basic velocity pressure.

Exposure factor $c_e(z)$ is a function of height above terrain and a function of terrain category. For the specific location of the building, i.e. outside of the city, low vegetation and no obstacles, the site can be classified as terrain category II. This category contains areas with low vegetation such as grass and isolated obstacles (trees, buildings) with separations of at least 20 obstacle heights. Figure 9.41 shows the variation of the exposure factor $c_e(z)$ as a function of height above terrain. For a height of 8.2 m, the exposure factor c_e is 2.2.

According to the Romanian National Annex, the value of q_b for the building site is:
$q_b = 0.5$ kN/m²

The external pressure coefficient c_{pe} for buildings depends on the size of the loaded area, A. EN 1991-1-4 gives two values: $c_{pe,1}$, intended for tributary areas less than 1 m², i.e. local cladding design; and $c_{pe,10}$ intended for major structural members. The overall pressure coefficient $c_{pe,10}$ depends on the position of the surface relative to the wind, see Figure 9.42. For windward and leeward surfaces (zones D and E), a single value is used, and it depends on the ratio between h and d (Table 9.9). For sidewalls, different values are used (zones A, B and C), depending on the point on the surface, excepting the case where the value of the edge distance e is larger than or equal to 5d, when a single value is used. In this case, for both cross-wind and along-wind, the separation is done in three zones, namely A, B and C because the edge distance e, which is the minimum from b and 2h amounts to 16.4 m, and therefore $e < d$. Special attention should be given to side-walls, where extreme local cladding pressures may be experienced.

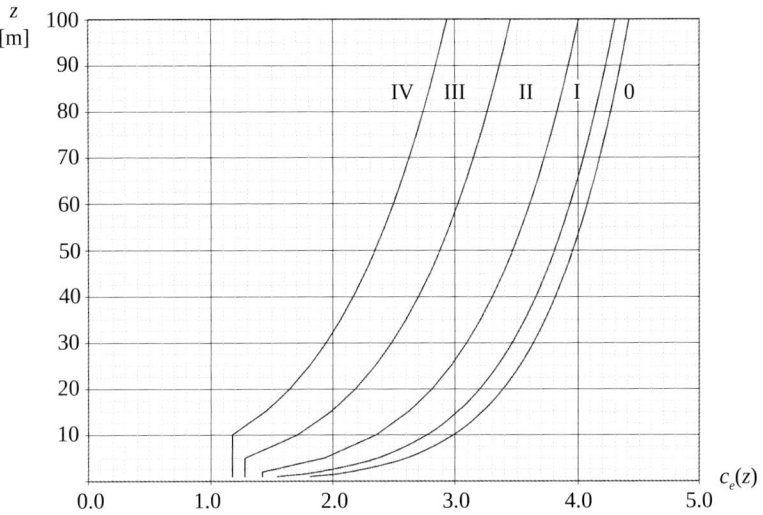

Figure 9.41 – Variation of exposure factor $c_e(z)$ for different terrain categories and flat terrain (orography coefficient $c_o = 1.0$, turbulence factor $k_l = 1.0$)

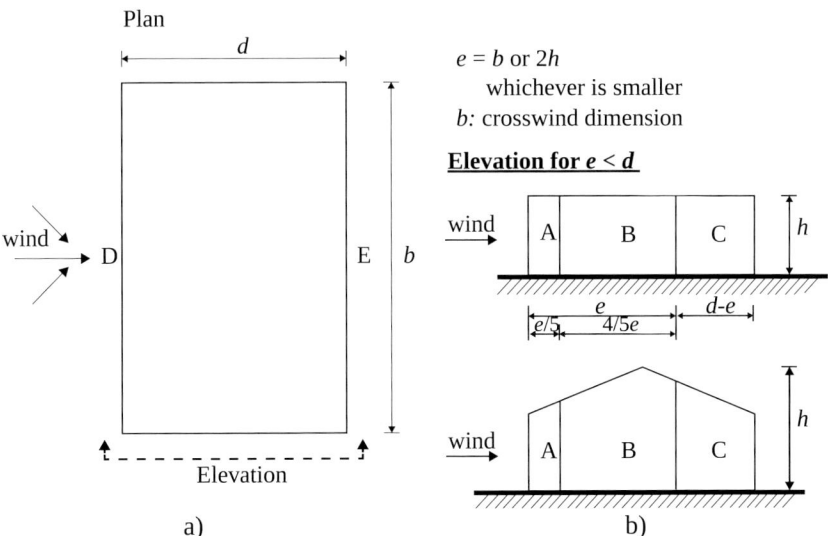

Figure 9.42 – Zones for vertical walls: a) plan view; b) elevation for $e < d$; elevation for $e \geq d$; elevation for $e \geq 5d$

9. Case Studies

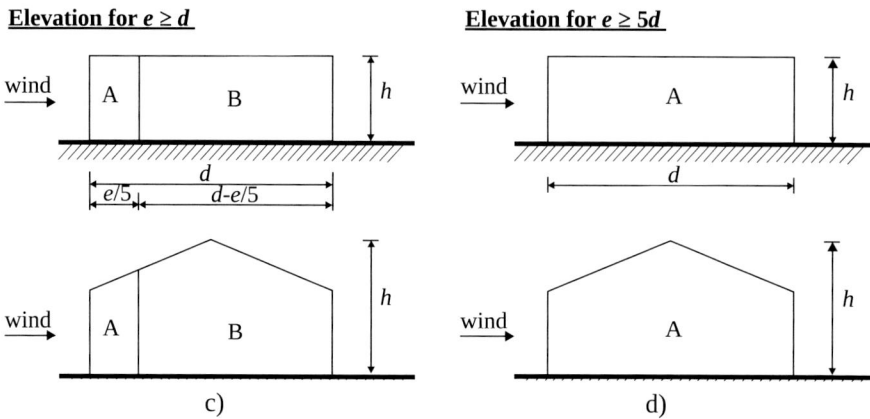

Figure 9.42 – Zones for vertical walls: a) plan view; b) elevation for $e < d$; elevation for $e \geq d$; elevation for $e \geq 5d$ (continuation)

Table 9.9 – Recommended values of external pressure coefficients for vertical walls of rectangular plan buildings

Zone	A		B		C		D		E	
h/d	$c_{pe,10}$	$c_{pe,1}$	$c_{pe,10}$	$c_{pe,1}$	$c_{pe,10}$	$c_{pe,1}$	$c_{pe,10}$	$c_{pe,1}$	$c_{pe,10}$	$c_{pe,1}$
5	−1.2	−1.4	−0.8	−1.1	−0.5		+0.8	+1.0	−0.7	
1	−1.2	−1.4	−0.8	−1.1	−0.5		+0.8	+1.0	−0.5	
≤ 0.25	−1.2	−1.4	−0.8	−1.1	−0.5		+0.8	+1.0	−0.3	

For vertical walls of rectangular plan buildings, the surface of the windward wall (zone D, see Figure 9.42) is split in different strips as a function of the height above terrain and for each strip the pressure distribution is assumed to be uniform. For the building under consideration (e.g. plan dimensions of 180 m × 78 m and height of 8.2 m), there is a single strip ($h = 8.2$ m $< L, H$), see Figure 9.43. The rules for the velocity pressure distribution for leeward wall and sidewalls (zones A, B, C and E, see Figure 9.42) may be given in the National Annex or be defined for the individual project. The recommended procedure is to take the reference height z_e as the height of the building, which means only one strip is defined.

9.3 SINGLE STOREY INDUSTRIAL WAREHOUSE IN BUCHAREST

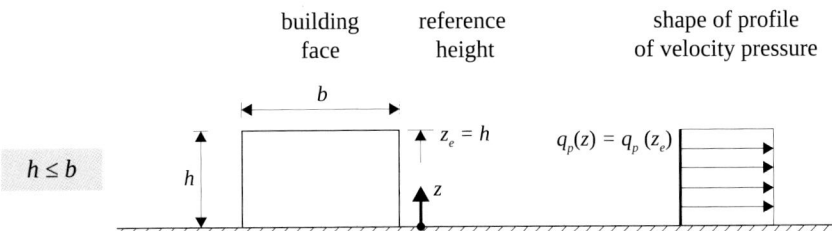

Figure 9.43 – The velocity pressure should be assumed to be uniform over each horizontal strip considered

The external pressure coefficients for roofs also vary with the point on the surface. When the slope of the roof, α, ranges between $-5° < \alpha < 5°$, the roof is considered flat and should be divided into zones, as shown in Figure 9.44. Table 9.10 shows the pressure coefficients for each zone and sharp eaves. It can be observed that roof corners (zone F) may experience very large pressures.

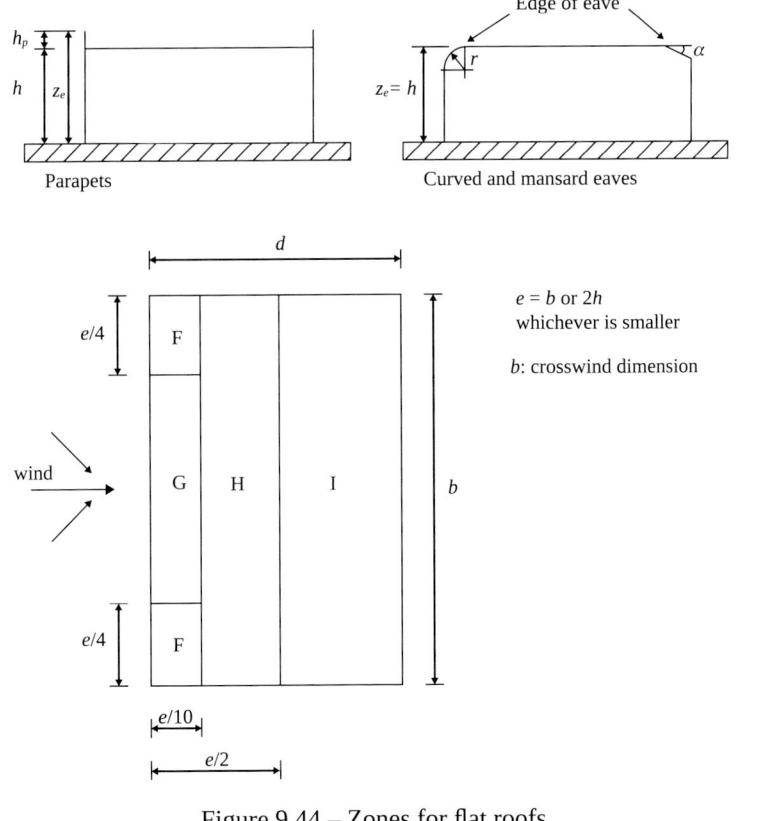

Figure 9.44 – Zones for flat roofs

9. Case Studies

Table 9.10 – External pressure coefficients for flat roofs and sharp eaves

Roof type	Zone							
	F		G		H		I	
	$c_{pe,10}$	$c_{pe,1}$	$c_{pe,10}$	$c_{pe,1}$	$c_{pe,10}$	$c_{pe,1}$	$c_{pe,10}$	$c_{pe,1}$
Sharp eaves	−1.8	−2.5	−1.2	−2.0	−0.7	−1.2	+0.2 / −0.2	

For roof sidewalls, different values are used (zones A, B and C), depending on the point on the surface, excepting the case where the value of the edge distance e is larger than or equal to 5d, when a single value is used. In our case, for both cross-wind and along-wind, the separation is done in three zones, namely A, B and C because the edge distance e, which is the minimum from b and 2h amounts to 16.4 m, and therefore e < d. Special attention should be given to side-walls, where extreme local cladding pressures may be experienced.

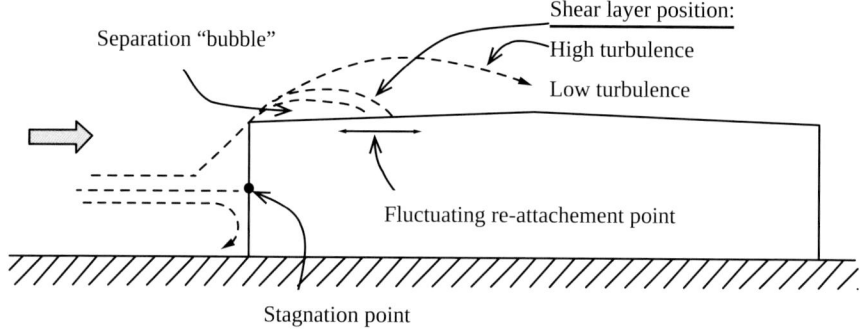

Figure 9.45 – Wind flow around a low-rise building

Seismic design load

This industrial building was designed according to the new Romanian seismic design code, P100/1-2013, which differs from the previous version issued in 2006 in several aspects. Regarding the seismic action, the reference acceleration $a_{g,R}$ increased from $0.24g$ to $0.30g$ for the ultimate limit state (10 % probability in 50 years event). The EC8-1 Type 1 spectrum constructed for soil type D was used for the design. The characteristic periods of the design spectrum are those indicated in Table

9.3 SINGLE STOREY INDUSTRIAL WAREHOUSE IN BUCHAREST

9.11. The building was classified as class II of importance, leading to an importance factor of $\gamma_I = 1.0$. Figure 9.46 shows the shape of the Type 1 spectrum, normalized by a_g, for 5 % damping.

Table 9.11 – Values of the parameters describing the Type 1 elastic response spectrum

Soil Type	S	T_B (s)	T_C (s)	T_D (s)
D	1.35	0.20	0.80	2.0

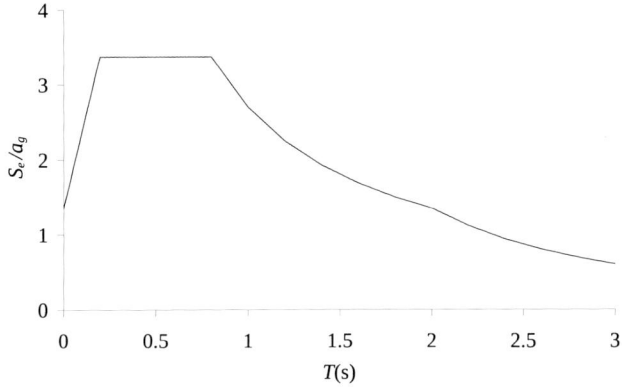

Figure 9.46 – Type 1 elastic response spectrum for ground types D (5 % damping)

Load combinations (according to EN 1990)

Partial safety factors are considered as follows:
- $\gamma_{M0} = 1.0$
- $\gamma_{M1} = 1.0$ (1.1 for seismic combinations)
- $\gamma_G = 1.35$ (permanent loads)
- $\gamma_Q = 1.50$ (imposed loads)
- $\psi_0 = 0.70$ (live, snow)
- $\psi_0 = 0.60$ (wind)
- $\psi_1 = 0.50$ (live)
- $\psi_1 = 0.20$ (wind)
- $\psi_2 = 0.30$ (live)
- $\psi_2 = 0.40$ (snow)
- $\varphi = 1.0$ - used in the computation of the modal masses

9. CASE STUDIES

ULS combinations
→ Fundamental combinations (persistent or transient design situations):

$$1.35\sum_{j=1}^{n} G_{k,j} + 1.5 Q_{k,1} + \sum_{i=2}^{m} 1.5 \psi_{0,i} Q_{k,i}$$

The following combinations have been considered:
C1:   1.35D + 1.5S
C2:   1.35D + 1.5WT
C3:   1.35D + 1.5WL
C4:   1.35D + 1.5S + 1.05WT
C5:   1.35D + 1.5S + 1.05WL
C6:   1.35D + 1.05S + 1.5WT
C7:   1.35D + 1.05S + 1.5WL
C8:   1.35D + 1.5L1
C9:   1.35D + 1.5L2
C10:   1.35D + 1.5L3
C11:   1.35D + 1.5L1 + 1.05S
C12:   1.35D + 1.5L1 + 1.05S + 1.05WT

→ Seismic design situations

$$\sum_{j=1}^{n} G_{k,j} + \gamma_1 A_{E,k} + \sum_{i=2}^{m} \psi_{2,i} Q_{k,i}$$

C13:   1.0D + 0.4S + E + 0.4L1

The inertial effects of the design seismic action were evaluated by taking into account the presence of the masses associated with all gravity loads appearing in the following combination of actions:

$$\sum_{j=1}^{n} G_{k,j} + \sum_{i=2}^{m} \psi_{E,i} Q_{k,i}$$

The combination coefficients ψ_{Ei} introduced in the previous formula for the calculation of the effects of the seismic actions were computed from the following expression:

$$\psi_{E,i} = \varphi \cdot \psi_{2,i}$$

SLS combinations:

→ Non-seismic combinations

$$\sum_{j=1}^{n} G_{k,j} + Q_{k,1} + \sum_{i=2}^{m} \psi_{0,i} Q_{k,i}$$

→ Seismic combination

$$\sum_{j=1}^{n} G_{k,j} + \gamma_I \cdot q \cdot \nu \cdot A_{E,k} + \sum_{i=2}^{m} \psi_{2,i} Q_{k,i}$$

ν is the reduction factor which takes into account the lower return period of the seismic action associated with the damage limitation requirement. Different values of ν may be defined for the various seismic zones of a country, depending on the seismic hazard conditions and on the protection of property objective. The recommended values of ν are 0.4 for importance classes III and IV and ν = 0.5 for importance classes I and II. In the present example, the value of ν is 0.5.

9.3.2.3 Global analysis

The analysis was done on two distinct 3D structural models for each of the two structures, i.e. A1 and A2, using SAP2000. Only building A1, located near the taller building, is presented.

Due to the reduced sensitivity of the structure to 2nd order effects (α_{cr} = 11.7 > 10), a first order elastic analysis with the explicit consideration of the global imperfections was performed according to EN 1993-1-1. It is noted that, under gravity load conditions, only the beams and the columns should be considered to resist such loads, without taking into account the bracing members.

Seismic action effects were evaluated using a modal response spectrum analysis and taking into account all modes that contribute significantly to the response. If the analysis is elastic, only the tension diagonals are taken into account in the structural model. The requirements regarding the number of modes to be considered in the analysis may be deemed to be satisfied if either of the following is fulfilled:

- the sum of the effective modal masses for the modes taken into account amounts to at least 90 % of the total mass of the structure;

9. Case Studies

- all modes with effective modal masses greater than 5 % of the total mass are taken into account.

In this example, the first criterion was verified. Because a spatial model was used, the condition was verified for each of the two main orthogonal horizontal directions. Table 9.12 shows the modal participating mass factors. It can be observed that the sum of the effective modal masses amounts to 98.2 % for the X direction (transverse) and 99.3 % for the Y direction (longitudinal). The first mode of vibration is translational along the Y direction with a period of 0.66 sec.

Table 9.12 – Modal participating mass ratio, structure A1

Mode	T_i (s)	$\Sigma(M_{i,X})$ (-)	$\Sigma(M_{i,Y})$ (-)
1	0.66	0.003	0.471
2	0.51	0.056	0.472
3	0.49	0.142	0.472
4	0.46	0.591	0.472
5	0.41	0.660	0.474
6	0.36	0.746	0.523
7	0.33	0.784	0.524
8	0.28	0.786	0.785
9	0.25	0.786	0.790
10	0.20	0.788	0.993
11	0.16	0.791	0.993
12	0.14	0.982	0.993

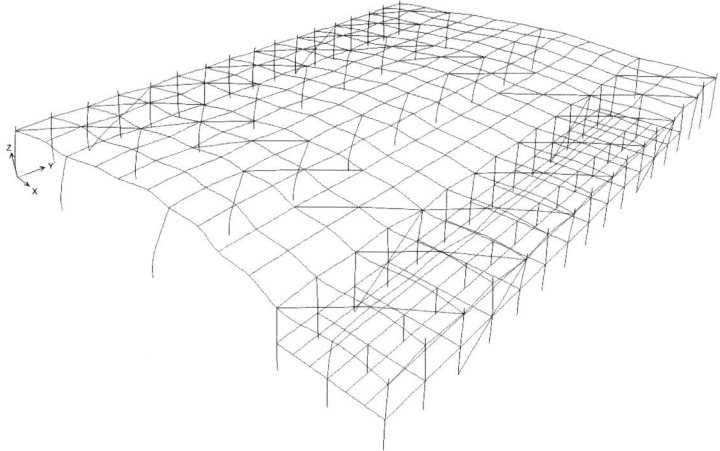

Figure 9.47 – First mode of vibration

9.3.2.4 Design verifications

The design for ultimate limit state and serviceability limit state, seismic design situation included, were done according to EN 1993-1-1 and EN 1993-1-8. Since the structure is designed using a non-dissipative concept, the additional rules from EC8-1 do not apply. These verifications are not shown here, but for more details can be found in the design manual about EC3-1-1 (Simões da Silva *et al*, 2016). It is further noted that combination 13 (seismic combination) was only critical for the design of the braces.

9.4 THE FIRE STATION OF NAPLES

9.4.1 General description

The erection of the new centre of the Fire Brigade of Naples started during the 1980's and was completed during the 1990's (see Figure 9.48). The Centre is composed of eight buildings with the following functions:

- Building A: garage, offices, lodgings and canteen for fire brigade;
- Building B: garage, storehouse, guesthouse, training;
- Building C: workshop;

9. CASE STUDIES

- Building D: scaffolding;
- Building E: heating and water station;
- Building F: guard house;
- Building G: head-quarters, offices and lodgings;
- Building H: gymnasium.

Figure 9.48 – The Fire Station of Naples

The preliminary design of the new Fire Station centre was developed in 1979-1980 and the structural typologies of all the buildings were selected in order to satisfy both functional and operational needs.

In particular, for building A (see Figure 9.49), the main functional requirement was identified as having wide covered spaces at the ground level completely free from columns in order to guarantee ease of circulation of the fireman trucks. Hence, a suspended structural scheme was chosen. The integration of this choice with the basic principle for exploiting the properties of the structural materials led to a composite system, where:

- reinforced concrete (r.c.) is used for vertical cores;
- steel is used for the members (beams and ties) of the suspended skeleton.

Only two months after the delivery of the general structural design of the Centre, the earthquake of 23 November 1980 caused a temporary stop in the design process. After this earthquake, the city of Naples was classified as seismic area for the first time, and a peak ground acceleration equal to $0.15g$ was assumed for the design earthquake (i.e. the seismic action having 10 % exceedance probability in 50 years, namely a mean return period equal to 475 years) at ultimate limit state (ULS). At the beginning of 1983, when the immediate emergency was over, the Public Works Department of Campania decided to continue the project of the new Fire Station, advancing to the phase of the detailed design and then to fabrication and erection. Hence, it was necessary to revise and improve the preliminary project in term of seismic resistance. The structural schemes of the seven steel buildings were characterized by different typologies of seismic resistant structures.

The need for building A to adapt the previous selected scheme of suspended structure to the new seismic requirements was faced and special damping devices were introduced, in order to seismically isolate the suspended steel skeleton from both horizontal and vertical movements at the top of the RC cores. As mentioned in sub-chapter 9.1, this application of seismic isolation was pioneering for that time, being the first in Europe, after very limited experience in the rest of the world.

The construction of building A started in July 1983 and its structure were completed in 1985 (Figure 9.49).

Altogether, the seven steel buildings of the Fire Station of Naples represent a catalogue of seismic resistant structural solutions, going from the most sophisticated passive control systems (Buildings A and B), to simple concentric braced frames (Buildings G and D), moment resisting frames (Buildings F and H), cantilever columns (Building C), resulting in very high earthquake protection in all cases.

Building B is based on the same structural mesh, 18 m × 18 m as Building A, but it uses an "all-steel" solution, with steel cores and Vierendeel type large span beams (Figure 9.50).

Figure 9.49 – The building A of the Fire Station of Naples under construction

9. Case Studies

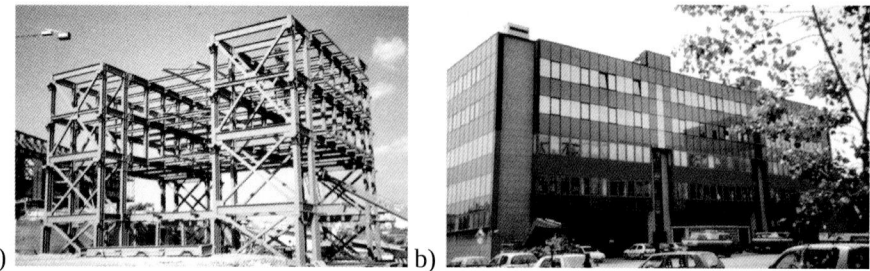

Figure 9.50 – The steel skeleton of building B (a) and its final configuration (b)

The cores resist as moment resisting frames in the longitudinal direction and as concentric braced frames in the transverse direction. Cores and floor structures are connected together by means of shock block transmitters (see Figure 9.51), which are special oleo-dynamic devices allowing free deformations without opposing resistance in serviceability conditions (i.e. slow movements of the structures as those produced by thermal changes), but creating a rigid link in case of dynamic displacements as those induced by earthquakes.

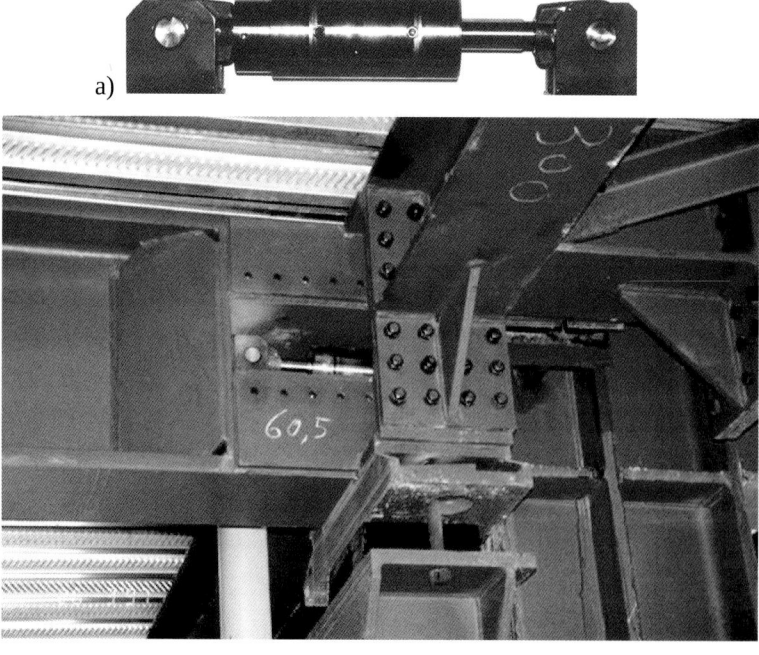

Figure 9.51 – a) The shock block transmitter; b) the connection between the beam end to the core column

In the latter case, the change of the structural scheme absorbs 83 % of the seismic input energy and limits the local damage to 17 % only. Nowadays, shock block transmitters are patented devices, which currently require qualification tests according to the European standard for anti-seismic devices (i.e. EN 15129 (CEN, 2009)). At the time of the project of the examined building no specific codes for testing seismic devices were available in Europe. Therefore, ad-hoc characterization tests were required to verify the response of the devices by imposing first a monotonic quasi-static increasing displacement, followed by an impulsive imposed displacement.

Building C is a simple single-storey industrial building with a partial mezzanine inside (Figure 9.52a). The main roof structure is made of single span steel trusses. The horizontal forces are resisted by columns in a cantilever configuration (Figure 9.52b).

a)   b)

Figure 9.52 – The structural scheme of building C (a) and its main entrance (b)

Building D, the Scaffolding, is a typical facility in all fire stations, where the firemen make their training exercises (Figure 9.53). The structural scheme is a concentric brace frame in both directions. The main façade is traditionally covered by a timber cladding.

Building F is the Guard House, whose structure is a simple two level moment resisting frame (Figure 9.54).

Building G hosts the Headquarters and its offices. The structural scheme is composed of concentrically braced frames for earthquake resistance in both directions, integrated by pinned frames. The perimetric braces are visible in the main façade, characterizing the architectural feature of the building (Figure 9.55). It is connected to building H by a complex stair-case system (Figure 9.56).

9. Case Studies

Figure 9.53 – The building D: the Scaffolding

a)   b)

Figure 9.54 – The Guard House at the entrance of the Centre

Figure 9.55 – The façade of the Head Quarter

9.4 THE FIRE STATION OF NAPLES

Figure 9.56 – The stair-case system connecting Buildings G and H

Building H contains the Gymnasium and its facilities, whose main structure is a portal frame with square box columns and tapered beams (Figure 9.57).

Figure 9.57 – Structure of the Gymnasium

The following sections are devoted to Building A, which has been selected for representing the characteristics of the Fire Station of Naples,

9. Case Studies

because it can be considered the most significant from the structural point of view, namely as an example of suitable seismic solution for a building playing a fundamental role for civil protection and emergency management during a catastrophic event.

9.4.2 Design considerations and constructional details

9.4.2.1 Design criteria

The Building A of the Fire Station of Naples is mainly formed by a steel skeleton suspended from a top grid, which is supported by reinforced concrete cores containing stairs and elevators. This choice was basically inspired by functional reasons, which required the ground floor to be completely free from columns in order to easily allow for the operational activity of the fireman trucks. From seismic point of view, such a building typology can be prone to significant demand in terms of displacements and accelerations, which can lead to unfeasible selection of members and details if conventional structural schemes have been adopted. Therefore, it was decided to decouple the dynamic behaviour for horizontal oscillations of the horizontal suspended steel structure from the vertical r.c. cores. This structural performance was possible by means of the introduction of special devices (see section 9.4.3) located both on the top of the r.c. towers and at each floor level (Mazzolani, 1986a,b,c), which were conceived to seismically isolate the whole steel suspended structure at the top of the r.c. cores. It should be noted that this design solution was pioneering at that time, but is still quite challenging today. Indeed, the current practice of seismic isolation is to use isolators at the base of the building (Naeim and Kelly, 1999), thus needing to account for mostly the displacements and the relevant effects of the overall superstructure. In the case of Building A it was necessary to properly characterize the interactions and the relative displacements among two macro sub-systems, namely the steel suspended structure and the r.c. towers. Of course the proper evaluation of the dynamic effects needed the use of refined models and adequate computational efforts that is nowadays easy affordable, but not at the beginning of 1980s'.

The idea of top seismic isolation had threefold objectives: 1) to allow free deformations and displacements due to the design loads at serviceability limit state; 2) to introduce supplementary damping elements to limit the displacements induced by the earthquake; 3) to introduce horizontal flexibility at the top of the vertical r.c. structure once the dissipative systems are activated by the earthquakes, thus lengthening the period of vibration. As it can be clearly recognized, increasing the fundamental period corresponds to reduce the seismic forces acting on the structure, but it can give rise to extremely large relative displacements and consequently P-Delta effects that can be unfeasible for the overall stability of the building. However, the dampers connected to the isolators were designed to control the relative displacements between the steel superstructure and r.c. cores. Of course, increasing the damping could further reduce the horizontal forces, thus giving an additional benefit because the building is less sensitive to the variability of ground motion features.

The design against vertical earthquake was another aspect accounted for. Indeed, although the introduction of horizontal deformability was crucial to isolate the building against horizontal accelerations, increasing the vertical deformability could not be acceptable. Therefore, the isolators were designed to guarantee adequate vertical rigidity and strength to resist the forces induced by vertical earthquakes. Constructional details of the isolators are shown in section 9.4.3.

In light of the considerations above described, the r.c. cores and their foundations were design to resist the total seismic actions, while the steel suspended structure was designed to resist gravity loads and the vertical earthquakes (Mazzolani, 1985a,b,c).

9.4.2.2 Geometry and typological features

The building A is composed of four upper floors and a ground floor completely free from columns (Figure 9.58). Its plan has an extended rectangular shape 26 m wide, which is longitudinally subdivided by transversal modules of 3 m (Figure 9.59). The whole structure is shown in Figure 9.60 during the erection phase. The total weight of steelwork is about 1500 tons. The vertical bearing structures are the r.c. cores, which contain the vertical links (stairs, elevators, poles). The cores are transversally coupled and each couple is longitudinally spaced 18 m apart, thus forming square meshes of 18×18 m (see Figure 9.61).

9. CASE STUDIES

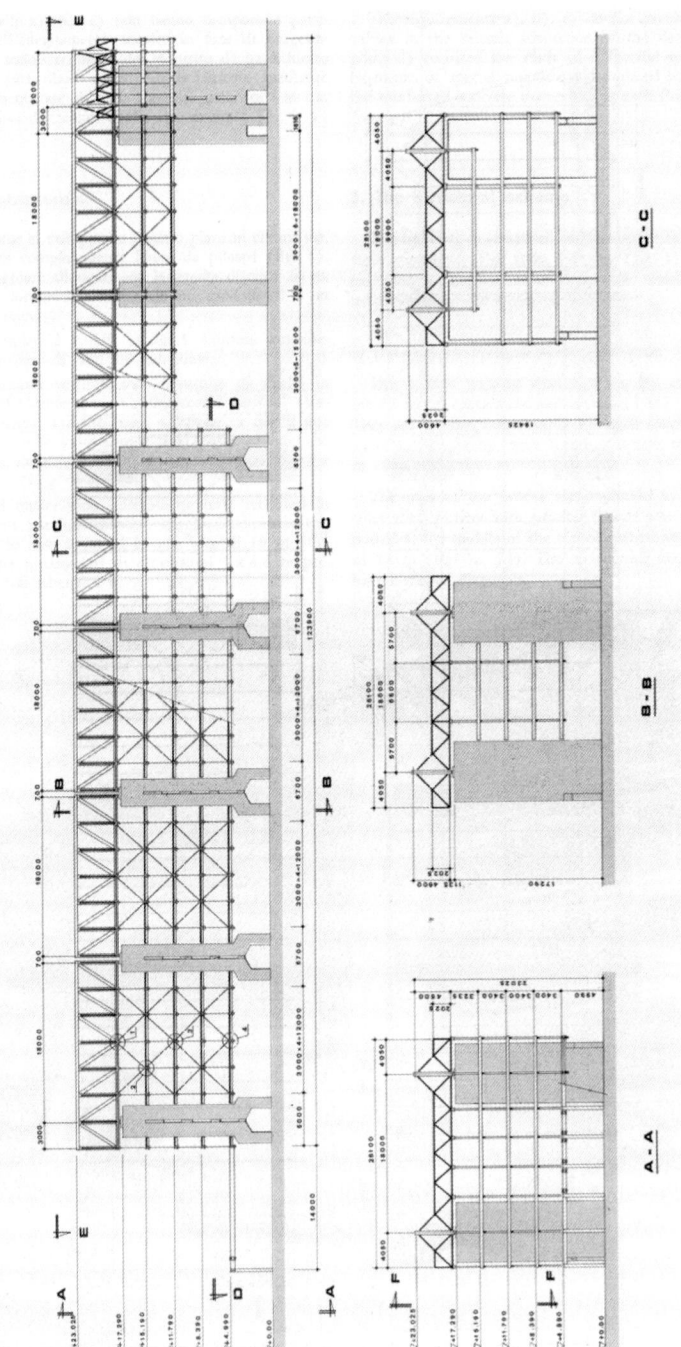

Figure 9.58 – Longitudinal and transverse sections of Building A

9.4 THE FIRE STATION OF NAPLES

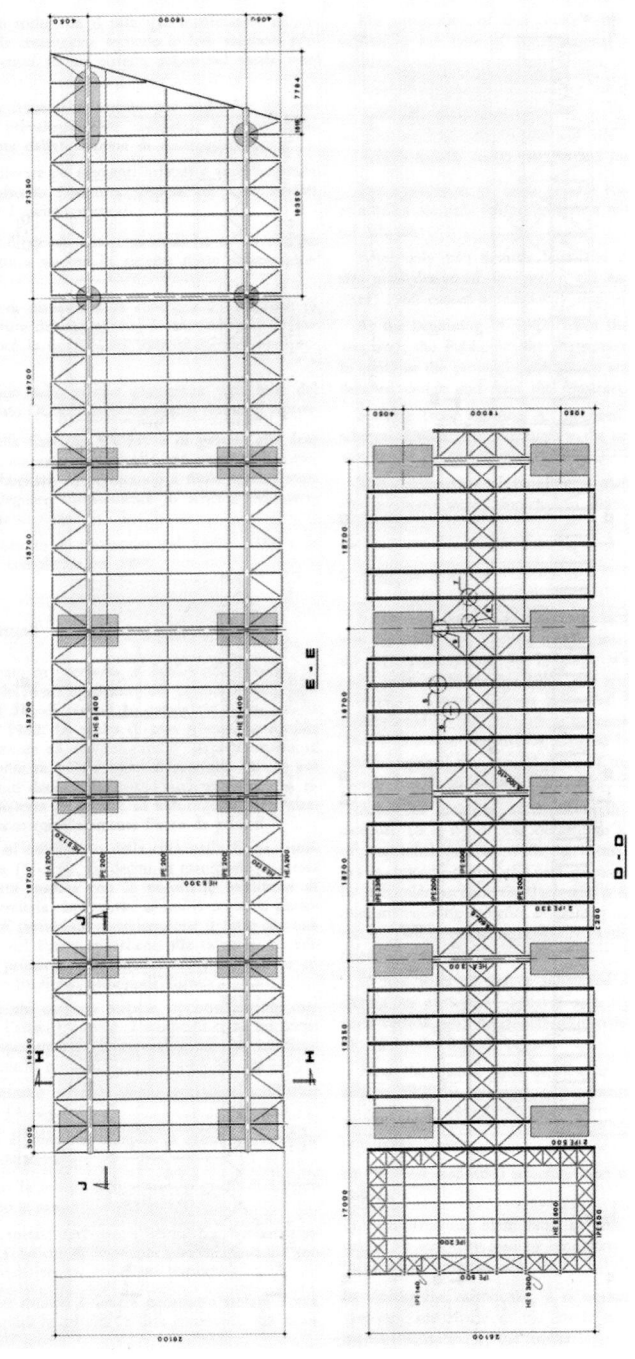

Figure 9.59 – Plan lay-out of Building A

9. CASE STUDIES

Figure 9.60 – The building A under constructions

Figure 9.61 – Erection phases of the top truss girders

9.4 THE FIRE STATION OF NAPLES

The tops of the cores are connected by longitudinal steel lattice girders with parallel chords, which are simply supported at the middle of the top of each tower and have a span of 18 m (see view "J-J" in Figure 9.62). The rectangular meshes of the trussed girders are 3 m × 4.6 m with k-shaped diagonals.

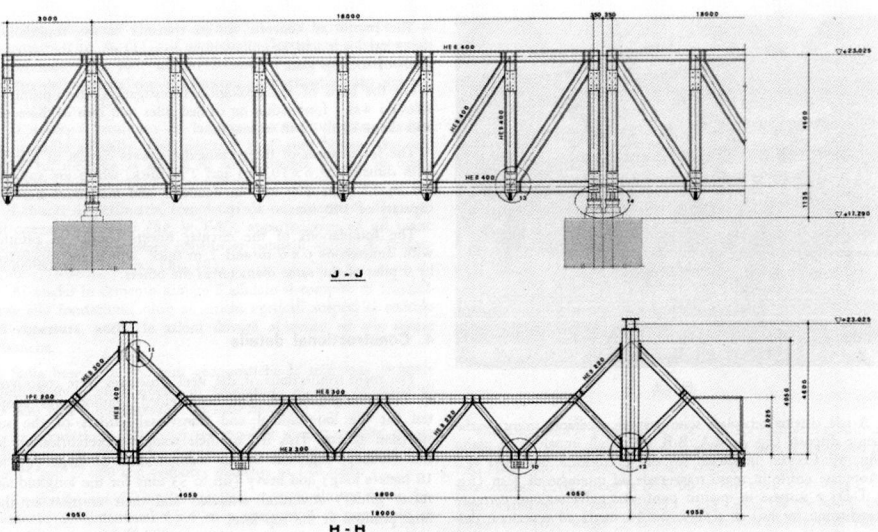

Figure 9.62 – Design sheets of longitudinal and transversal girders

The transverse top girders are spaced 3 m apart, being supported at each vertical member of the longitudinal girder. They also are trusses with a span of 18 m and lateral cantilevers of about 4 m (see view "H-H" in Figure 9.62). Their wall members have a V scheme and the parallel chords are shaped at the ends in order to accommodate their height of about 2 m to the one of the longitudinal girders that have height equal to 4.6 m (see Figures 9.62 and 9.63).

Longitudinal and transverse girders together form an orthogonal grid at the roof level, supported by the r.c. cores, with a horizontal mesh of 18 m × 18 m, which is integrated by horizontal bracings along the longitudinal perimeter (see view "E-E" in Figure 9.59). By means of ties in various positions (see views "A-A", "B-B" and "C-C" in Figure 9.58) the floor decks are suspended to the top grid system (see Figure 9.64). The main structure of each suspended floor is composed by couples of transverse IPE beams, 3 m spaced (see view "D-D" in Figure 9.59). The four suspension points at each floor identify three spans of 8.10-9.90-8.10 meters, with the exception

9. Case Studies

of the upper floor which is about 2 m × 4 m less wide than the perimeter. Each floor structure is integrated by horizontal bracings (see Figure 9.65), which are located in the middle of the plane along the longitudinal axis, with expansions near the cores (see view "D-D" in Figure 9.59).

Figure 9.63 – The connection between the top girders

Figure 9.64 – Floor decks sustained by steel ties to the top transversal trusses

9.4 THE FIRE STATION OF NAPLES

The floor structures are covered by trapezoidal steel sheets, connected by studs (Nelson type) to the upper flanges of the IPE beams, which act compositely with the concrete cast slab (see Figure 9.66).

The perimeter ties are integrated by bracing diagonals, which are located in a vertical external plane before the curtain-walls (see Figure 9.67).

Figure 9.65 – Floor bracings

Figure 9.66 – Steel trapezoidal sheeting for covering of the floor structures

9. CASE STUDIES

Figure 9.67 – The suspended skeleton composed of steel ties and bracing on the facade

The r.c. cores transmit to the foundation the wind loads and seismic actions in addition to the gravity loads of the suspended structure, which are transferred by the top grid. On the basis of the geological investigation, the selected solution was a foundation on drilled piles with 800 mm of diameter and about 18 m to 20 m of length.

The foundations of the rectangular cores are plinths with dimensions 6 m × 10.8 m and 2 m thick, which are supported by 15 piles of 800 mm and maximum bearing capacity of 140 tons each.

The foundations of the circular cores are footings with dimensions 6 m×6 m and 2 m thick, which are supported by 9 piles of the same diameter as the others.

9.4.2.3 Constructional details

Full strength connections were designed for the main structural components. In particular, butt-welded splices were adopted for the longitudinal and transverse girders of the suspension system. This solution required their complete prefabrication in the workshop, followed by the transportation of large (up to 18 meters long) and heavy (up to 33 tons for the longitudinal girders) structural elements and their erection in the final position at the top level.

The longitudinal girders (see view "J-J" in Figure 9.62) are completely built-up by welding. The unified cross section of the bars is composed of two HEB 400 profiles, which are coupled in order to obtain a box section. After the building-up in the workshop (Figure 9.68a), the longitudinal girders were transported to the yard (Figure 9.68b) and lifted on the core tops (Figures 9.68 c, d), where they were located on special supports (Figures 9.68 e, f). The transverse girders (see view "H-H" in Figure 9.62) are also fully welded by using HEB 300 profiles as chords and HEB 220 profiles as wall bars. Its central part with a length of 18 m and the two external cantilever parts were prefabricated in the workshop separately.

Figure 9.68 – Workshop fabrication (a), transportation (b), lifting on the top of the r.c. cores (c, d) and connection of the steel girder to special supports (e, f) (continuation)

9. Case Studies

The connections between transversal and longitudinal girders (Figure 9.69a) are made by means of end-plate joints with high strength steel bolts in calibrated holes in order to strictly limit the slip due to the bolt-hole clearance (Figure 9.69b).

Figures 9.70 and 9.71 show some additional details of the joints as they were built in the real structure.

Figure 9.69 – Connection of transversal to longitudinal girders: global view a) and detail b)

Figure 9.70 – The detail of the top suspension of ties

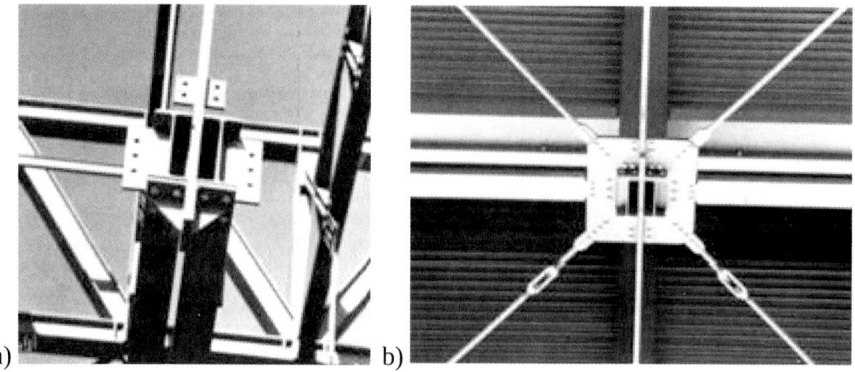

Figure 9.71 – Some details of the structural joints: a) intermediate and b) lateral node

9.4.3 The anti-seismic devices

The special devices are located at the top of the r.c. cores, where the reticular suspension grid is supported. As previously discussed in section 9.4.2.1, these special systems, were designed to provide twofold performance objectives, namely:

a) to allow free deformations and displacements under the design loads in serviceability conditions;
b) to dampen the displacements of the structure produced by the seismic actions, but to lengthen the period of vibration of the structure sufficiently to reduce the seismic forces.

The first performance objective is guaranteed by the support bearings, which provide fixed and mono- or multi-directional moving hinges by means of teflon-neopren sliders. The suspension structure is subdivided in plan into separate square meshes of 18 m × 18 m, with one fixed corner (spherical hinge) and three corners (cylindrical hinges) free to expand, according to the scheme of Figure 9.72.

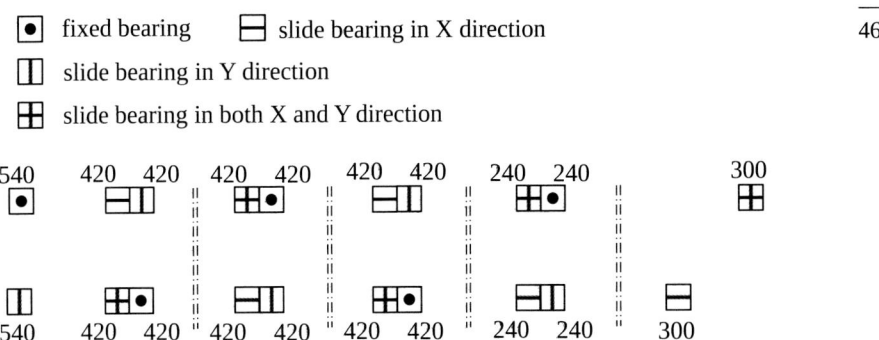

Figure 9.72 – Plan location of the top supports and maximum design loads (in tons)

The second performance objective complies with the following requirements (Mazzolani, 1986c, 1990):

- the anti-seismic device must solely operate during the earthquake, and it must behave as an internal restraint during the service conditions;

- the bearing must be fully resist the vertical earthquakes, because vertical accelerations can significantly impair the performance of suspended structures;
- the energy absorption is based on the yielding of dissipating elements, which are activated by horizontal actions;
- the dissipative elements must be simple, cheap and easy to be replaced;
- the support must allow the horizontal displacements, which are necessary to plastically deform the dissipative elements;
- the period of vibration of the structures after the yielding of the hysteretic devices increases in order to reduce the seismic action.

On the basis of the results of the collected experience at international level in that period, multi-layer sandwich made of steel plate and neoprene was used as soft support. Such a type of arrangement is intervenes when the stress level transmitted from the core to the suspension system exceeds the level due to the service actions. When the serviceability actions are exceeded, the relevant displacements are resisted by dissipative elements that are design to yield in bending and shear. The elastic rigidity of the dissipating elements must be higher than that of the neoprene support, so that these elements play the role of restraints during the behaviour of the structure under serviceability limit state. During this phase the structure behaves as a system of high frequency (i.e. small period of vibration) for small excitations. On the contrary, when the seismic excitation increases the dissipative elements in the devices start yielding, by playing the role of "hysteretic fuse". The frequency of the system decays (i.e. the period of vibration increases) and the device no longer amplifies the incoming acceleration, but acts as an isolator. The seismic performance under this behavioural phase was deeply investigated by means of nonlinear dynamic analyses that demonstrated the beneficial effects due both to lengthening of the period of vibration and large energy dissipated by the system, with significant reduction of the displacements at the supports.

The application of the above criteria led to the construction of the support device that is shown in Figure 9.73 in vertical section and in plan.

The upper part develops the function a) of a support at serviceability limit state. The neoprene is restrained into an external ring (1) and allows the rotations of the upper element, which is connected to the structure (2), with respect to the lower plate (3). This plate (3) can slip horizontally, thanks to the teflon layer below

9.4 THE FIRE STATION OF NAPLES

(4), which is counter faced with a stainless steel plate. The presence of special locks or rails (5) characterizes the support as a spherical or a cylindrical hinge.

The lower part of the device represents the seismic support, made of a multi-layer steel-neoprene sandwich (6), which is compressed during the vertical quake for both up and down movements by means of ties (7), which raise the lower plate (8).

Figure 9.74 illustrates the constructional details of two support devices coupled together at the intermediate joint between two adjacent structural meshes in the middle top of each core. Figure 9.75 shows the real view of a couple of support devices, as they can be observed from inside of the upper floor through a special window, which also can be used for maintenance operations.

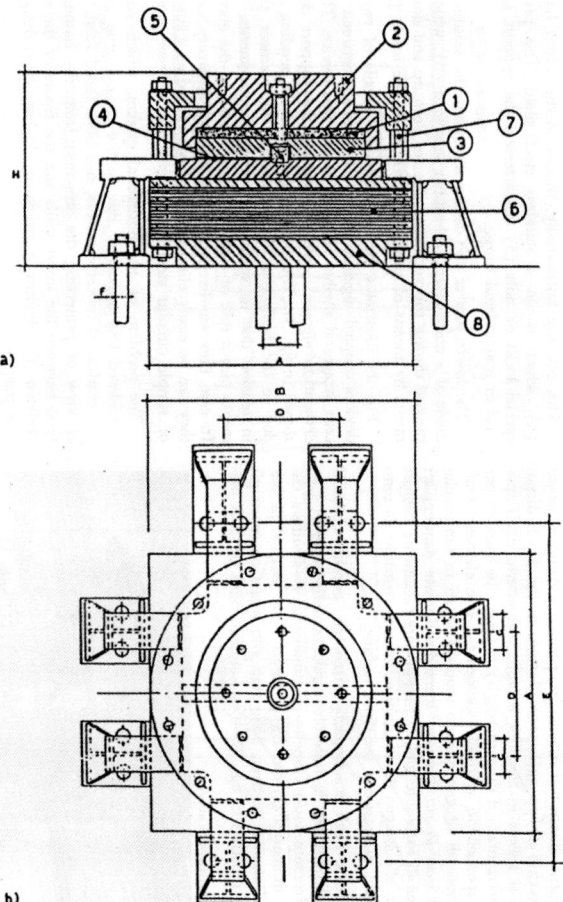

Figure 9.73 – (a) vertical section and (b) plan of the special support device

9. CASE STUDIES

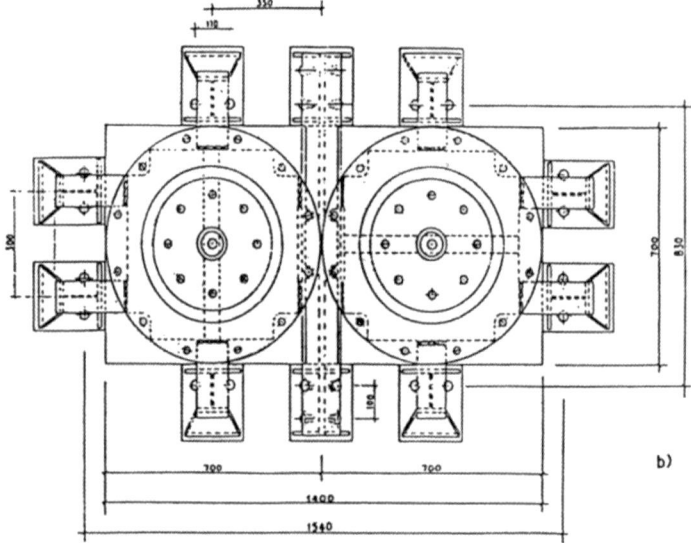

Figure 9.74 – Constructional details of the support device between two adjacent structural meshes: (a) vertical section; (b) plan

The dissipative elements, which complete the support device, are shown in Figure 9.76. The element (a) yields in bending and shear due to the horizontal displacements; the element (b) yields in tension due to vertical ascending displacements in case of vertical earthquakes. Both elements play a role of restraint when loaded in the elastic range. At ultimate limit state, these elements are expected to be severely damaged, but they can be easily removed and substituted as "mechanical fuses", being made of simple circular low carbon steel bars.

9.4 THE FIRE STATION OF NAPLES

The problem of damping the horizontal earthquakes, which are anyway less important due to the considerable rigidity of the r.c. cores, has been faced from the point of view of protecting the structure from the pounding effects between the core walls and the floors of the suspended skeleton. The devices shown in Figure 9.77 were designed with this purpose.

Figure 9.75 – Real view of the device between two adjacent structural meshes

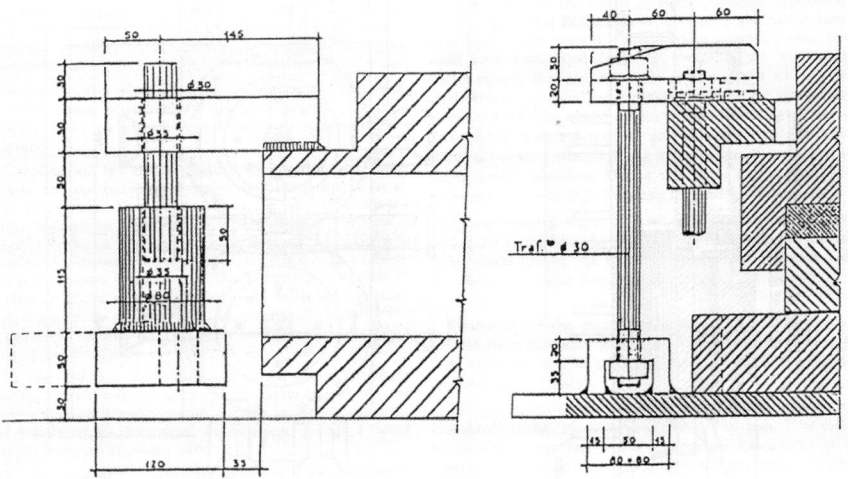

Figure 9.76 – Energy dissipation elements of the device

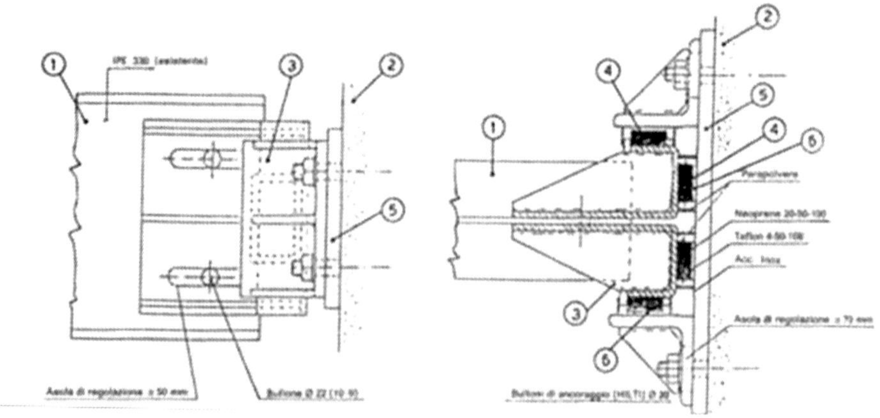

9. CASE STUDIES

Figure 9.77 – Details of the device used for avoiding pounding effects between the r.c. walls and the suspended floors

The edge of each floor beam (1), which is adjacent to a r.c. core (2), is closed by a special head (3) where four cellular rubber pads are connected to (4). Due to the horizontal movements, such pads can come in touch with a stainless steel plate attached to the r.c. wall (5). Contrary, vertical displacements are freely allowed by the contact between teflon and stainless steel (6).

This device is illustrated in Figure 9.78, while Figure 9.79 shows the lower view of the device, which is included between the steel floor structure and the wall of the r.c. core.

Figure 9.78 – Lateral (a) and plane (b) views of the device used to avoid pounding effects between the cores and the suspended floors

9.4 THE FIRE STATION OF NAPLES

Figure 9.79 – Lower view of the device used to avoid pounding effects between the cores and the suspended floors

REFERENCES

AISC 341-05 (2005). Seismic Provisions for Structural Steel Buildings, *American Institute of Steel Construction*, Inc. Chicago, Illinois, USA.

AISC 341-10 (2010). Seismic Provisions for Structural Steel Buildings, *American Institute of Steel Construction*, Inc. Chicago, Illinois, USA.

AISC 358-10 (2010). Prequalified Connections for Special and Intermediate Steel Moment Frames for Seismic Applications, *American Institute of Steel Construction*, Inc. Chicago, Illinois, USA.

AISC 341-16 (2016). Seismic Provisions for Structural Steel Buildings, *American Institute of Steel Construction*, Inc. Chicago, Illinois, USA.

AISC 358-16 (2016). Prequalified connections for special and intermediate steel moment frames for seismic applications, *American Institute of Steel Construction*, Inc. Chicago, Illinois, USA.

Akiyama H (1985). *Earthquake-Resistant Limit-State Design for Buildings*, University of Tokyo Press.

Araújo M, Macedo L, Marques M, Castro JM, (2016). Code-based record selection methods for seismic performance assessment of buildings, *Earthquake Engineering & Structural Dynamics,* 45:129-148.

ASCE 7-10 (2013). Minimum Design Loads for Buildings and Other Structures, *American Society of Civil Engineers*, Third Printing, incorporating errata, Supplement 1, and revised commentary, Reston, Virginia, USA.

Astaneh-Asl A, Goel SC, Hanson RD (1982). *Cyclic Behavior of Double Angle Bracing Members with End Gusset Plates*, Research Report UMEE 82R7, Department of Civil Engineering, University of Michigan, Ann Arbor, MI.

Astaneh-Asl A, Goel SC (1984). Cyclic In-Plane Buckling of Double Angle Bracing, *Journal of Structural Engineering*, American Society of Civil Engineers, 110(9):2036-2055 (www.asce.org).

Astaneh-Asl A, Goel SC, Hanson RD (1985). Cyclic Out-of-Plane Buckling of Double Angle Bracing, *Journal of Structural Engineering*, American Society of Civil Engineers, 111(5):1135-1153 (www.asce.org).

REFERENCES

Astaneh-Asl A, Goel SC, Hanson RD (1986). Earthquake-resistant design of double angle bracings, *Engineering Journal*, AISC, Chicago (www.aisc.org).

Astaneh-Asl A, Cochran ML, Sabelli R (2006). Seismic Detailing of Gusset Plates for Special Concentrically Braced Frames, *Structural Steel Educational Council - Steel TIPS*.

Ayers JM, Sun TY (1973). *Nonstructural damages, San Fernando, California, earthquake of February 9, 1971*, U.S. Dept. of Commerce, National Oceanic and Atmospheric Admin.

Baker JW, Cornell AC (2006). Spectral shape, epsilon and record selection, *Earthquake Engineering & Structural Dynamics*, 35:1077-1095.

Benioff H (1934). The physical evaluation of seismic destructiveness, *Bulletin of the Seismological Society of America*, 24(2):398-403.

Beresnev IA, Atkinson GM (1998). FINSIM-A FORTRAN program for simulating stochastic acceleration time histories from finite faults, *Seismological Research Letters* 69(1):27-32.

Bertero VV (1996). The need for multi-level seismic design criteria, 11WCEE – *Proceedings of the 11th World Conference on Earthquake Engineering*, Acapulco, Mexico, 23-28 June.

Bertero VV (1997). Earthquake Engineering, *Lecture Notes, University of California*, Berkeley, 1997.

Biot MA (1941). A mechanical analyzer for the prediction of earthquake stresses, *Bulletin of the Seismological Society of America*, 31(2):151-171.

Bisch P (2009). Ch 1 Introduction: Seimic design and Eurocode 8. In *Seismic Design of Buildings to Eurocode 8*, Elghazouli AY (ed.), Spon Press.

Bommer JJ, Acevedo AB (2004). The use of real earthquake accelerograms as input to dynamic analysis, *Journal of Earthquake Engineering*, 8(1):43-91.

Bommer JJ, Pinho R (2006). Adapting earthquake actions in Eurocode 8 for performance-based seismic design, *Earthquake Engineering & Structural Dynamics*, 35(1):39-55.

Boore DM (2003). Simulation of ground motion using the stochastic method, *Pure and Applied Geophysics*, 160:635{676.

Bordea S (2010). *Dual frame systems of Bukling Restrained Braces*, PhD Thesis, University Politehnica Timisoara, Romania, Series 5 No. 53.

Burmeister S, William PJ (2008). Horizontal floor diaphragm load effects on composite beam design, December 2008, *Modern Steel Construction*.

Calvi GM, Timothy S, Roldán R, O'Reilly G, Simões da Silva L, Rebelo C, Castro M, Agusto H, Landolfo R, Della Corte G, Terraciano G, Salvatore W, Morelli F (2015). *Displacement based seismic design of steel moment resisting frame structures (DISTEEL)*, Directorate-General for Research and Innovation, EUR 27157 EN, 2015.

CEN (1998). ENV 1993-1-1:1998 – Eurocode 3: Design of steel structures – Part 1-1: General rules and rules for buildings, *European Committee for Standardization*, Brussels.

CEN (2002a). EN 1990-2002 – Eurocode 0: Basis of structural design, *European Committee for the Standardization*, Brussels.

CEN (2002b). EN 1991-1-1:2002 – Eurocode 1: Actions on structures – Part 1-1: General actions -Densities, self-weight, imposed loads for buildings, *European Committee for the Standardization*, Brussels.

CEN (2003). EN 1991-1-3:2003 – Eurocode 1: Actions on structures – Part 1-3: General actions - Snow loads, *European Committee for the Standardization*, Brussels.

CEN (2004a). EN 1998-1-1:2004 – Eurocode 8: Design of structures for earthquake resistance – Part 1: General rules, seismic actions and rules for buildings, *European Committee for Standardization*, Brussels.

CEN (2004b). EN 1997-1:2004 – Eurocode 7: Geotechnical design - Part 1: General rules, *European Committee for the Standardization*, Brussels.

CEN (2004c). EN 1998-5:2004 – Eurocode 8: Design of structures for earthquake resistance Part 5: Foundations, retaining structures and geotechnical aspects, *European Committee for the Standardization*, Brussels.

CEN (2004d). EN 10025-1:2004 – Hot rolled products of structural steels. Part 1: General technical delivery conditions, *European Committee for Standardization*, Brussels.

CEN (2004e). EN 1994-1-1:2004 – Eurocode 4: Design of composite steel and concrete structures – Part 1.1: General rules and rules for buildings, *European Committee for Standardization*, Brussels.

CEN (2004f). EN 10164:2004 – Steel products with improved deformation properties perpendicular to the surface of the product - Technical delivery conditions, *European Committee for the Standardization*, Brussels.

CEN (2005a). EN 1992-1-1: 2005 – Eurocode 3: Design of concrete structures – Part 1-1: General rules and rules for buildings, *European Committee for the Standardization*, Brussels.

References

CEN (2005b). EN 1993-1-1: 2005 – Eurocode 3: Design of steel structures – Part 1-1: General rules and rules for buildings, *European Committee for the Standardization*, Brussels.

CEN (2005c). EN 1998-3:2005 – Eurocode 8: Design of structures for earthquake resistance – Part 3: Assessment and retrofitting of buildings, *European Committee for the Standardization*, Brussels.

CEN (2005d). EN 1993-1-8:2005 – Eurocode 3: Design of steel structures – Part 1-8: Design of joints, *European Committee for Standardization*, Brussels.

CEN (2005e). EN 1993-1-10:2005 – Eurocode 3: Design of steel structures – Part 1-10: Material toughness and through-thickness properties, *European Committee for Standardization*, Brussels.

CEN (2005f). EN 1991-1-4:2005 – Eurocode 1: Actions on structures – Part 1-4: General actions - Wind actions, *European Committee for Standardization*, Brussels.

CEN (2006a). EN 10210-1:2006 – Hot finished structural hollow sections of non-alloy and fine grain steels – Part 1: Technical delivery conditions, *European Committee for Standardization*, Brussels.

CEN (2006b). EN 10219-1:2006 – Cold formed welded structural hollow sections of non-alloy and fine grain steels – Part 1: Technical delivery conditions, *European Committee for Standardization*, Brussels.

CEN (2006c). EN 1993-1-5:2006 – Eurocode 3: Design of steel structures – Part 1-5: Plated structural elements, *European Committee for Standardization*, Brussels.

CEN (2009). EN 15129:2009 – Anti-seismic devices, *European Committee for Standardization*, Brussels.

CEN (2011). EN 1090-1:2009+A1:2011 – Execution of steel structures and aluminium structures – Part 1: Requirements for conformity assessment of structural components, *European Committee for Standardization*, Brussels.

CEN (2017). EN 1992-4:2017 – Eurocode 2 – Design of concrete structures – Part 4: Design of fastenings for use in concrete, European Committee for Standardization, Brussels.

Chi WM, Deierlein GG, Ingraffea AR (1997). Finite element fracture mechanics investigation of welded beam-column connections, Rep. No. SAC/BD-97/05, *SAC Joint Venture*, Sacramento, California.

Chopra AK (2011). *Dynamics of Structures*, Prentice-Hall International Series in Civil Engineering and Engineering Mechanics, 4th Edition.

Chopra AK, Chintanapakdee C (2001). Comparing response of SDF systems to near-fault and far-fault earthquake motions in the context of spectral regions, *Earthquake Engineering and Structural Dynamics*, 30:1769-1789.

Ciutina AL, Dubina D (2008). Column Web Stiffening of Steel Beam-to-Column Joints Subjected to Seismic Actions, *Journal of Structural Engineering*, American Society of Civil Engineers, 134(3):505-510.

Ciutina A, Dubina D, Danku G (2013). Influence of steel-concrete interaction in dissipative zones of frames: I – Experimental study, *Steel and Composite Structures*, 15(3):299-322.

Clifton GC, El Sarraf R (2005). *Composite Floor Construction Handbook*, HERA Report R4-107. N. Z. HERA, Manukau City, New Zealand.

Clough RW, Penzien J (1993). *Dynamics of Structures*, McGraw-Hill International Editions, 2nd Edition.

Cochran ML (2003). Seismic design and steel connection detailing, NASCC 2003 – *Proceedings of North American Steel Construction Conference*, Baltimore, MD, April 2-5.

Cowie K, Hicks S, MacRae G, Clifton GC, Fussell A (2013). Seismic design of composite metal deck and concrete-filled diaphragms – a discussion paper, *Proceedings of Steel Innovations Conference 2013*, Christchurch, New Zealand, 21-22 February.

D'Aniello M, Landolfo R, Piluso V, Rizzano G (2012). Ultimate Behaviour of Steel Beams under Non-Uniform Bending, *Journal of Constructional Steel Research*, 78:144-158.

D'Aniello M, La Manna Ambrosino G, Portioli F, Landolfo R (2013). Modelling aspects of the seismic response of steel concentric braced frames, *Steel and Composite Structures*, 15(5):539-566.

Davaran A, Hoveidae N (2009). Effect of mid-connection detail on the behaviour of cross-bracing systems, *Journal of Constructional Steel Research*, 65(4):985-990.

Dekker RWA, Snijder HH, Maljaars J (2015). Experimental study into bending-shear interaction of rolled I-shaped sections, NSCC-2015 – *The 13th Nordic Steel Construction Conference*, 23-25 September, Tampere, Finland . Heinisuo M & Mäkinen J (eds.), Tampere: Tampere University of Technology, 1-10.

REFERENCES

Dinu F, Dubina D, Neagu C, Vulcu C, Both I, Herban S (2012). Experimental and numerical evaluation of an RBS coupling beam for moment-resisting steel frames in seismic areas, *Steel Construction – Design and Research*, 6(1):27-33.

Dubina D (2008). Structural analysis and design assisted by testing of cold-formed steel structures, *Thin-Walled Structures*, 46(7-9):741-764.

Dubina D (2010). Dual-steel frames for multistory buildings in seismic areas, Keynote lecture, in *Proceedings of SDSS'Rio 2010, International Colloquium Stability and Ductility of Steel Structures*, Batista E, Vellasco P, de Lima L (eds.), 1:59-80 , 08-10 September 2012, Rio de Janeiro, Brazil.

Dubina D, Fülöp LA, Aldea A, Demetriu S, Nagy Zs (2006). Seismic performance of cold-formed steel framed houses. STESSA 2006 – *Proceedings of the 5th International Conference on Behaviour of Steel Structures in Seismic Areas*, Yokohama, Japan, 14-17 August, 429-435.

Dubina D, Stratan A, Dinu F (2009a). Design and Performance based Evaluation of Tower Centre International building in Bucharest, Part I: Structural design, *Steel Construction – Design and Research*, 2:4/2009, Ernst&Sohn, a Wiley Company.

Dubina D, Stratan A, Dinu F (2009b). Design and Performance based Evaluation of Tower Centre International building in Bucharest , Part II: Performance-based Seismic Evaluation and Robustness, *Steel Construction – Design and Research*, 3(1):2010, Ernst&Sohn, a Wiley Company.

Dubina D, Stratan A, Muntean N, Dinu F (2008). Experimental program for evaluation of moment beam-to-column joints of high strength steel components, CONNECTIONS VI – *Proceedings of the 6th International Workshop Connections in Steel Structures VI*, June 22–25, Chicago, U.S.A., Bjorhovde R, Bijlaard FSK, Geschwindner LF (eds.), 355-366.

Dubina D, Ungureanu V, Ciutina A, Mutiu M, Grecea D (2010). Innovative sustainable steel framing based affordable house solution for continental seismic areas, ICSA 2010 – *Structures and Architecture - Proceedings of the 1st International Conference on Structures and Architecture*, 21–23 July, Guimarães, Portugal, 1341-1348.

Dubina D, Ungureanu V, Landolfo R (2012). *Design of Cold-formed Steel Structures*, ECCS Eurocode Design Manuals, ECCS Press/Ernst&Sohn.

Dubina D, Vulcu C, Stratan A, Ciutina A, Grecea D, Loan A, Tremeea A, Braconi A, Fulop L, Jaspart J-P, Demonceau J-F, Hoang VL, Comeliau L, Kuhlmann

U, Kleiner A, Rasche C, Landolfo R, D'Aniello M, Portioli F, Beg D, Cermelj B, Moze P, Simões da Silva L, Rebelo C, Tenchini A, Kesti J, Salvatore W, Caprili S, Ferrini M (2015). *High Strength Steel in Seismic Resistant Building Frames* - HSS-SERF. EC, RFCS, EUR 26933, Luxembourg, EU.

Ebadi P, Sabouri-Ghomi S (2012). Conceptual study of X-braced frames with different steel grades using cyclic half-scale tests, *Earthquake Engineering and Engineering Vibration*, 11:313-329.

ECCS (1986). Recommended Testing Procedure for Assessing the Behaviour of Structural Steel Elements under Cyclic Loads, P045, ECCS Technical Committee 1 – Structural Safety and Loadings, Technical Working Group 1.3 – Seismic Design, European Convention for Constructional Steelwork, Brussels.

ECCS (1988). *European Recommendations for Steel Structures in seismic Zones*, P054, ECCS Technical Committee 13 – Seismic Design, European Convention for Constructional Steelwork, Brussels.

Elghazouli AY (ed.) (2009). *Seismic design of buildings to Eurocode 8*, Taylor and Francis/Spon Press, London.

Engelhardt MD, Popov EP (1989). On Design of Eccentrically Braced Frames, *Earthquake Spectra*, EERI, 5(3):495-511.

Engelhardt MD, Popov EP (1992). Experimental Performance of Long Links in Eccentrically Braced Frames, *Journal of Structural Engineering*, American Society of Civil Engineers, 118(11):3067-3088.

EU (2013). Commission Implementing Regulation No 1062/2013 of 30 October 2013 on the format of the European Technical Assessment for construction products, *Official Journal of the European Union*.

European Strong-Motion Database (http://www.isesd.cv.ic.ac.uk).

Fajfar P (1999). Capacity spectrum method based on inelastic demand spectra, *Earthquake Engineering and Structural Dynamics*, 28(9):979-993.

Fajfar P (2000). A nonlinear analysis method for performance-based seismic design, *Earthquake Spectra* 16(3):573-592.

Fardis MN, Carvalho E, Elnashai A, Faccioli E, Pinto P, Plumier A (2005). *Designer's Guide to EN 1998-1 and EN 1998-5 . Eurocode 8:Design of structures for earthquake resistance. General Rules. Seismic actions, design rules for buildings, foundations and retaining structures*, Thomas Telford , London.

FEMA 350 (2000). Recommended Seismic Design Criteria for New Steel Moment-Frame Buildings, *Federal Emergency Management Agency*, Washington, DC.

REFERENCES

FEMA 356 (2000). Prestandard and commentary for the seismic rehabilitation of buildings, *Federal Emergency Management Agency*, Washington, DC.

FEMA 445 (2006). Next-Generation Performance-Based Seismic Design Guidelines, *Federal Emergency Management Agency*, Washington, DC.

FEMA 461 (2007). Interim Testing Protocols for Determining the Seismic Performance Characteristics of Structural and Nonstructural Components, *Federal Emergency Management Agency*, Washington, DC.

FEMA 695 (2009). Quantification of Building Seismic Performance Factors, *Federal Emergency Management Agency*, Washington, DC.

FEMA 750 (2009). Recommended Seismic Provisions for New Buildings and Other Structures, *Federal Emergency Management Agency*, Washington, DC.

Fiorino L, Iuorio O, Landolfo R (2014). Designing CFS structures: The new school BFS in Naples, *Thin-Walled Structures* 78:37-47.

Fiorino L, Iuorio O, MacIllo V, Landolfo R (2012). Performance-based design of sheathed CFS buildings in seismic area, *Thin-Walled Structures* 61: 248-257.

Fulop LA, Dubina D (2007). *The importance of earhquake loads in the design of single- storey portal frames*, Technical Report (unpublished), Department of steel Structures and Structural Mechanics of University Poitehnica Timisoara, Romania.

Gulvanessian H, Calgaro JA, Holicky M (2002). *Designers' guide to EN 1990: Eurocode 0: basis of structural design*, Thomas Telford.

Gasparini DA, Vanmarcke EH (1979). *Simulated earthquake motions compatible with prescribed response spectra. Evaluation of Seismic Safety of Buildings Report No. 2*, Department of Civil Engineering, MIT, Cambridge, Massachusetts, 99.

Gioncu V, Mazzolani FM (2002). *Ductility of Seismic Resistant Steel Structures*, Spon Press, London.

Gioncu V, Pectu D (1997a). Available rotation capacity of wide-flange beams and beam-columns. Part 1: theoretical approaches, *Journal of Constructional Steel Research*, 43(1–3):161-217.

Gioncu V, Pectu D (1997b). Available rotation capacity of wide-flange beams and beam-columns. Part 2: experimental and numerical tests, *Journal of Constructional Steel Research*, 43(1-3):219-44.

Grecea D, Dinu F, Dubina D (2004). Performance criteria for MR steel frames in seismic zones, *Journal of Constructional Steel Research* 60: 739-749.

Greiner R, Kettler M, Lechner A, Freytag B, Linder J, Jaspart J-P, Boissonnade N, Bortolotti E, Weynand K, Ziller C, Oerder R (2009). *SEMI-COMP: Plastic member capacity of semi-compact steel sections - a more economic design*, EC, RFCS, EUR 23735, Luxembourg, EU.

Gupta A, Krawinkler H (1999). *Influence of column web stiffening on the seismic behaviour of beam-to-column joints*, Stanford University, Stanford, California.

Hakuno M, Shidawara M, Hara T (1969). Dynamic Destructive Test of a Cantilever Beam, Controlled by an Analog-Computer, *Transactions of the Japan Society of Civil Engineers*, 171-179.

Hancock J, Bommer JJ, Stafford PJ (2008). Numbers of scaled and matched accelerograms required for inelastic dynamic analyses, *Earthquake Engineering & Structural Dynamic* 37:1585-1607.

HAZUS (1997). Earthquake Loss Estimation Methodology, Technical Manual, *National Institute of Building for the Federal Emergency Management Agency*, Washington, DC.

Hjelmstad KD, Popov EP (1983). Cyclic Behavior and Design of Link Beams, *Journal of Structural Engineering*, American Society of Civil Engineers, 109(10):2387-2403.

Hjelmstad KD, Popov EP (1984). Characteristics of Eccentrically Braced Frames, *Journal of Structural Engineering*, American Society of Civil Engineers, 110(2):340-353.

Hjelmstad KD, Lee SG (1989). Lateral Buckling of Beams in Eccentrically Braced Frames, *Journal of Constructional Steel Research*, 14(4):251-272.

Hong J-K, Uang C-M, Okazaki T, Engelhardt MD (2015). Link-to-Column Connection with Supplemental Web Doublers in Eccentrically Braced Frames, *Journal of Structural Engineering*, American Society of Civil Engineers, 141(8):04014200.

Horne MR (1949). Contribution to The design of steel frames by Baker JF, *Structural Engineer*, 27:421.

Housner GW (1956). Limit design of structures to resist earthquakes, *Proceedings of First World Conference on Earthquake Engineering*.

Housner GW (1959). Behavior of structures during earthquakes, *American Society of Civil Engineers*, EM4.

References

INCERC database (http://www.incerc2004.ro/accelerograme.htm and http://www.incerc.ro/download.htm)

ITACA database (http://itaca.mi.ingv.it).

Jaspart J-P, Weynand K (2016). *Design of joints in steel and composite structures*, ECCS Eurocode Design Manuals, ECCS Press/Ernst&Sohn.

Jaspart J-P, Demonceau J-F, D'Aniello M, Tartaglia R, Landolfo R, Stratan A, Maris C, Dubina D, Elghazouli A, Nunez EM (2017). *EQUALJOINTS – Design Manual of prequalified beam-to-column joints*, European pre-QUALified steel JOINTS, Report on design procedure, prequalification charts and examples of application.

Johnson S (2005). *Improved seismic performance of special concentrically braced frames*, Master thesis, University of Washington, Seattle.

Kasai K, Popov EP (1985). On Seismic Design of Eccentrically Braced Frames, 8WCEE – *Proceedings of the 8th World Conference on Earthquake Engineering*, 5:387-394, IAEE.

Kasai K, Popov EP (1986a). Cyclic Web Buckling Control for Shear Link Beams, *Journal of Structural Engineering*, American Society of Civil Engineers, 112(3):505-523.

Kasai K, Popov EP (1986b). General Behavior of WF Steel Shear Link Beams, *Journal of Structural Engineering*, American Society of Civil Engineers, 112(2):362-382.

Krawinkler H (1978). Shear in beam-column joints in seismic design of steel frames, *Engineering Journal*, American Institute of Steel Construction, 5(3):82-91.

Landolfo R (2011). Cold-formed steel structures in seismic area: research and applications, VIII CMM – *Proceedings of the 8th Conference in Steel and Composite Structures*, Guimarães, Portugal, pp. I-3–I-22, 24–25 November, cmm Press.

Landolfo R (ed.) (2013). *Assessment of EC8 provisions for seismic design of steel structures*, Technical Committee 13 – Seismic Design, No 131/2013, European Convention for Constructional Steelwork.

Landolfo R (2016). European qualification of seismic resistant steel bolted beam-to-column joints: the EQUALJOINTS project, CONNECTIONS VIII – *Proceedings of 8th International Workshop on Connections in Steel Structures*, Carter CJ, Hajjar JF (eds.), Boston, U.S.A, 579-588, 24-26 May, AISC.

Lee KH, Stojadinovic B, Goel SC, Margarian AG, Choi J, Wongkaew A, Rayher BP, Lee DY (2000). *Parametric tests on unreinforced connections*, Rep. No. SAC/BD-00/01, SAC Joint Venture, Sacramento, California.

Lehman DE, Roeder CW, Herman D, Johnson S, Kotulka B (2008). Improved Seismic Performance of Gusset Plate Connections, *Journal of Structural Engineering*, American Society of Civil Engineers, 134(6), June 1.

Mazzolani FM (1985a). Le strutture sismo-resistenti della nuova Caserma VV.FF. di Napoli, *Proceedings of the Italian Conference on Steel Constructions* – CTA, Montecatini, Italy, October.

Mazzolani FM (1985b). Applicazione di apparecchiature dissipative antisismiche in un edificio civile, *Proceedings of the CTE Conference on Special Problems for Industrializad Construction in seismic zones*, Perugia, Italy, November.

Mazzolani FM (1985c). Seismic resistant system for a composite steel concrete building, *Proceedings of the IABSE-ECCS Symposium*, Luxembourg, September.

Mazzolani FM (1986a). Base isolation system for steel seismic-resistant structures, IAEG1986 – *Proceedings of the International Symposium on Engineering Geology Problems in Seismic Areas*, Bari, Italy, April.

Mazzolani FM (1986b). L'applicazione del Base isolation system a strutture sismo-resistenti in acciaio, *Acciaio*, 4.

Mazzolani FM (1986c). The seismic resistant structures of the New Fire Station of Naples, *Costruzioni Metalliche*, 38(6):343-362.

Mazzolani FM (1990). The seismic resistant structures of the building A of the New Fire Station of Naples (in Italian), *L'Ingegnere*, 12:4-23.

Mazzolani FM, Piluso V (1996). *Theory and Design of Seismic Resistant Steel Frames*, E & FN Spon, an imprint of Chapman & Hall, London.

Mazzolani FM, Piluso V (1997). Plastic design of seismic resistant steel frames, *Earthquake Engineering and Structural Dynamics*, 26167-191.

Mazzolani FM, Gioncu V (eds.) (2000). Seismic resistant steel structures, *CISM (International Center for Mechanical Sciences) Courses and Lectures* No 420, Springer Wien NewYork.

Mazzolani FM, Landolfo R, Della Corte G, Faggiano B (2006). *Edifici con Struttura di Acciaio in Zona Sismica*, IUSS press.

MOC-2008 (2008). Manual de diseno de obras civiles. Diseno por sismo. Recomendaciones y Comentarios, *Instituto de Investigaciones Electricas, Comision Federal de Electricidad* (in Spanish).

References

MRDPA (2013). P100: Part 1/2013 – Code for earthquake resistant design. Design provisions for buildings, *Ministry of Regional Development and Public Administration*, Romania.

MTCT (2006). Seismic Design Code P100: Part I, P100-1/2006: Rules for Buidings (in Romanian), Ministry's Order No 1711/2006, Officeal Monitor of Romania No. 803 bis, *Ministry of Transport, Constructions and Tourism*.

Naeim F, Kelly JM (1999). *Design of Seismic Isolated Structures: From Theory to Practice*, John Wiley and Sons, Inc., New York.

Nakashima M, Wakabayashi M (1992). *Analysis and design of steel braces and braced frames in building structures. Stability and ductility of steel structures under cyclic loading*, Fukumoto Y and Lee GC, (eds.), CRC Press, Boca Raton, Fla., 309-321.

NBCC (2005). National Building Code of Canada 2005. Ottawa, Canada: Institute for Reseach Construction, *National Research Council of Canada*.

NEHRP (2003). NEHRP Recommended Provisions for Seismic Regulations for New Buildings and Other Structures, Part I – Provisions, *Federal Emergency Management Authority*.

NEHRP (2009). NEHRP Recommended Seismic Provisions for New Buildings and Other Structures, Part I – Provisions, *Federal Emergency Management Authority*.

Newmark NM, Hall WJ (1969). Seismic Design Criteria for Nuclear Reactor Facilities, 4WCEE – *Proceedings of the 4th World Conference on Earthquake Engineering*, Santiago, Chile, 2(B-4):37-50.

Newmark NM, Hall WJ (1982). Earthquake Spectra and Design, *Earthquake Engineering Research Institute*, Okland, U.S.A.

NTC (2008). *Norme tecniche per le costruzioni*, DM 14 gennaio 2008, pubblicato sulla Gazzetta Ufficiale n. 29 del 4 febbraio 2008 – Suppl. Ordinario n. 30.

Nussbaumer A, Borges L, Davaine L (2011). *Fatigue design of steel and composite structures*, ECCS Eurocode Design Manuals, ECCS Press/Ernst&Sohn.

NZS 1170.5 (2004). Structural Design Actions Part 5: Earthquake actions, *New Zealand Standard*.

Obama B (2016). *Executive Order: Establishing a Federal Earthquake Risk Management Standard*. https://www.whitehouse.gov/the-press-office/2016/02/02/executive-order-establishing-federal-earthquake-risk-management-standard.

References

Ohi K, Takanashi K (1998). Seismic diagnosis for rehabilitation and upgrading of steel gymnasiums, *Engineering Structures*, 20(4-6):533-539, 1998.

Okazaki T, Arce G, Ryu H-C, Engelhardt MD (2005). Experimental Study of Local Buckling, Overstrength, and Fracture of Links in Eccentrically Braced Frames, *Journal of Structural Engineering*, American Society of Civil Engineers, 131(10):1526-1535.

Okazaki T, Engelhardt MD, Nakashima M, Suita S (2006). Experimental Performance of Link-to-Column Connections in Eccentrically Braced Frames, *Journal of Structural Engineering*, American Society of Civil Engineers, 132(8):1201-1211.

Okazaki T, Engelhardt MD, Drolias A, Schell E, Uang CM (2009). Experimental Investigation of Link-to-Column Connections in Eccentrically Braced Frames, *Journal of Constructional Steel Research*, ASCE, 65(7):1401-1412.

Palmer KD (2012). *Seismic behavior, performance and design of steel concentrically braced framed systems*, PhD Thesis, University of Washington, Seattle.

Palmer KD, Roeder CW, Lehman DE, Okazaki T, Shield C (2013). Experimental Performance of Steel Braced Frames Subjected to Bidirectional Loading, *Journal of Structural Engineering*, American Society of Civil Engineers, 139(8):1274-1284.

Paulay T, Priestley MJN (1992). *Seismic Design of Reinforced Concrete and Masonry Buildings*, John Wiley & Sons Inc, Chichester.

Picard A, Beaulieu D (1989a). Theoretical study of the buckling strength of compression members connected to coplanar tension members, *Canadian Journal of Civil Engineering*, 16(3):239-248.

Picard A, Beaulieu D (1989b). Experimental study of the buckling strength of compression members connected to coplanar tension members, *Canadian Journal of Civil Engineering*, 16(3):249-257.

Pinto A, Taucer F, Dimova S (2007). Pre-normative research needs to achieve improved design guidelines for seismic protection in the EU. EUR 22858 EN – 2007, *European Commission - Joint Research Centre*, ISPRA.

Plumier A, Doneux C (ed.) (2001). *Seismic Behaviour and Design of Composite Steel Concrete Structures*, LNEC Edition, Lisbon.

Priestley MJN (2000). Performance-Based seismic design, 12WCEE – *Proceedings of the 12th World Conference on Earthquake Engineering*, Auckland, New Zealand, 30 January – 4 February.

REFERENCES

Pseudo Dynamic (PsD) Tests – Non Linear Structural Dynamics Techniques, http://www.aboutcivil.org/pseudo-dynamic-tests.html

Rai DC (2000). Future trends in earthquake-resistant design of structures, *Current Science*, 79(9).

Ricles JM, Popov EP (1989). Composite Action in Eccentrically Braced Frames, *Journal of Structural Engineering*, American Society of Civil Engineers, 115(8):2046-2065.

Ricles JM, Mao C, Lu LW, Fisher JW (2000). Development and evaluation of improved details for ductile welded unreinforced flange connections, Rep. No. SAC BD 00-24, *SAC Joint Venture*, Sacramento, California.

Rihal SS (1992). Performance and behavior of non-structural building components during the Whittier Narrows, California (1987) and Loma Prieta, California (1989) earthquakes: selected case studies, Report No. ATC-29, *Proc. Seminar and Workshop on Seismic Design and Performance of Equipment and Nonstructural Elements in Buildings and Industrial Structures*, Applied Technology Council, Redwood City, California, 119-143.

Roeder CW (2002). General Issues Influencing Connection Performance, *Journal of Structural Engineering*, American Society of Civil Engineers, 128(4):420-428.

Sabau GA, Poljansek M, Taucer F, Pegon P, Molina Ruiz FJ, Tirelli D, Viaccoz B, Stratan A, Dubina D, Ioan-Chesoan A (2014). DUAREM – *Full-Scale experimental validation of dual eccentrically braced frame with removable links*, TA Project Final Report in Seismic Engg. Resc. Infrastructures for European Synergies – SERIES 227887, EU, JRC, JRC 93136, EUR 27030, doi 10.2788/539418, Luxembourg, EU.

Sabelli R, Hohbach D (1999). Design of Cross-Braced Frames for Predictable Buckling Behavior, *Journal of Structural Engineering*, American Society of Civil Engineers, 125(2):163-168.

Sabelli R, Sabol TA, Easterling WS (2011). *Seismic Design of Composite Steel Deck and Concrete-filled Diaphragms: A Guide for Practicing Engineers*, NEHRP Seismic Design Technical Brief No. 5, NIST GCR 11-917-10.

Sandu L, Degeratu M, Hasegan L, Georgescu A, Cosoiu C (2004). *Design Report: Modelling of wind action against Bucharest Tower Centre International (in Romanian)*, Contract No. 427 / 2004, Technical University of Construction Bucharest, UTCB.

SEAOC (1995). A framework for performance-based design, *Structural Engineers Association of California*, Vision 2000 Committee, Sacramento, California.

Sedlacek G, Müller C (2006). The European standard family and its basis. *Journal of Constructional Steel Research*, 62:1047-1059.

Seismosoft (2014). *SeismoStruct v7.0 – A computer program for static and dynamic nonlinear analysis of framed structures*, available from http://www.seismosoft.com.

SHARE (2013). *Seismic Hazard Harmonization in Europe, 7th Framework Program of the European Commission*, http://www.share-eu.org (accessed on June 23, 2017).

Simões da Silva L, Rebelo C, Nethercot D, Marques L, Simões R, Vila Real P (2009). Statistical evaluation of the lateral-torsional buckling resistance of steel I-beams – Part 2: Variability of steel properties, *Journal of Constructional Steel Research*, 65(4):832-849.

Simões da Silva L, Simões R, Gervásio H (2016). *Design of steel structures*, 2nd Edition, ECCS Eurocode Design Manuals, ECCS Press/Ernst&Sohn.

da Silva LS, Tankova T, Marques L, Kuhlmann U, Kleiner A, Spiegler J, Snijder HH, Dekker R, Taras A, Popa N (2017). Safety Assessment across Modes Driven by Plasticity, Stability and Fracture, *ce/papers*, 1:3689-3698.

Somerville PG (2003). Magnitude scaling of the near fault rupture directivity pulse, *Physics of the Earth and Planetary Interiors*, 137:201-212.

Somerville PG, Smith NF, Graves R, Abrahamson NA (1997). Modification of Empirical Strong Ground Motion Attenuation Relations to Include the Amplitude and Duration Effects of Rupture Directivity, *Seismological Research Letters*, 68(1):199-222.

Solomos G, Pinto A, Dimova S (2008). *A review of the seismic hazard zonation in national building codes in the context of Eurocode 8*. JRC scientific and technical report. European Commission, Joint Research Centre - EUR 23563.

Stoman SH (1989). Effective Length Spectra for Cross Bracings, *Journal of Structural Engineering*, American Society of Civil Engineers, 115(12):3112-3122.

Sumner EA, Murray TM (2002). Behavior of Extended End-Plate Moment Connections Subject to Cyclic Loading, *Journal of Structural Engineering*, American Society of Civil Engineers, 128(4):501-508.

References

Tanabashi R (1935). Tests to determine the behavior of riveted joints of steel structures under alternate bending moments, *Memoirs of the College of Engineering*, Kyoto Imperial University, VIII(4):164.

Takanashi K (1975). Non-linear Earthquake Response Analysis of Structures by a Computer Actuator on-line System, *Transactions of the Architectural Institute of Japan*, 229:77-83.

Takanashi K, Nakashima M (1987). Japanese Activities on On-Line Testing, *Journal of Engineering Mechanics*, 113(7):1014-1032.

Tankova T, Simões da Silva L, Marques L, Rebelo C, Taras A (2014). Towards a standardized procedure for the safety assessment of stability design rules, *Journal of Constructional Steel Research*, 103:290-302.

Taranath SB (2005). *Wind and Earthquake resistant buildings*, CRC, Taylor &Francis.

Tenchini A, D'Aniello M, Rebelo C, Landolfo R, Simões da Silva L, Lima L (2014). Seismic performance of dual-steel moment resisting frames, *Journal of Constructional Steel Research*, 101.

Tremblay R, Archambault M, Filiatrault A (2003). Seismic Response of Concentrically Braced Steel Frames Made with Rectangular Hollow Bracing Members, *Journal of Structural Engineering*, American Society of Civil Engineers, 129(12):1626-1636.

Uniform Building Code (1997). *International Council of Building Officials*, Whittier, California.

Vamvatsikos D, Cornell CA (2002). Incremental Dynamic Analysis, *Earthquake Engineering and Structural Dynamics*, 31(3):491-514.

Van Cranenbroeck J (2005). *A New Total Station Tracking GPS Satellites in a Network RTK Infrastructure Perspective*, From Pharaohs to Geoinformatics, Working Week and GSDI-8, Cairo Egypt, April 16-21.

Vayas I, Dinu F, (2000). *Evaluation of the response of moment frames in respect to various performance criteria*, STESSA 2000, Balkema, 643-648.

Veletsos AS, Newmark NM, (1960). Effect of inelastic behavior on the response of simple systems to earthquake motions, 2WCEE – *Proceedings of the 2nd World Conference on Earthquake Engineering*, Tokyo, Japan, 2: 895-912.

Wakabayashi M, Nakamura T, Yoshida N (1977). Experimental studies on the elastic–plastic behavior of braced frames under repeated horizontal loading, *Bull. Disaster Prevention Research Institute*, Kyoto University, 27(251):121-154.

Wald F, Hofmann J, Kuhlmann U, Bečková S, Gentili F, Gervásio H, Henriques J, Krimpmann M, Ožbolt A, Ruopp J, Schwarz I, Sharma A, Simoes da Silva L, van Kann J (2014). *INFASO + – "Valorisation of knowledge for innovative fastening solutions between steel and concrete"*, RFS2-CT-2012-00022, Brussels.

Whitmore RE (1952). *Experimental Investigation of Stresses in Gusset Plates*, Bulletin No. 16, Engineering Experiment Station, University of Tennessee, Knoxville, May.

Whitman RV, Hong S-T, Reed J (1973). *Damage statistics for high-rise buildings in the vicinity of the San Fernando Earthquake*, Report No. 7, Massachusetts Institute of Technology, 1973.

Wilson EL, Der Kiureghian A, Bayo EP (1981). A replacement for the SRSS method in seismic analysis, *Earthquake Engineering and Structural Dynamics*, 9:187-194.

Wilson E. *Dynamic Analysis By Numerical Integration*, CSI, Technical Paper. http://www.comp-engineering.com/downloads/technical_papers/CSI/20.pdf.

World Geodetic System website of the NGA (2012). National Geospatial-Intelligence Agency, April 2.

Yoo JH (2006). *Analytical investigation on the seismic performance of special concentrically braced frames*, PhD Thesis, University of Washington, Seattle.

Zaharia R, Dubina D (2014). Fire design of concrete encased columns: Validation of an advanced calculation model, *Steel and Composite Structures*, 17(6):835-850.

Zeng Y, Anderson JG, Yu G (1994). A composite source model for computing realistic synthetic strong ground motions, *Geophysical Research Letters*, 21(8):725-728.